ACID-BASE AND POTASSIUM HOMEOSTASIS

CONTEMPORARY ISSUES
IN NEPHROLOGY
VOL. 2

Series Editors
BARRY M. BRENNER MD
JAY H. STEIN MD

Published January 1978
Vol. 1. Sodium and Water Homeostasis

Forthcoming volumes in series

Vol. 3 Immunological Mechanisms of Renal Disease
    Curtis B. Wilson MD Guest Editor
Vol. 4 Hormonal Function and the Kidney
Vol. 5 Nephrolithiasis
    Fredric L. Coe MD Guest Editor

ACID-BASE AND POTASSIUM HOMEOSTASIS

Edited by **Barry M. Brenner,** M.D.

Samuel A. Levine Professor of Medicine, Harvard Medical School; Director, Laboratory of Kidney and Electrolyte Physiology, and Physician, Peter Bent Brigham Hospital, Boston, Massachusetts

and **Jay H. Stein,** M.D.

Professor of Medicine and Physiology and Chairman, Department of Medicine, The University of Texas Health Science Center at San Antonio, San Antonio Texas

CHURCHILL LIVINGSTONE
NEW YORK EDINBURGH AND LONDON 1978

CHURCHILL LIVINGSTONE
Medical Division of Longman Inc.

Distributed in the United Kingdom by Churchill Livingstone, 23 Ravelston Terrace, Edinburgh EH4 3TL and by associated companies, branches and representatives throughout the world.

© Longman Inc. 1978

All rights reserved. No part of this publication may be reproduced, stored in a retrieval system, or transmitted in any form or by any means, electronic, mechanical, photocopying, recording or otherwise, without prior permission of the publishers (Longman Inc., 19 West 44th Street, New York, N.Y. 10036).

First published 1978

ISBN 0 443 08017 8

Library of Congress Cataloging in Publication Data

Main entry under title:
Acid-base and potassium homeostasis.
  (Contemporary issues in nephrology; v. 2)
  Includes index.
  1. Acid-base imbalances. 2. Potassium deficiency diseases. 3. Homeostasis. I. Brenner, Barry M.
II. Stein, Jay H., 1937–   III. Series.
RC630.A25   616.3'9   78-8839
ISBN 0-443-08017-8

Manufactured in the United States of America

Preface

This volume of Contemporary Issues in Nephrology focuses on disorders of acid-base balance and potassium homeostasis. A pre-eminent group of investigators and clinicians have been asked to integrate basic physiologic concepts with pathogenesis and pathophysiology of disease.

The initial chapter by O'Connor and Kunau summarizes the current concepts of hydrogen and potassium handling by the kidney. The important evidence demonstrating that hydrogen ion secretion is the primary mediator of bicarbonate reabsorption is presented. The effects of extracellular volume, the carbon dioxide tension, calcium and phosphorous metabolism, and the role of potassium and mineralocorticoids in the regulation of bicarbonate transport are also considered. Similarly, the transport characteristics of potassium and the primary factors that affect renal potassium transport (tubular fluid flow rate, acid-base balance, mineralocorticoids, potassium intake, and sodium delivery) are described so as to provide a modern basis for understanding the alterations in renal potassium transport that occur in various clinical disorders.

Narins, in his chapter on the renal acidoses, briefly reviews the pertinent aspects of bicarbonate transport as it applies to both the acidosis of renal failure and to renal tubular acidosis. The importance of the anion gap in the differential diagnosis of the metabolic acidoses is emphasized. The author also clearly defines the pathophysiologic basis of uremic acidosis, renal tubular acidosis, and classifies the latter into its various clinical and pathogenetic subgroups.

Relman begins his chapter on lactic acidosis by exploring the various biochemical pathways that lead to lactate accumulation. He follows with elegant descriptions of the various clinical entities and their therapies.

Sebastian and his colleagues provide a scholarly exposition of the pathogenesis of metabolic alkalosis. From this overview, one can easily grasp the basis for the development of this acid-base abnormality in a wide variety of disorders. Cohen and Madias follow with a review of the various acid-base disorders of respiratory origin. The initial discussion on the physiologic regulation of carbon dioxide tension is of importance for the understanding of the basis of respiratory acidosis and respiratory alkalosis. The interrelationships among these primary acid-base disturbances of respiratory origin and their renal compensation are discussed in detail.

The final four chapters deal with various aspects of potassium homeostasis. Kliger and Hayslett categorize the disorders associated with hypo- and hyperkalemia. The pathophysiologic mechanisms involved, the clinical characteristics, and the means to differentiate among these disorders are nicely outlined. Cronin and Knochel present an excellent description of the consequences of potassium deficiency. This is an area which has not been widely emphasized in the past. Schambelan and associates then elucidate the various syndromes associated with either mineralocorticoid excess or deficiency. This group of investigators have had extensive clinical and experimental experience with these disorders and share their numerous insights with great clarity.

Finally, Bardgette and Stein discuss the recent flurry of activity in the evaluation of the pathophysiology of Bartter's Syndrome. This disorder can be utilized as a model for studying alterations of the renin-angiotensin system, as well as water and potassium balance. Recently, patients with this syndrome have also been shown to have markedly enhanced production of renal prostaglandins. The authors suggest that the basic defect in the disorder is renal potassium wasting and that other manifestations of the disease are the predictable consequences of this abnormality.

In summary, the material assembled in Volume 2 of this Series aims to provide a broad overview of various acid-base and potassium disturbances, with emphasis on pathogenesis and pathophysiology. We feel strongly that this approach will lead to a clear and practical understanding of these conditions and, therefore, sound therapeutic approaches.

May 1978   JHS
                                                                  BMB

Contributors

JOHN J. BARDGETTE, M.D.
Fellow, Division of Renal Diseases, University of Texas Health Science Center at San Antonio, San Antonio, Texas

JORDAN J. COHEN, M.D.
Professor of Medicine, Chief, Nephrology Division, Tufts-New England Medical Center, Boston, Massachusetts

ROBERT E. CRONIN, M.D.
Medical Service, Veteran's Administration Hospital, Assistant Professor of Medicine, University of Texas Southwestern Medical School, Dallas, Texas

JOHN P. HAYSLETT, M.D.
Associate Professor of Medicine and Pediatrics, Chief, Section of Nephrology, Yale University School of Medicine, New Haven, Connecticut

HENRY N. HULTER, M.D.
Assistant Clinical Professor of Medicine, University of California, Chief, Renal Service, U.S. Public Health Service Hospital, San Francisco, California

ALAN S. KLIGER, M.D.
Department of Internal Medicine, Renal Section, Yale University School of Medicine, New Haven, Connecticut

JAMES P. KNOCHEL, M.D.
Associate Chief of Staff for Research and Chief, Renal Section, Veteran's Administration Hospital, Professor of Internal Medicine, University of Texas Southwestern Medical School, Dallas, Texas

ROBERT KUNAU, M.D.
Associate Professor of Medicine, University of Texas Health Science Center at San Antonio, San Antonio, Texas

NICOLAOS E. MADIAS, M.D.
Assistant Professor of Medicine, Assistant Physician, Tufts-New England Medical Center, Boston, Massachusetts

ROBERT G. NARINS, M.D.
Associate Professor of Medicine, Acting Chief, Division of Nephrology, University of California, Los Angeles, Center for the Health Sciences, Los Angeles, California

GERALD O'CONNOR, M.D.
Renal Division, Department of Medicine, University of Texas Health Science Center at San Antonio, San Antonio, Texas

ARNOLD S. RELMAN, M.D.
Editor, New England Journal of Medicine, Professor of Medicine, Harvard Medical School, Boston, Massachusetts

FLOYD C. RECTOR, Jr. M.D.
Professor of Medicine, Director, Division of Nephrology, University of California, San Francisco, California

MORRIS SCHAMBELAN, M.D.
Associate Professor of Medicine, University of California, Assistant Director, Clinical Study Center, San Francisco General Hospital Medical Center, San Francisco, California

ANTHONY SEBASTIAN, M.D.
Associate Professor of Medicine, Assistant Director, General Clinical Research Center, University of California, San Francisco, California

JAY H. STEIN, M.D.
Professor of Medicine and Physiology, Chairman, Department of Medicine, University of Texas Health Science Center at San Antonio, San Antonio, Texas

Contents

1. *Renal transport of hydrogen and potassium* — 1
   Gerald O'Connor, M.D. and Robert T. Kunau, M.D.

2. *The renal acidoses* — 30
   Robert G. Narins, M.D.

3. *Lactic acidosis* — 65
   Arnold S. Relman, M.D.

4. *Metabolic alkalosis* — 101
   Anthony Sebastian, M.D., Henry N. Hulter, M.D., and
   Floyd C. Rector, Jr., M.D.

5. *Acid-base disorders of respiratory origin* — 137
   Jordan J. Cohen, M.D. and Nicolaos E. Madias, M.D.

6. *Disorders of potassium balance* — 168
   Alan S. Kliger, M.D. and John P. Hayslett, M.D.

7. *The consequences of potassium deficiency* — 205
   Robert E. Cronin, M.D. and James P. Knochel, M.D.

8. *Mineralocorticoid excess and deficiency syndromes* — 232
   Morris Schambelan, M.D., Anthony Sebastian, M.D., and
   Henry N. Hulter, M.D.

9. *Pathophysiology of Bartter's syndrome* — 269
   John J. Bardgette, M.D. and Jay H. Stein, M.D.

Index — 289

1

Renal transport of hydrogen and potassium

GERALD O'CONNOR
ROBERT T. KUNAU

Renal transport of hydrogen ions
Evidence for the presence of hydrogen ion secretion in the renal tubule
Quantitative role of hydrogen secretion in urinary acidification
Characteristics of the hydrogen ion secretory mechanism
    Rate
    Active versus passive secretion
    Alkali removal
    Relationship between hydrogen ion secretion and the transport of other ions
Factors influencing hydrogen ion secretion
    Extracellular fluid (ECF) volume
    Carbon dioxide tension

    Calcium, phosphorous, and parathyroid hormone
    Potassium and mineralocorticoids
Renal transport of potassium
The proximal convoluted tubule
The pars recta
The descending limb of Henle's loop
The ascending limb of Henle's loop
The distal convoluted tubule
The collecting tubule and duct
Factors affecting renal potassium transport
    Sodium
    Potassium intake
    Mineralocorticoids
    Acid-base balance
    Tubular fluid and urine flow rate

RENAL TRANSPORT OF HYDROGEN IONS

The ability of the kidney to regulate urinary acidification is essential to normal acid-base homeostasis. In particular, precise control of the plasma bicarbonate concentration is largely dependent upon the kidney. First, by reabsorbing all of the filtered bicarbonate below a certain threshold level, inappropriate loss of bicarbonate into the urine is avoided. Alternatively, if the plasma bicarbonate level is raised above a threshold level, reabsorption is incomplete, permitting the excretion of the excess bicarbonate into the urine. Second, the kidneys are responsible for regenerating, and returning to the extracellular fluid, that bicarbonate decomposed by reaction with nonvolatile acids (e.g., H_2SO_4, HCl, H_3PO_4) and with organic acids in the extracellular fluid. The reaction of these acids with bicarbonate results in the formation of a sodium salt. As these salts are presented to the kidney in the glomerular filtrate, the anion is excreted with hydrogen whereas the sodium,

2 ACID BASE AND POTASSIUM HOMEOSTASIS

together with newly formed bicarbonate, is returned to the extracellular fluid. Inasmuch as little free hydrogen is present in the final urine, urinary hydrogen is excreted either as titratable acid or bound to NH_3 and excreted as NH_4^+. The quantity of hydrogen excreted as titratable acid and NH_4^+ is equivalent to the amount of bicarbonate regenerated by the renal tubules.

At this point, it seems appropriate to review several generally held concepts regarding bicarbonate reabsorption and acid excretion and then to discuss certain aspects of this subject in greater detail. Hydrogen ions may be produced within the cell by the hydration of CO_2, forming carbonic acid (H_2CO_3) which disassociates into hydrogen and bicarbonate (Fig. 1.1.a). Alternatively, a redox mechanism may split water into hydrogen and

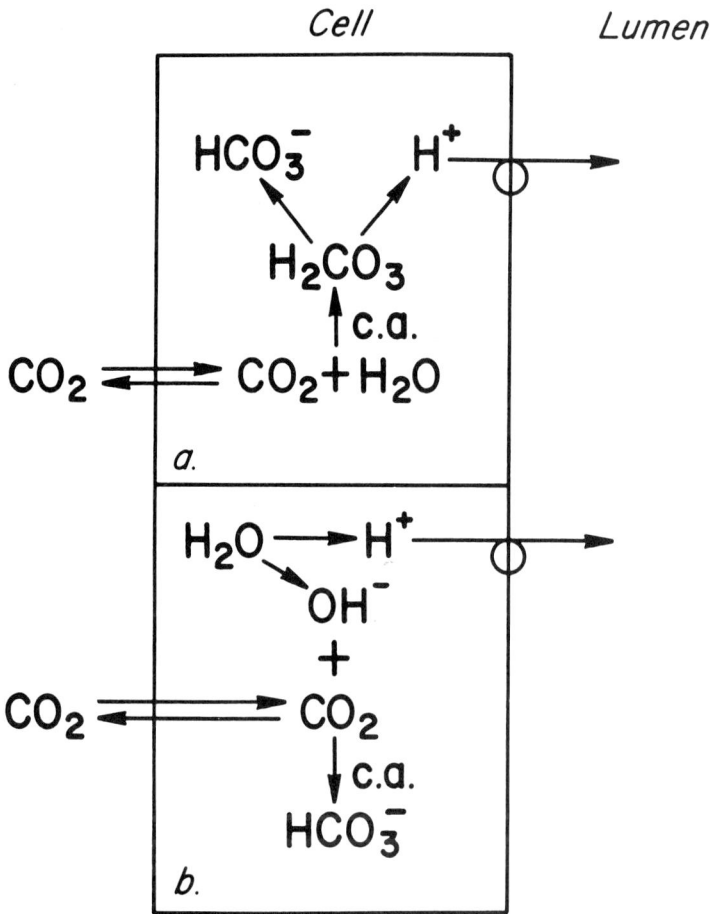

Figure 1.1 Intracellular hydrogen ion generation through hydration of carbon dioxide (a) or redox mechanism involving splitting of water molecules (b).

RENAL TRANSPORT 3

hydroxyl (OH⁻) ions with the resultant OH⁻ being neutralized by CO_2 (Fig. 1.1.b). In any case, the hydrogen ions provided are actively transported across the luminal membrane where they react with filtered bicarbonate to form H_2CO_3, which dehydrates to H_2O and CO_2 (Fig. 1.2.a). The hydrogen ions may also react with certain salts (e.g., A⁻) such as phosphate, to form the acid salt (HA) or with NH_3 to form NH_4^+ (Fig. 1.2.b). In the

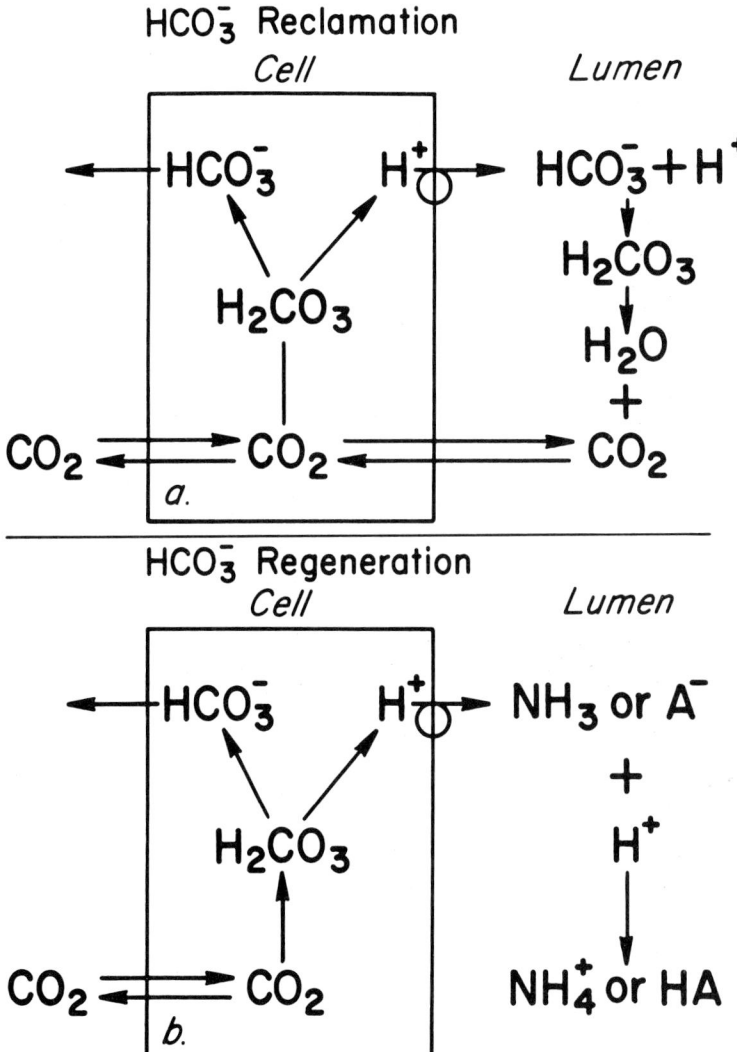

Figure 1.2 Reclamation of filtered bicarbonate (a) and regeneration of bicarbonate (b). Quantitatively, hydrogen ions utilized for reclamation far exceed those utilized for bicarbonate regeneration.

latter instances, new bicarbonate is generated and returned to the extracellular fluid whereas in the first circumstance, filtered bicarbonate is reabsorbed. On a quantitative basis, the hydrogen ions that are utilized for bicarbonate reabsorption greatly exceed the amount that serve to regenerate new bicarbonate. As indicated by Rector, the relative distribution of the reaction of hydrogen ions between filtered bicarbonate and nonbicarbonate buffers depends upon the pH of the tubular urine. As bicarbonate reabsorption nears completion, the luminal bicarbonate concentration decreases, and an increasingly larger fraction of the secreted hydrogen reacts with nonbicarbonate buffers to form titratable acid and NH_4^+ (Rector, 1973).

Carbonic anhydrase (c.a.) may serve several functions in this model. The hydration of CO_2 and dehydration of H_2CO_3 are steps of the following reaction which are catalyzed by carbonic anhydrase ($CO_2 + H_2O \underset{c.a.}{\rightleftarrows}$ $H_2CO_3 \rightleftharpoons H^+ + HCO_3^-$)* (Maren, 1967). As shown in Figure 1.1, the catalytic effect of carbonic anhydrase within the cell would be to facilitate a supply of hydrogen ions to a secretory pump. In addition, in certain portions of the nephron, the dehydration of luminal H_2CO_3 is catalyzed by carbonic anhydrase present at this site (Rector, Carter and Seldin, 1965; Vieira and Malnic, 1968). This prevents the intraluminal H_2CO_3 concentration from increasing above its equilibrium state and thereby lowers the pH gradient against which hydrogen ions are secreted. The rate constants of the uncatalyzed reaction are also of physiologic importance as a determinant of the reaction rate after inhibition of carbonic anhydrase. At 37°C, the ratio of the dehydration constant ($H_2CO_3 \xrightarrow{K_{-1}} CO_2 + H_2O$) to the hydration constant ($H_2O + CO_2 \xrightarrow{K_1} H_2CO_3$) is roughly 300:1, indicating that under equilibrium conditions the concentration of CO_2 is about 300 times higher than that of H_2CO_3 (Garg and Maren, 1972).

Evidence for the presence of hydrogen ion secretion in the renal tubule

Although it is implied above that tubular secretion of hydrogen ions is responsible for bicarbonate reabsorption and urinary acidification, direct evidence in favor of the existence of this mechanism and of its quantitative importance remain areas of intensive investigation and controversy.

Pitts and Alexander (1945) noted that the quantity of titratable acid present in the urine of an acidotic dog infused with sodium phosphate could not be accounted for by the reabsorption of the alkaline fraction of filtered buffers from tubular urine. Thus, it was suggested that hydrogen ions were being added to the urine. Net addition of acid might occur if hydrogen ions

---

* Current evidence supports the view that the reaction $OH^- + CO_2 \rightleftarrows HCO_3^-$ is catalyzed by carbonic anhydrase. However, as the end products are the same and hydroxyl ions may be negligible at a physiologic pH, the reaction may be viewed as indicated (Maren, 1967).

are transported directly from cell to lumen (as in Fig. 1.1) or if bicarbonate is absorbed after its intraluminal production from CO_2 and H_2O, the hydrogen ion so generated being bound to intraluminal buffers (Fig. 1.3).

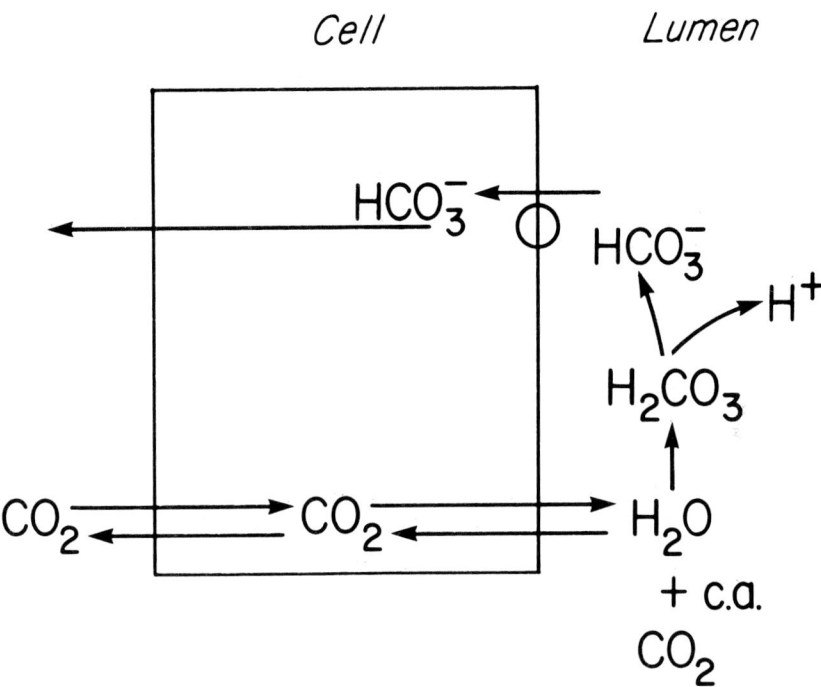

Figure 1.3 Schematic portrayal of ionic bicarbonate reabsorption. See text for details.

It is apparent that the equilibrium reaction ($CO_2 + H_2O \rightleftarrows H_2CO_3 \rightleftarrows H^+ + HCO_3^-$) will be shifted in opposite directions by bicarbonate reabsorption from the lumen or hydrogen ion secretion into it. Secretion of hydrogen ions into bicarbonate containing tubular fluid will generate H_2CO_3 at a rate equal to the reabsorptive rate of bicarbonate. However, as the uncatalyzed rate of H_2CO_3 dehydration is insufficient to account for the quantity of bicarbonate reabsorbed if equilibrium between H_2CO_3 and plasma pCO_2 is maintained, the intraluminal concentration of H_2CO_3 will increase until the rate of dehydration equals its rate of formation (Walser and Mudge, 1960). Under these circumstances, the pH of the tubular fluid in situ will be more acid than the pH of the same fluid after its removal from the lumen and after equilibrium between the H_2CO_3 and plasma pCO_2 of the animal has occurred (i.e., an acid disequilibrium exists). Thus, the existence of an acid-disequilibrium pH indicates hydrogen ion secretion. On the other hand, under conditions in which the luminal bicarbonate concentration is low and nonbicarbonate buffers are present, bicarbonate reabsorption lowers the intraluminal concentration of H_2CO_3 (Fig. 1.3). The concentration of

H_2CO_3 decreases until the rate of CO_2 hydration generates H_2CO_3 at a rate equal to its consumption. Under these conditions, the pH measured in situ is higher than that measured following the removal of tubular fluid from the lumen and equilibration of the fluid with the plasma pCO_2 allowed to occur, that is, an alkaline disequilibrium is present (Rector, 1973).

Rector et al. (1965) were the first to compare directly the in-situ pH with the pH calculated from the tubular fluid bicarbonate concentration and the plasma pCO_2. In the normal rat undergoing a $NaHCO_3$ diuresis, no disequilibrium was present in the proximal convoluted tubule whereas a significant disequilibrium pH of -0.88 pH units was observed in the distal convoluted tubule. Following the intravenous administration of carbonic anhydrase, the distal disequilibrium was obliterated. On the other hand, the administration of an inhibitor of carbonic anhydrase resulted in a disequilibrium pH of -0.85 units in the proximal convolution. Although the magnitude of the disequilibrium pH in both the proximal and distal tubules was less, comparable results have been presented by Vieira and Malnic (1968). Demonstration of an acid disequilibrium pH under these circumstances is consistent with the presence of H_2CO_3 in the lumen and its production as a result of the interaction of secreted hydrogen ions and intraluminal bicarbonate.

In the studies described above, it was assumed that the pCO_2 of tubular fluid was in diffusion equilibrium across the tubular wall and that the tubular fluid pCO_2 was equal to the plasma pCO_2 of the animal. Malnic and Mello-Aires (1971) have suggested, however, that the proximal tubular wall may be a barrier to CO_2 diffusion and that CO_2 diffusion is adversely affected by inhibition of carbonic anhydrase. If this is the case, the acid disequilibrium pH may, in part, reflect an increase in tubular pCO_2 as well as an increase in luminal H_2CO_3 concentration. Consistent with this finding is the observation that the proximal tubular pCO_2 may exceed the systemic plasma pCO_2 by approximately 16 mmHg (Sohtell and Karlmark, 1976). On the other hand, DuBose and associates (1977) have recently reported that the pCO_2 in the normal rat renal cortex is considerably higher than the level in systemic arterial plasma but that the pCO_2 is at diffusion equilibrium throughout the renal cortex. The effect of these recent observations, if confirmed, on interpretations based upon the acid disequilibrium pH will require further study.

Further confirmatory evidence in favor of a hydrogen secretory process being involved in urinary acidification is derived from in-vitro studies in the urinary bladder of the turtle, *Pseudemys scripta* (Schwartz et al., 1974). Two observations based upon these studies appear to be particularly pertinent. First, when examined under appropriate circumstances, the quantity of CO_2 produced and available at the luminal membrane is insufficient to account for the rate of urinary acidification if bicarbonate reabsorption is the primary event leading to acidification. Secondly, when examined in the presence of exogenous CO_2 and bicarbonate, pCO_2 measurements with a pCO_2 electrode demonstrated the pCO_2 at the mucosal border to be 9.4 mmHg higher than the serosal pCO_2. This increase in pCO_2 is larger than can be attributed to metabolic CO_2 production and must have resulted from the interaction of secreted hydrogen ion and bicarbonate at the mucosal border.

Quantitative role of hydrogen secretion in urinary acidification

The above data suggest that hydrogen ion secretion plays a role in urinary acidification and bicarbonate reabsorption. The magnitude of the contribution of hydrogen ion secretion, however, is unclear. As mentioned, carbonic anhydrase (c.a.) may serve several functions in the acidification process, a major component being its effect on facilitating hydrogen ion delivery to a luminal secretory site. In the absence of carbonic anhydrase, the hydrogen ion secretory mechanism continues to function because of the uncatalyzed reaction, albeit at a slower rate (Maren, 1967). It is not surprising, therefore, that the administration of a carbonic anhydrase inhibitor does not completely abolish bicarbonate reabsorption. In fact, approximately 50 percent of the filtered load of bicarbonate may be reabsorbed by the kidney after inhibition of carbonic anhydrase. Maren (1967) and Rector (1973) have analyzed the possible contribution of the uncatalyzed reaction to the observed rate of bicarbonate reabsorption in the dog kidney and proximal tubule of the rat, respectively. Their calculations indicate that the contribution of the uncatalyzed reaction is insufficient to account for the quantity of bicarbonate reabsorbed after carbonic anhydrase inhibition. Thus, the hydrogen ion provided by both the catalyzed and uncatalyzed reactions can only account for a portion of renal bicarbonate reabsorption and urinary acidification.

With the use of stationary microperfusion techniques, Malnic and Steinmetz (1976) have used kinetic data to subdivide proximal and distal convoluted tubule acidification into its various components. The results shown in Table 1.1 are compiled from their studies performed in normal

Table 1.1 Components of renal tubular acidification in cortical tubules of rat kidney as calculated from stopped-flow microperfusion data[a,b]

Fraction	Convoluted tubules in control rats	
	Proximal	Distal
1. Total HCO_3^- reabs.	13.3 (100%)	14.8 (100%)
2. After Diamox	7.9 (59.3)	2.94 (20.0)
3. V_{unc} (D + P)	1.75 (13.1)	1.10 (7.5)
4. V_{cat} (1 − 2)	5.4 (40.6)	11.8 (80.0)
5. Remainder (2 − 3)	6.15 (46.2)	1.84 (12.5)
6. Total H (3 + 4)	7.15 (53.8)	12.9 (87.5)

[a] Ionic flows in mmoles/liter · sec. Numbers in parentheses are percentages.
[b] 1 to 6 components of bicarbonate reabsorption V_{unc} (D + P), uncatalyzed rate of H ion secretion obtained during Diamox infusion, and microperfusion with phosphate. V_{cat}, catalyzed rate of H ion secretion. Remainder: uncatalyzed HCO_3^- reabsorption minus uncatalyzed H ion secretion. Total H: total H ion secretion by tubular epithelium.

animals. The rates are expressed in mmol/liter·sec. Fraction 1, the total rate of bicarbonate reabsorption, was determined from the acidification rates observed during perfusion of the tubule with $NaHCO_3$. Fraction 2 represents the rate of bicarbonate reabsorption observed after carbonic anhydrase inhibition; fraction 3, the rate of phosphate acidification in the absence of bicarbonate in the perfusion fluid after inhibition of carbonic anhydrase, the latter being an estimate of the uncatalyzed rate of hydrogen ion secretion. The other fractions (4, 5, 6) are the sum of, or the difference between, observed values. In both the proximal and distal convoluted tubules, the fraction of urinary acidification catalyzed by carbonic anhydrase, fraction 4, only accounts for 40 and 80 percent of bicarbonate absorbed, respectively. The uncatalyzed component, fraction 3, may account for only 13.1 and 7.5 percent, respectively. A significant portion, fraction 5, remains and cannot be accounted for either by catalyzed or by uncatalyzed hydrogen ion generation. Fraction 5 represents 46 percent of the bicarbonate absorption (and hydrogen ion secretion) in the proximal and 12.5 percent in the distal tubule. Fraction 6 represents the quantity of bicarbonate absorption which can be attributed to both the catalyzed and uncatalyzed fractions (i.e., 3 and 4). Maren (1967) has suggested that the component represented by fraction 5 is transported as ionic bicarbonate. On the other hand, Rector (1973) has proposed another means to account for this component. He suggests that following carbonic anhydrase inhibition, the luminal concentration of H_2CO_3 would increase and result in a concentration gradient of the acid from the tubular lumen to the cell. As a result, this H_2CO_3 would "recycle" from the lumen into the cell and could provide hydrogen for the secretory process or neutralize intracellular OH^-. In any case, this proposal suggests a means by which hydrogen ion secretion could continue without the necessity of invoking a second mechanism to account for urinary acidification. This concept is schematically presented in Figure 1.4.

Characteristics of the hydrogen ion secretory mechanism

Rate

Historically, an evaluation of the maximum rates of tubular hydrogen ion secretion has been made by the infusion of $NaHCO_3$ and by the subsequent determination of tubular bicarbonate reabsorption. It is now recognized, however, that such estimates of the tubular maximum (Tm) of bicarbonate may not measure the maximum rates of bicarbonate reabsorption. In particular, the reabsorptive rates of bicarbonate are affected by changes in extracellular fluid volume, a factor that was not considered in earlier studies (Purkerson, Lubovitz and White, 1969; Kurtzman, 1970b).

Malnic, De Mello-Aires and Giebisch (1971) have utilized the stationary microperfusion technique to measure rates of bicarbonate reabsorption (hydrogen ion secretion). Droplets of alkaline buffers (e.g., 100 mm $NaHCO_3$), equilibrated together with raffinose at a physiologic pCO_2 to

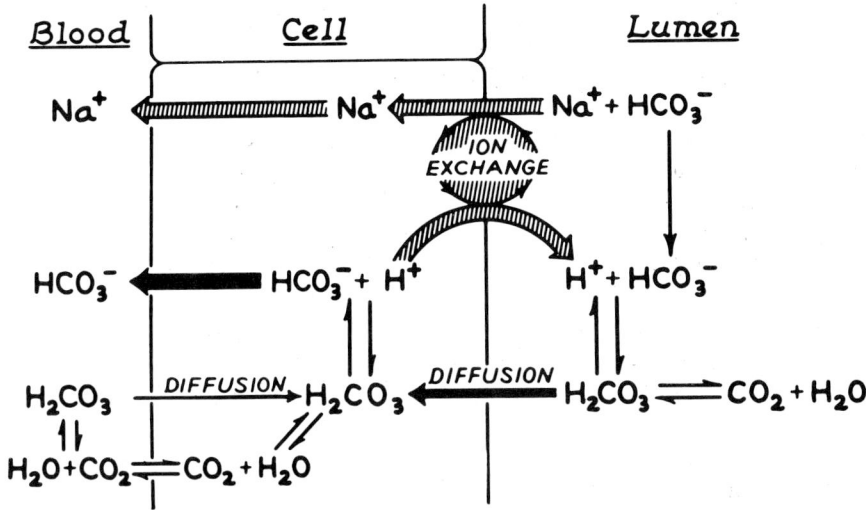

Figure 1.4 Proposed model for carbonic acid recycling. Following carbonic anhydrase inhibition, luminal concentration of carbonic acid increases, creating a concentration gradient and allowing recycling of carbonic acid, lumen to cell. This "recycled" carbonic acid, then, provides hydrogen ion for the secretory process. (From Rector, 1973, with permission of the publisher)

retard volume reabsorption, were placed in tubular segments and the change in pH/unit time recorded with a micro pH electrode. From the measured pH, the Henderson-Hasselbalch equation and, assuming that tubular and plasma pCO_2 are in equilibrium, the tubular bicarbonate concentration can be calculated and the initial to steady state value plotted semilogarithmically against time. The fall in tubular bicarbonate under these circumstances is described by a single exponential, Figure 1.5. When measured by this technique, acidification rates are roughly similar in the proximal and distal tubules (Malnic and Giebisch, 1972). These studies also indicate that the rate of tubular acidification is proportional to the chemical gradient (i.e., the concentration of the alkaline buffer in the lumen) and that the tubular hydrogen ion secretory mechanism is normally highly unsaturated. The proportionality between hydrogen ion secretion and luminal buffer concentration suggests either that the hydrogen secretory pump varies its rate in response to the pH gradient against which hydrogen ions are pumped or that the system corresponds to a pump-leak system in which a constant pump is opposed by a variable leak of hydrogen ions from the lumen, the latter being dependent upon the luminal concentration of hydrogen ions (Malnic, De Mello-Aires and Giebisch, 1972).

Active versus passive secretion

The direction and/or magnitude of the potential differences across various nephron segments are insufficient to account for the pH gradients observed

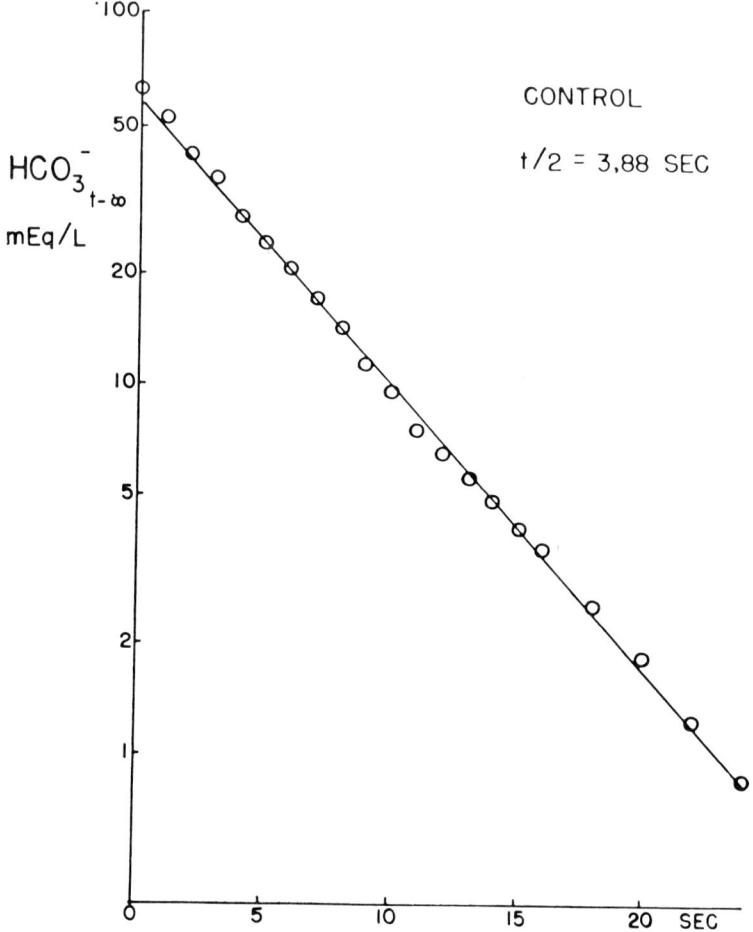

Figure 1.5 Change in luminal bicarbonate concentration with time (Malnic et al., 1971, with permission of the publisher). See text for details.

(Barratt et al., 1974; Schaefer, Troutman and Andreoli, 1974; Vieira and Malnic, 1968; Rau and Frömter, 1974). This suggests that hydrogen ion secretion throughout the nephron occurs via an active process.

The hydrogen ion secretory process in certain segments of the nephron may not only occur by an active means but may also be electrogenic. Frömter and Gessner (1974) observed that perfusion of both the peritubular capillaries and lumen of the early proximal tubule (<1 mm from the glomerulus) with glucose-free solutions containing both chloride and bicarbonate generated an active transport potential of 1 mV, lumen positive. In contrast, similar studies in the middle and late proximal convoluted tubules in which $NaHCO_3$ Ringer's, sodium-free choline HCO_3 Ringer's, or bicarbonate-free NaCl Ringer's was used indicated that these segments do not generate a significant transport potential (Frömter, Sato and Gessner, 1976). The difference in findings between the early and late segments of the proximal convoluted tubule was attributed to the presence of an electroneutral sodium-hydrogen exchange in the distal parts of the proximal tubules, whereas the positive potential differences in the early segments were thought to be due to the presence of a hydrogen ion secretory system, which increased bicarbonate over active sodium transport. In the cortical collecting tubule, the lumen may become electropositive when sodium transport is inhibited (Stoner, Burg and Orloff, 1974). Moreover, this positive potential difference was obliterated by acetazolamide administration. Thus, it appears that hydrogen ion secretion is not only an active process but may be electrogenic as well.

Recently, McKinney and Burg (1977b) have demonstrated that isolated segments of the rabbit cortical collecting tubule may demonstrate bicarbonate secretion as well as absorption, depending upon previous dietary treatment. Whereas cortical collecting tubules from untreated rabbits demonstrated no net bicarbonate transport, following pretreatment of the rabbit with NH_4CL or $NaHCO_3$, net bicarbonate absorption and secretion, respectively, were observed. Thus, under conditions in which a bicarbonate load must be excreted, a bicarbonate secretory process in the cortical collecting tubule may contribute to the urinary bicarbonate.

Alkali removal

Essential to the process of urinary acidification is the means by which alkali is transferred from the cell to the peritubular blood. The electrophysiologic studies of Frömter and associates (1976) indicate that the peritubular cell membrane is very permeable to bicarbonate and suggest that bicarbonate can move freely and passively across this membrane, possibly in the form of OH^-. Of interest is that acetazolamide decreases the permeability of this membrane for bicarbonate exit (Garcia Filho and Malnic, 1976).

Relationship between hydrogen ion secretion and the transport of other ions

In order to maintain electrical neutrality, the secretion of hydrogen ions must be related in some manner to the transport of another ion. There is some uncertainty, however, about the character of this linkage. When $NaHCO_3$ is replaced by choline HCO_3 in the tubular lumen, acidification in the proximal convoluted tubule proceeds normally (Malnic et al., 1971). Similarly, when the concentration of sodium in both the lumen and peritubular capillaries are lowered below the concentration in the cell by replacement with choline, acidification remains close to normal levels whereas sodium reabsorption is almost abolished (Giebisch and Malnic, 1976). If sodium removal from the cell is inhibited at the level of the peritubular membrane exchange pump, sodium reabsorption is dramatically reduced whereas the rate of acidification is minimally affected (Giebisch and Malnic, 1976). These findings in the proximal convoluted tubules are in agreement with the observations in a number of in-vitro studies of the urinary bladder of the turtle (Steinmetz, Omachi and Frazier, 1967) and of the cortical collecting tubule (Stoner et al., 1974). These studies suggest a loose or indirect coupling of hydrogen ion secretion to the reabsorption of sodium.

Yet, there is some evidence to suggest a more direct coupling between sodium transport and hydrogen secretion. Thus, the addition of sodium to the bathing medium results in proton movement from vesicles previously obtained from rabbit proximal tubular brush border into the media (Murer, Hopfer and Kinne, 1976). In studies in the isolated proximal straight tubule of the rabbit, McKinney and Burg (1977a) observed that removal of sodium from both the bath and the perfusate markedly decreases, but does not abolish, bicarbonate absorption. Similar findings were noted when a potassium-free bath, or ouabain, was used. They concluded that the absorption of bicarbonate in this nephron segment is strongly dependent upon sodium, but not completely so.

Another possibility is for hydrogen ion secretion to be coupled to the transport of some anion. A number of years ago, Kashgarian, Warren and Levitin (1965) observed that the chloride concentration measured in raffinose droplets previously placed into the proximal tubular lumen is significantly greater than plasma in the presence of hypercapnia but equal to plasma in normal rats. Conceivably, the higher chloride concentration in the former group can be related to the hypercapnia-induced stimulation of hydrogen and chloride secretion. Giebisch and Malnic (1976) reported that when sodium reabsorption is inhibited by peritubular and luminal perfusion with solutions with a low sodium concentration, the fall in luminal bicarbonate is associated with a concomitant increase in chloride concentration. Giebisch and Malnic summarized the possible relationship between hydrogen ion secretion and the transport of other ions by concluding that the coupling of hydrogen ion secretion and the transport of sodium and chloride may be indirect, flexible, nonspecific, and a consequence of the magnitude of the electrogenic potentials developed by hydrogen ion secretion and of the conductance of the transport pathways involved.

Factors influencing hydrogen ion secretion

Extracellular fluid (ECF) volume

The state of extracellular fluid volume expansion may have a profound influence on renal bicarbonate transport. Purkerson and associates (1969) observed that when the plasma bicarbonate concentration in the rat is increased by hypertonic $NaHCO_3$ in a manner that minimizes expansion of the ECF volume, tubular bicarbonate reabsorption rises with little evidence of a maximum reabsorptive rate. At a plasma bicarbonate concentration of 57.5 mEq/l, bicarbonate reabsorption is 57.1 mEq/l GFR. In contrast, when the plasma bicarbonate concentration is similarly increased after expansion of the ECF volume with isotonic NaCl, a tubular maximum bicarbonate reabsorption between 40–45 mEq/l GFR was observed. Kurtzman (1970b) has reported experiments of a comparable nature in the dog, studies that indicate tubular bicarbonate reabsorption in the dog is even more sensitive to variations in ECF volume than in the rat.

Micropuncture studies in the rat have shown that isotonic expansion of the ECF volume decreases the reabsorption of $NaHCO_3$ as well as of NaCl in the proximal convoluted tubule (Kunau et al., 1966; Levine et al., 1976). At least part of the influence of ECF volume on renal bicarbonate transport, therefore, appears to be mediated at this site.

Carbon dioxide tension

Even when allowances are made for changes in the effective ECF volume, an increase in plasma pCO_2 has been shown to enhance renal bicarbonate reabsorption in the bicarbonate-loaded dog (Kurtzman, 1970b).

Although hyperventilation reduces bicarbonate reabsorption in the proximal convoluted tubule in the absence of bicarbonate loading, it is difficult to demonstrate an effect of hypercapnia on bicarbonate reabsorption by free-flow micropuncture studies in the nonbicarbonate-loaded animal (Malnic et al., 1972). However, following elevation of the plasma bicarbonate concentration by $NaHCO_3$ infusion, reabsorption in both sites was shown to vary directly with the plasma pCO_2. These observations can perhaps best be explained by the interrelationship of the effects of hydrogen ion secretion and intratubular bicarbonate concentration on bicarbonate reabsorption. In the studies of Malnic and coworkers, a small but significant acid-disequilibrium pH was observed in the proximal convoluted tubule following acute hypercapnia in both the normal and bicarbonate-loaded rat whereas the magnitude of the disequilibrium pH in the distal convoluted tubule was diminished. The proximal disequilibrium pH was felt to develop as a result of a simultaneous increase in the tubular bicarbonate concentration and hydrogen ion secretion. Under these conditions, the activity of luminal carbonic anhydrase is insufficient to catalyze the increased quantity of H_2CO_3 formed. In the animal not loaded with bicarbonate, the development of an acid disequilibrium in the proximal tubule will produce a less favorable gradient for hydrogen ion secretion and thereby prevent an increase in bicarbonate reabsorption. In contrast, in the bicarbonate-loaded state, the higher intraluminal bicarbonate concentration will permit a proximal effect to be observed despite the development of an acid disequilibrium.

The decrease of the acid disequilibrium pH in the distal tubule of the hypercapnic nonbicarbonate-loaded rat, on the other hand, was related to the lower intratubular bicarbonate concentration present. This bicarbonate concentration (approximately 5.75 mEq/l) could limit effective hydrogen secretion as well as excess H_2CO_3 formation. Consequently, an increase in bicarbonate reabsorption would not be observed under these conditions.

After bicarbonate loading, adequate bicarbonate was present in the distal tubule to permit an influence of hypercapnia on bicarbonate reabsorption to be demonstrated.

Similar reasoning was set forth for the reduction in the distal tubular disequilibrium pH observed during hypocapnia. In this situation, although the intraluminal bicarbonate concentration is elevated, the decreased hydrogen ion secretion that results from the lowered pCO_2 serves to diminish intraluminal H_2CO_3 formation. These micropuncture studies indicate that both hypercapnia and hypocapnia may affect hydrogen secretion in the proximal and the distal convoluted tubules. Whether hypercapnia affects bicarbonate reabsorption depends not only on the increase in hydrogen ion secretion but on the intraluminal pH and bicarbonate concentration. Under physiologic circumstances, it is possible that renal bicarbonate reabsorption is not significantly enhanced by an elevation in the plasma pCO_2 tension until such time that the intraluminal bicarbonate concentration is spontaneously increased.

Recent microperfusion experiments by De Mello-Aires and Malnic (1975) have indicated that changes in the peritubular environment may have a direct influence on net acidification rates, observations particularly relevant to the effect of pCO_2. In these experiments, acidification of an intraluminal buffer solution was examined while the peritubular capillary network was controlled by perfusion with various solutions. Perfusion with a solution with a high pCO_2, at a constant pH, enhanced acidification in the proximal but reduced it in the distal convoluted tubule. A perfusate with a low pCO_2 did not alter acidification when the pH was maintained at physiologic levels. These results suggest that hypercapnia may have a direct influence on hydrogen ion secretion whereas the effect of hypocapnia may be mediated through the associated increase in pH.

Calcium, phosphorous, and parathyroid hormone

A number of clinical observations have suggested that parathyroid hormone (PTH) may so inhibit renal hydrogen ion secretion that high endogenous levels may induce a metabolic acidosis. For the most part, however, it has been difficult to characterize clearly the effect of PTH on renal hydrogen ion secretion in the experimental animal. Crumb and associates (1974) observed that the intravenous administration of a purified parathyroid extract results in a modest decrease in renal bicarbonate reabsorption from 24.6 to 22.5 mM/l GFR in the normal dog and from 26.9 to 22.6 mM/l GFR in the thyroparathyroidectomized dog. Other researchers have also observed bicarbonate reabsorption to be higher in thyroparathyroidectomized than in normal dogs and to be reduced to approximately normal values by the administration of PTH or cyclic AMP (Karlinsky et al., 1974). Micropuncture studies, both in the dog (Puschett and Zurbach, 1976) and in the rat (Bank and Aynedjian, 1976) have indicated that PTH administration in the thyroparathyroidectomized animal results in an increase in bicarbonate delivery out of the proximal convoluted tubule. In neither study, however, was urinary bicarbonate excretion increased. While it appears that PTH can affect bicarbonate transport, its physiologic role, if any, remains to be defined.

Gold and associates (1973) induced phosphate depletion in the dog by feeding a low phosphate diet and an aluminum hydroxide gel. As phosphate depletion was generated, the plasma bicarbonate decreased from 24.5 to 21 mEq/l. Bicarbonate titration studies revealed both a decrease in the bicarbonate threshold and in the bicarbonate Tm. As hypo- rather than hyperparathyroidism was likely present in these studies, high endogenous levels of parathyroid hormone can not be responsible for the acidification defect observed. These studies suggest, therefore, that phosphorus depletion per se can impair renal hydrogen ion secretion. Nevertheless, as phosphorus depletion may be associated with hyperparathyroidism, an interrelationship between the two is possible. In this regard, the bicarbonaturia induced in the patient with hereditary fructose intolerance by PTH may be partially reversed by the administration of sodium phosphate (Morris, Sebastian and McSherry, 1972).

Extraparathyroid hypercalcemia can result in a modest increase in the serum bicarbonate concentration. This effect may not be related to changes in PTH secretion but possibly represents a more direct effect of hypercalcemia on renal hydrogen ion secretion (Crumb et al., 1974).

Potassium and mineralocorticoids

The metabolic alkalosis that has been assumed to be a common companion of potassium deficiency is in truth a complex and poorly understood disorder. In the rat, the experimental induction of potassium depletion is frequently accomplished with the use of mineralocorticoids or associated with the dietary provision of large quantities of a poorly absorbed anion while NaCl is restricted, thus making an interpretation of the singular effect of potassium deficiency on renal hydrogen ion secretion difficult. In this species, the alkalosis and hypokalemia that occur after mineralocorticoid administration and dietary potassium restriction is corrected by KCl administration to the nephrectomized animal (Orloff, Kennedy and Berliner, 1953). This suggests that the extracellular alkalosis of this type is largely the consequence of transcellular shifts. On the other hand, it seems that an increase in renal hydrogen ion secretion is essential to the maintenance of the alkalosis. If this increase were not essential, the alkalosis should be corrected. Nevertheless, it has not been convincingly demonstrated that the renal contribution to the maintenance of alkalosis is a selective effect of potassium depletion.

In the dog, simple potassium depletion causes a modest metabolic acidosis as a consequence of an impaired ability to acidify the urine maximally (Burnell, Teubner and Simpson, 1974). As a result, urinary NH_4^+, titratable acid, and net hydrogen ion excretion are diminished. The metabolic acidosis so generated results in an increase in NH_3 production and, given time, NH_4^+

excretion returns to normal. Of interest, however, is the finding that when the potassium deficiency induced reduction in mineralocorticoid production is overcome by exogenous aldosterone administration, an acidosis does not develop in the potassium-deficient dog (Hulter, Sigala and Sebastian, 1977). In addition, when infused with $NaHCO_3$, the potassium-deficient dog appears to reabsorb more bicarbonate than either the normal or acutely potassium-loaded dog at any level of ECF volume expansion, a finding which suggests that the capacity of the potassium-deficient dog to reabsorb bicarbonate at elevated plasma levels may be enhanced (Kurtzman, White and Rogers, 1973).

The difference in response between the dog and the rat to potassium deficiency has not been adequately defined. It is conceivable that the magnitude of the contribution of transcellular shifts in the genesis of the extracellular alkalosis which develops in the potassium deficient state, variations in mineralocorticoid activity, or the role of associated factors such as the renal chloride wastage may differ in the two species.

Seldin, Welt and Cort (1956) observed that deoxycorticosterone induces a metabolic alkalosis in the rat only when potassium deficiency is permitted to develop. Similar findings in the dog were earlier reported by Giebisch, Macloed and Pitts (1955). Hulter, Sigala and Sebastian (1977) have recently reported that desoxycorticosterone increases net acid excretion and the plasma bicarbonate concentration when the dogs studied are placed on a diet deficient in potassium, but not when ample dietary potassium is provided. These observations suggest that the ability of mineralocorticoids to enhance renal hydrogen ion secretion may depend upon a concomitant potassium deficiency.

Mineralocorticoid deficiency appears to be capable of inducing a metabolic acidosis even when allowances are made for the hyperkalemia that may develop because of the lack of these steroids (Hulter et al., 1977). Despite previous evidence to the contrary (Kurtzman, Martin and Rogers, 1971), mineralocorticoid deficiency does not impair the ability of the urinary epithelium to establish a maximum pH gradient (Hulter et al., 1977; Al-Awqati et al., 1976). Rather, the decrease in hydrogen ion secretion is likely the consequence of a decrease in secretory rate.

RENAL TRANSPORT OF POTASSIUM

It is generally appreciated that the kidney is capable of both potassium secretion and reabsorption. As the character of renal potassium transport depends upon the nephron segment involved, this section will review segmental potassium transport in the mammalian nephron.

The proximal convoluted tubule

Potassium has been demonstrated to be freely filtered at the glomerulus (LeGrimellec, Poujeol and Roffignac, 1975). The concentration of potassium in fluid obtained from the mammalian proximal convoluted tubule is the same as, or slightly lower than, the plasma concentration (Bennett, Clapp and Berliner, 1967; Malnic, Klose and Giebisch, 1964). Under hydropenic conditions, approximately 50 percent of the filtered potassium is reabsorbed along the portion of the proximal tubule accessible to micropuncture study, which includes the proximal two-thirds of this nephron segment. Fractional potassium reabsorption in the proximal convolution is, for the most part, dependent upon those factors which influence filtrate reabsorption along this nephron site. As it has recently been shown that under normal circumstances the lumen is electronegative only in the initial one-third of the proximal convoluted tubule, but is electropositive in the later two-thirds. (Barratt et al., 1974), it is conceivable that potassium reabsorption in the initial electronegative portion may be active, whereas the positive lumen potential in the later segments may facilitate reabsorption of potassium by passive means. The presence of active potassium reabsorption in the proximal convolution has also been inferred from studies performed in certain unique conditions, that is, following acetazolamide infusion (Beck, Senesky and Goldberg, 1973) and stationary microperfusion in the presence of a poorly absorbable solute (Malnic, Klose and Giebisch, 1966).

The pars recta

Grantham, Qualizza, and Irwin (1974) reported that net fluid reabsorption occurs along the length of the pars recta of the rabbit when studied under in vitro conditions and when the tubules are bathed in rabbit serum. However, when para-aminohippurate (PAH) is added to the bath, net secretion of potassium, sodium, and water occurs, presumably moving into the lumen through paracellular channels. Thus, the pars recta appears to be a site where either potassium reabsorption or secretion may occur.

The descending limb of Henle's loop

The potassium concentration at the bend of Henle's loop of juxtamedullary nephrons of the rat kidney exceeds the plasma concentration by several fold (Jamison et al., 1976; Jamison, 1970). In fact, the quantity of potassium present at this site may equal or actually exceed that filtered. Jamison and associates (1976) have suggested that potassium is excreted into the descending limb of Henle's loop, the quantity of potassium secreted depending on the concentration gradient of potassium from medullary interstitium to tubule lumen and on the flow rate of tubular fluid.

Rocha and Kokko (1973), however, have reported that the isolated descending limb of the loop of Henle of the rabbit is highly impermeable to potassium and thereby a barrier to the transfer of potassium from the medullary interstitium into the lumen. At present, it is unclear how to reconcile these findings with the suggestion of Jamison and associates (1976).

The ascending limb of Henle's loop

The quantity of potassium present at the early distal tubule usually is about 10 percent of that filtered (Giebisch, Klose and Malnic, 1967). Although the concentration of potassium at the bend of the loop of Henle is considerably higher than plasma, the potassium concentration at the early distal tubule is usually less than one-half that of plasma (Malnic, Klose and Giebisch, 1964; Malnic et al., 1966). Recent in-vitro studies have shown that the thick ascending limb of Henle's loop actively transports chloride (Burg and Green, 1973; Rocha and Kokko, 1973a). As a result, an electropositive potential is developed in the lumen and a portion of potassium as well as sodium reabsorption may occur by passive means.

The distal convoluted tubule

The distal convoluted tubule consists of several different anatomic and functional components (Woodhall and Tisher, 1973; Morel, Chabardes and Imbert, 1976). For an extensive discussion of these various components, the reader is referred to the recent review by Wright (1977). In the rat, both reabsorption and secretion of potassium have been observed to occur in this nephron segment (Malnic et al., 1964; Malnic et al., 1966). The potassium concentration usually increases progressively along this segment, from a value usually less than one-half the plasma concentration in the early to a level that exceeds the plasma concentration by severalfold in the late distal tubule (Malnic et al., 1964; Malnic et al., 1966).

The nature of potassium transport along the distal convoluted tubule and a determination of those factors that influence this transport have been areas of intensive study. The following model of distal tubular cellular potassium transport (Figure 1.6) has been proposed by Giebisch and coworkers (Giebisch et al., 1967; Giebisch, 1971).

The basic components of this model consist of: (1) a pump at the peritubular border which moves potassium into and sodium out of the cell; (2) asymmetrical electrical potentials across the peritubular and luminal membranes because of different permeability properties of these membranes; and (3) a potassium reabsorptive pump at the luminal border of the cell. In this model, potassium is assumed to be actively transported into the cell but to flow passively from the cell, across either the luminal or peritubular membrane, depending upon the magnitude of the electrical and chemical concentration gradients across these membranes and the permeabilities of these membranes to potassium. Potassium movement into the lumen of the distal tubule does not require an active mechanism but can be accounted for by passive forces. It also follows that the net transfer of potassium across the luminal border is controlled by the relative magnitude of those passive forces that move potassium from cell to lumen and of the quantity of potassium that is actively reabsorbed at the luminal border.

DISTAL TUBULE CELL

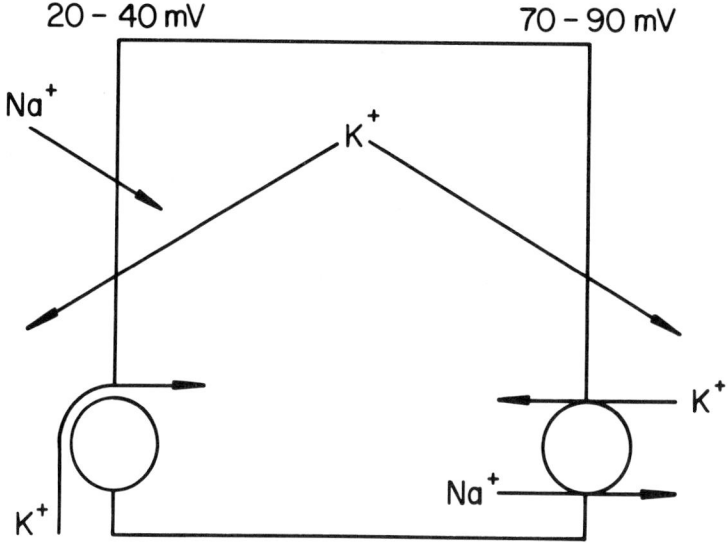

Figure 1.6 Schematic model of some properties of distal tubular cells. (Giebisch, 1971, with permission of the publisher)

It is possible that a number of aspects of this model may require modification. Wright (1977), for example, has recently proposed that potassium in the late distal tubule is secreted against an electrochemical gradient. In his studies, the transepithelial potential difference (PD) was recorded as tubular fluid was collected from a late distal segment. The observed tubular fluid to plasma potassium ratios exceeded those predicted on the basis of the transepithelial PD and were felt to indicate transport against an electrochemical gradient. Further, a number of studies have demonstrated situations in which the relationships between the transepithelial PD and the tubular fluid to plasma potassium ratio or changes in these parameters do not reflect a tight coupling between the transepithelial PD and tubular potassium (Malnic et al., 1966a and b; Wright et al., 1971).

The collecting tubule and duct

An assessment of the characteristics of potassium transport along the collecting system has frequently been made from a comparison of the quantity of potassium present at the superficial late distal tubule and in the final urine.

Direct study of the transport characteristics of the cortical collecting tubule can be performed only with the use of in-vitro perfusion techniques (Grantham, Burg and Orloff, 1970). When so examined against an electrochemical gradient, active potassium secretion has been demonstrated in tubular segments isolated from rabbits maintained on a normal diet. Removal of sodium from the lumen decreases the amount of potassium secreted. This effect of sodium was not felt to be related to a coupled exchange of the two ions but to an interdependence that is more indirect (Grantham, 1976).

Recent in-vivo studies by Reineck, Osgood and Stein (1977) suggest that the cortical collecting tubule may be a significant contributor to urinary potassium excretion under certain conditions. In these experiments, fractional potassium delivery rates to the base of the exposed papilla and to the superficial late distal tubule were compared. The authors demonstrated that under hydropenic conditions, no net transfer of potassium occurs between late distal and the papillary base whereas after Ringer's loading, net addition of potassium occurs. Pharmacologic maneuvers designed to reduce potassium delivery to the bend of juxtamedullary loops during Ringer's loading do not affect this relationship, a result that suggests the net addition between distal tubule and papillary base is not due to heterogeneity of loop function.

Both Jamison (1970) and Diezi et al. (1973) have directly examined potassium transport in the papillary portion of the collecting duct by in-vivo micropuncture techniques. In normal animals, both potassium reabsorption from and potassium secretion into the terminal ducts were observed. Although the character of the potassium secretory process has yet to be defined, the reabsorption observed presumably is an active process since it can proceed against an electrochemical gradient (Diezi et al., 1973).

Factors affecting renal potassium transport

Sodium

It is generally assumed that potassium is secreted into the distal nephron in exchange for reabsorbed sodium, that is, potassium secretion depends on the availability of sodium in the distal tubule for the potassium-sodium exchange to occur. The first micropuncture studies to examine directly the relationship between sodium reabsorption and potassium secretion in the distal convoluted tubule were those of Malnic and associates (1966a, 1966b). In rats maintained on a sodium-deficient diet that depressed both sodium and potassium excretion, they noted the quantity of potassium secreted in the distal tubule to be approximately one-tenth of the quantity of sodium reabsorbed. Clearly, the quantity of sodium available at this site was not so rate limiting for a one-to-one exchange with potassium to occur, an observation confirmed by others (Kunau, Webb and Borman, 1974). The depressed potassium excretion noted by Malnic and associates in these sodium-deficient rats was due to events beyond the superficial late distal tubule, possibly in the cortical collecting tubule.

The studies of Malnic and coworkers (1966a, 1966b), however, do not exclude the possibility that sodium may have a more indirect effect on potassium transport in the distal tubule (e.g., by influencing the transepithelial potential difference). In addition, it has recently been demonstrated that the infusion of sodium-containing solutions can increase the flow rate of distal tubular fluid (see below) and thereby influence potassium transport. Of these two factors, the influence of sodium on distal tubular flow rate appears to be more important than the effect of sodium on the transtubular PD in affecting distal potassium transport. (Khuri et al., 1975).

Potassium intake

The renal response to an increase in potassium intake is to accelerate the rate of urinary potassium excretion. This response is in part influenced by the previous level of potassium ingestion, that is, the kaliuretic response to an acute potassium load being greater in animals that have been fed a high potassium diet chronically (Wright et al., 1971). Chronic high potassium ingestion in the rat is associated with enhancement of potassium secretion along the superficial distal convoluted tubule (Kunau et al., 1974). In the papillary collecting duct of rats fed a high potassium diet, both potassium secretion and reabsorption may be seen (Diezi et al., 1973).

DeMello-Aires, Giebisch and Malnic (1973) have suggested that the accelerated distal tubular potassium secretion seen in the potassium-loaded state is due largely to an increase in the intracellular transport pool of potassium as a result of an increase in the peritubular uptake of potassium. In addition, a chronic increase in potassium intake can result in a more favorable electrical gradient for distal tubular potassium secretion (Wright et al., 1971). Although micropuncture studies suggest that changes in potassium transport in the superficial distal tubule are sufficient to account for the enhanced kaliuresis seen in the potassium-loaded animal (Wright et al., 1971), a supplementary role of the collecting system cannot be excluded. Grantham has proposed that the collecting system, presumably in large part the cortical collecting tubule, may serve, by increasing potassium secretion, to minimize the loss of potassium from the lumen, which might occur because of a more favorable gradient of potassium between the lumen and the cell (Grantham, 1976).

Recently, an increase in renal $Na^+K^+ATPase$ activity has been implicated as playing a role in the kaliuresis associated with chronic (Finkelstein and Hayslett, 1975; Silva, Hayslett and Epstein, 1973; Silva et al., 1975) but not acute (Katz and Lindheimer, 1975) potassium loading. Finkelstein and Hayslett noted that an increase in the activity of the enzyme in the outer medulla but not in the cortex or the inner medulla-papillary regions (Finkelstein and Hayslett, 1975) correlated with the ability of an intact kidney to excrete potassium after acute potassium administration. In contrast, Silva et al. (1975) reported that an increase in the activity of this enzyme in the inner medulla-papillary region was of importance in the ability of the isolated perfused rat kidney to demonstrate net potassium secretion. Neither study, however, permits a ready explanation for the observation that the superficial cortical distal tubule is the major adaptive site for the enhanced potassium excretion seen after either chronic or acute potassium loading (Wright et al., 1971).

In rats maintained on a potassium-deficient diet, potassium reabsorption, rather than secretion, can be observed along the distal convoluted tubule (Malnic et al., 1964). Reabsorption of potassium along the papillary collecting duct is also a consistent finding in the potassium-deficient rat, presumably related to active reabsorption at this site (Diezi et al., 1973).

Mineralocorticoids

The influence of mineralocorticoids on renal potassium transport takes place in the distal nephron; potassium transport in the proximal nephron appears to be unaffected (Wiederholt and Wiederholt, 1968). Adrenalectomized rats do not demonstrate a progressive increase in potassium concentration along the length of the distal tubule as do normal rats (Wiederholt and Wiederholt, 1968; Hierholzer et al., 1965). The altered distal tubular potassium transport seen in the mineralocorticoid-deficient state seems to be related primarily to two factors. First, aldosterone administration to adrenalectomized rats can increase and thereby normalize the permeability of the luminal membrane of the distal tubule to potassium (Wiederholt et al., 1973). This effect is not mediated through steroid-induced protein synthesis (Wiederholt et al., 1973). Second, chronic (>5 days), but not acute (<2 days), aldosterone administration to adrenalectomized rats results in a restoration of the distal tubular intracellular potassium content (Wiederholt, Agulian and Khuri, 1974). Obviously, both effects of aldosterone would have a favorable influence on distal tubular potassium secretion.

Although mineralocorticoids may increase renal Na^+K^+ stimulated ATPase activity, it is not certain to what extent this is an important aspect of their effect on potassium transport in the distal nephron.

Mineralocorticoids also appear to be capable of exerting an influence on potassium transport in the collecting system. Gross, Imai and Kokko (1975) observed that the potential difference measured in an isolated segment of rabbit cortical collecting tubule could vary depending upon the mineralocorticoid level. In tubules from rabbits fed a regular diet, the potential difference was $+3.7$ mV (lumen positive). After mineralocorticoid stimulation, the potential difference was greater than -30 mV, presumably as a result of an enhancement of electrogenic transport. This latter potential could favor either the entrapment of luminal potassium delivered from more proximal sites, or, if a passive element of potassium secretion were present, promote the movement of potassium from cell to lumen.

It is clear that the kaliuretic effect of mineralocorticoids can be disassociated from their antinatriuretic effect (deGraeff and Schuur, 1960; Swingle, et al., 1954; Williamson, 1963). Nevertheless, when animals are maintained on a diet free of sodium, the kaliuretic response to mineralocorticoid administration can be markedly diminished, if not abolished (Seldin, Welt and Cort, 1956). There appear to be several reasons for the ability of sodium-free diets to minimize the kaliuretic effect of mineralocorticoids. First, under certain circumstances sodium deficiency may enhance potassium removal from the lumen of the distal tubule (De Mello-Aires et al., 1973). Second, sodium-deficient diets may be associated with a reduction in the flow rate of tubular fluid. Third, as suggested by Grantham, low concentrations of sodium in the cortical collecting tubule in sodium-deficient states may impair potassium secretion at this site (Grantham, 1976).

Acid-base balance
A relationship between urine pH and potassium excretion has been frequently observed. For example, both the infusion of sodium bicarbonate and the administration of a carbonic anhydrase inhibitor render the urine alkaline and enhance potassium excretion. Respiratory and metabolic acidosis can have the opposite effect. Studies of this nature have suggested the possibility that potassium and hydrogen compete for a common secretory site (Berliner, 1959–1960).

Recently, Malnic et al. (1971) have presented data that have done much to clarify the effect of acute acid-base variations on renal potassium transport. Figure 1.7, derived from their studies, illustrates the quantity of potassium, which is expressed as a fraction of the filtered load, present along the distal convoluted tubule of the rat under a variety of acute acid-base disturbances. Acute alkalosis either of metabolic or respiratory origin results in a marked increase in potassium secretion along the distal convoluted tubule. In contrast, both respiratory and metabolic acidosis result in a diminution of potassium secretion along this segment. The ability of acute acid-base disturbances to alter potassium secretion in the distal convoluted tubule is largely related to variations in the chemical gradient of potassium between cell and lumen (De Mello-Aires et al., 1973). For example, acute metabolic alkalosis, which can be induced by sodium bicarbonate infusion, increases the concentration of potassium in the intracellular transport pool by accelerating the active uptake of potassium across the peritubular border. Acidosis presumably has the opposite effect.

In-vitro studies in the isolated cortical collecting tubule of the rabbit have shown that the intraluminal pH can have an influence on potassium transport at this site (Boudry, Stoner and Burg, 1976). As the luminal pH was decreased by lowering the pH of the perfusate from 7.4 to 6.8, potassium secretion decreased by almost 50 percent. The intraluminal pH directly influences potassium secretion at this site by altering an active component of potassium secretion (Boudry et al., 1976).

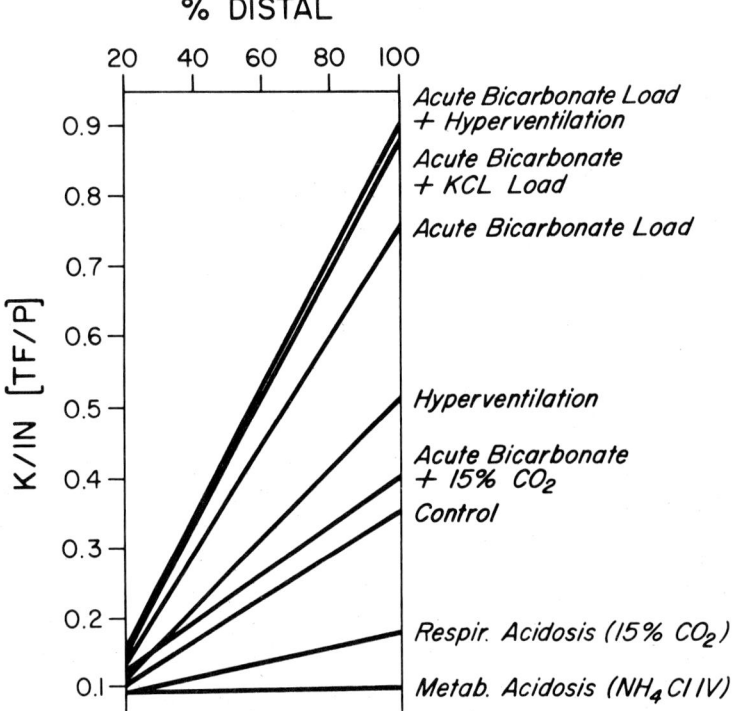

Figure 1.7 Addition of potassium along distal tubule in various acid-base disturbances. (Malnic et al., 1971, with permission of the publisher)

Most of the above studies have dealt with acute acid-base disturbances. Under more chronic conditions, the effects of acid-base disturbances on renal potassium transport are more complex and less well understood. Chronic respiratory (Carter, Seldin and Teng, 1959) and metabolic (De-Sousa et al., 1974) acidosis, for example, can be associated with a kaliuresis for reasons that are still unclear.

Tubular fluid and urine flow rate

Recently, it has been demonstrated that the flow rate of tubular fluid through the distal convoluted tubule can be an important modulator of potassium transport at this site (Khuri et al., 1975; Kunau et al., 1974; Reineck et al., 1975). The extent to which distal tubular potassium transport is affected by changes in the flow rate of tubular fluid depends upon the degree to which the net driving forces for cell-to-lumen potassium movement change as the flow rate varies. If the movement of potassium from cell to tubular lumen increases in proportion to an increase in flow rate, the transtubular potassium gradient* will remain constant, and the quantity of potassium present at the late distal tubule, expressed as a fraction of the filtered load, will vary directly with the flow rate of tubular fluid. Conversely, should the quantity of potassium that moves from cell to tubular lumen remain constant, the transtubular potassium gradient will decrease as the flow rate is enhanced, and the quantity of potassium at the late distal tubule will be unaffected by the increase in flow rate.

Following chronic potassium loading (Khuri et al., 1975; Kunau et al., 1974; Reineck et al., 1975) and the infusion of hypertonic solutions (Kunau et al., 1974), the late distal transtubular potassium gradient remains constant over a wide range of tubular fluid flow rates. As a result, potassium secretion in the distal tubule varies in parallel with changes in flow rate. On the other hand, in rats previously fed a low potassium diet (Khuri et al., 1975), the late distal transtubular potassium gradient decreases progressively as the flow rate is enhanced and potassium secretion along the length of the distal tubule does not increase as the flow rate is increased.

Reineck and associates (1975) have recently suggested that potassium transport in the collecting system may also be influenced by the urinary flow rate. When studied in a number of models in the rat kidney, net reabsorption occurred along the collecting system during low urinary flow rates whereas no net transport occurred at higher flow rates.

ACKNOWLEDGMENT

Portions of this work were supported by NIH Program Grant AM-17387, Grant HL-18875, and Training Grant AM-07103.

REFERENCES

Al-Awqati, Q., Norby, L. H., Mueller, A. & Steinmetz, P. R. (1976). Characteristics of stimulation of H^+ transport by aldosterone in turtle urinary bladder. *Journal of Clinical Investigation*, 58, 351–358.

Bank, N. & Aynedjian, H. S. (1976). A micropuncture study of the effect of parathyroid hormone on renal bicarbonate reabsorption. *Journal of Clinical Investigation*, 58, 336–344.

Barratt, L. J., Rector, F. C., Jr., Kokko, J. P. & Seldin, D. W. (1974). Factors governing the transepithelial potential difference across the proximal tubule of the rat kidney. *Journal of Clinical Investigation*, 53, 454–464.

Beck, L. H., Senesky, D. & Goldberg, M. (1973). Sodium independent active potassium reabsorption in proximal tubule of the dog. *Journal of Clinical Investigation*, 52, 2641–2645.

Bennett, C. M., Clapp, J. R. & Berliner, R. W. (1967). Micropuncture study of the proximal and distal tubule in the dog. *American Journal of Physiology*, 213, 1254–1262.

Berliner, R. W. (1959–1960). Renal mechanisms for potassium excretion. *Harvey Lecture Series*, 55, 141–171.

Boudry, J. F., Stoner, L. C. & Burg, M. B. (1976). Effect of acid lumen pH on potassium transport in renal cortical collecting tubules. *American Journal of Physiology*, 230, 239–244.

---

* The transtubular potassium gradient is expressed by a ratio of the concentration of potassium in the tubular fluid to that in plasma (TF/P K^+). Potassium delivery as a percent of the filtered load can be expressed as (TF/P K^+/TF/P Inulin) x 100%.

Burg, M. & Green, N. (1973). Function of thick ascending limb of Henle's loop. *American Journal of Physiology*, **224**, 659–668.

Burnell, J. M., Teubner, E. J. & Simpson, D. P. (1974). Metabolic acidosis accompanying potassium deprivation. *American Journal of Physiology*, **227**, 329–333.

Carter, N. W., Seldin, D. W. & Teng, H. C. (1959). Tissue and renal response to chronic respiratory acidosis. *Journal of Clinical Investigation*, **39**, 949–960.

Crumb, C. K., Martinez-Maldonando, M., Eknoyan, G. & Suki, W. N. (1974). Effects of volume expansion, purified parathyroid extract and calcium on renal bicarbonate absorption in the dog. *Journal of Clinical Investigation*, **54**, 1287–1294.

deGraeff, J. & Schuur, M. A. M. (1960). Severe potassium depletion caused by the abuse of laxatives. *Acta Medica Scandinavica*, **166**, 407–422.

De Mello-Aires, M. Giebisch, G. & Malnic, G. (1973). Kinetics of potassium transport across single data tubules of rat kidney. *Journal of Physiology* (London), **232**, 47–70.

De Mello-Aires, M. & Malnic, G. (1975). Peritubular pH and pCO_2 in renal tubular acidification. *American Journal of Physiology*, **228**, 1766–1774.

De Sousa, R. C., Harrington, J. T., Ricanati, E. S., Shelkrot, J. W. & Schwartz, W. B. (1974). Renal regulation of acid-base equilibrium during chronic administration of mineral acid. *Journal of Clinical Investigation*, **53**, 465–476.

Diezi, J., Michoud, P., Aceves, J. & Giebisch, G. (1973). Micropuncture study of electrolyte transport across papillary collecting duct of the rat. *American Journal of Physiology*, **224**, 623–634.

DuBose, T. D., Carter, N. W., Percacco, L. R. & Kokko, J. P. (1977). Determination of in-situ pCO_2 in the rat nephron. *Proceedings of the American Society of Nephrology*, **9**, 103A.

Finklestein, F. O. & Hayslett, J. P. (1975). Role of medullary Na-K-ATPase in renal potassium adaptation. *American Journal of Physiology*, **229**, 524–528.

Frömter, E. & K. Gessner, K. (1974). Active transport potentials, membrane diffusion potentials and streaming potential across rat kidney proximal tubular epithelium. *Pfluegers Archiv*, **351**, 85–98.

Frömter, E., Sato, K. & Gessner, K. (1976). Electrical studies on the mechanism of H^+/HCO_3^- transport across rat kidney proximal tubule. In *Proceedings of the 6th International Congress of Nephrology, Florence 1975*, pp. 108–112, Karger: Basel.

Garcia Filho, E. M. & Malnic, G. (1976). pH in cortical peritubular capillaries of rat kidney. *Pfluegers Archiv*, **363**, 211–217.

Garg, L. C. & Maren, T. H. (1972). The rates of hydration of carbon dioxide and dehydration of carbonic acid at 37°. *Biochim. Biophys. Acta*, **261**, 70–76.

Giebisch, G., Macloed, M. B. & Pitts, R. F. (1955). Effects of adrenal steroids on renal tubular reabsorption of bicarbonate. *American Journal of Physiology*, **183**, 377–386.

Giebisch, G., Klose, R. M. & Malnic, G. (1967). Renal tubular potassium transport. *Bulletin of the Swiss Academy of Medical Science*, **23**, 287.

Giebisch, G. (1970). Renal potassium excretion. In *The Kidney*, Rouiller, C. & Muller, A. F. (Editors). Vol. 3, p. 329. New York and London: Academic Press.

Giebisch, G. & Malnic, G. (1976). Studies on the mechanism of tubular acidification. *The Physiologist*, **19**, 511–524.

Gold, L. W., Massry, S. G., Arieff, A. I. & Coburn, J. W. (1973). Renal bicarbonate wasting during phosphate depletion. A possible cause of altered acid-base homeostasis in hyperparathyroidism. *Journal of Clinical Investigation*, **52**, 2556–2562.

Grantham, J. J., Burg, M. B. & Orloff, J. (1970). The nature of the transtubular Na and K transport in rabbit collecting tubules. *Journal of Clinical Investigation*, **49**, 1815–1826.

Grantham, J. J., Qualizza, P. B. & Irwin, R. L. (1974). Net fluid secretion in proximal straight tubules in vitro. *American Journal of Physiology*, **226**, 191–197.

Grantham, J. J. (1976). Renal transport and excretion of potassium. In *The Kidney*, Brenner, B. M. and Rector, F. C. (Editors). pp. 299–317. Philadelphia: Saunders.

Gross, J. B., Imai, M. & Kokko, J. P. (1975). A functional comparison of the cortical collecting tubule and distal convoluted tubule. *Journal of Clinical Investigation*, **55**, 1284–1294.

Haddy, F. J. & Scott, J. B. (1971). Biosassy and other evidence for participation of chemical factors in local regulation of blood flow. *Circulation Research* (*suppl. 1*), **29**, 186–192.

Hierholzer, K., Wiederholt, W., Holzgreve, H., Giebisch, G., Klose, R. M. & Windhager, E. E. (1965). Micropuncture study of renal transtubular concentration gradients of sodium and potassium in adrenalectomized rats. *Pfluegers Archiv*, **285**, 193–210.

Hulter, H. N., Ilnicki, L. P., Harbottle, J. A. & Sebastian, A. (1977a). Impaired renal H+ secretion and NH_3 production in mineralocorticoid-deficient glucocorticoid-replete dogs. *American Journal of Physiology*, **232**, F136–F146.

Hulter, H. N., Sigala, J. F. & Sebastian, A. (1977b). Effect of pre-existing dietary restriction on the renal action of mineralocorticoid hormones. *Clinical Research*, **25**, 436A.

Jamison, R. L. (1970). Micropuncture of superficial and juxtamedullay nephrons in the rat. *American Journal of Physiology*, **218**, 46–55.

Jamison, R. L., Lacy, F. B., Pennell, J. P. & Sanjana, V. M. (1976). Potassium secretion by the descending limb or pars recta of the juxtamedullary nephron in vivo. *Kidney International*, **9**, 323–332.

Karlinsky, M. L., Sager, D. S., Kurtzman, N. A. & Pillay, V. K. G. (1974). Effect of parathormone and cyclic adenosine monophosphate on renal bicarbonate reabsorption. *American Journal of Physiology*, **227**, 1226–1231.

Kashgarian, M., Warren, Y. and Levitin, H. (1965). Micropuncture study of proximal renal tubular chloride transport during hypercapnia in the rat. *American Journal of Physiology*, **209**, 655–658.

Katz, A. I. & Lindheimer, M. D. (1975). Relation of Na-K-ATPase to acute changes in renal tubular sodium and potassium transport. *Journal of General Physiology*, **66**, 209–222.

Khuri, R. N., Wiederholt, M., Strider, N. & Giebisch, G. (1975). Effects of flow rate and potassium intake on distal tubular potassium transfer. *American Journal of Physiology*, **228**, 1249–1261.

Kunau, R. T., Jr., Frick, A., Rector, F. C., Jr. & Seldin, D. W. (1966). Effect of extracellular fluid volume expansion, potassium deficiency and pCO_2 on bicarbonate reabsorption in the rat kidney. *Clinical Research*, **14**, 380.

Kunau, R. T., Jr., Webb, H. L. & Borman, S. C. (1974). Characteristics of the relationship between the flow rate of tubular fluid and potassium transport in the distal tubule of the rat. *Journal of Clinical Investigation*, **54**, 1488–1495.

Kurtzman, N. A. (1970a). Relationship of extracellular volume and CO_2 tension to renal bicarbonate reabsorption. *American Journal of Physiology*, **219**, 1299–1304.

Kurtzman, N. A. (1970b). Regulation of renal bicarbonate reabsorption by extracellular volume. *Journal of Clinical Investigation*, **49**, 586–595.

Kurtzman, N. A., Martin, G. W. & Rogers, P. W. (1971). Aldosterone deficiency and renal bicarbonate reabsorption. *Journal of Laboratory and Clinical Medicine*, **77**, 931–940.

Kurtzman, N. A., White, M. G. & Rogers, P. W. (1973). The effect of potassium and extracellular volume on renal bicarbonate reabsorption. *Metabolism, Clinical and Experimental*, **22**, 481–492.

LeGrimellec, C., Poujeol, P., and Rouffignac, C. de (1975) ^{3}H-inulin and electrolyte concentrations in Bowman's capsule in rat kidney. *Pfluegers Archiv*, **354**, 117–131.

Levine, D. Z., Nash, L. A., Chan, T. & Dubrovskis, A. H. E. (1976). Proximal bicarbonate reabsorption during ringer and albumin infusion in the rat. *Journal of Clinical Investigation*, **57**, 1490–1497.

Malnic, G., Klose, R. M. & Giebischz, G. (1964). Micropuncture study of renal potassium excretion in the rat. *American Journal of Physiology*, **206**, 674–686.

Malnic, G., Klose, R. M. & Giebisch, G. (1966a). Microperfusion study of distal tubular potassium and sodium transfer in rat kidney. *American Journal of Physiology*, **211**, 548–559.

Malnic, G., Klose, R. M. & Giebisch, G. (1966b). Micropuncture study of distal tubular potassium and sodium transport in rat nephron. *American Journal of Physiology*, **211**, 529–547.

Malnic, G. & De Mello-Aires, M. (1971). Kinetic study of bicarbonate reabsorption in the proximal tubule of the rat. *American Journal of Physiology*, **220**, 1759–1767.

Malnic, G., De Mello-Aires, M. & Giebisch, G. (1971). Potassium transport across renal distal tubular during acid-base disturbances. *American Journal of Physiology*, **221**, 1192–1208.

Malnic, G., De Mello-Aires, M. & Giebisch, G. (1972). Micropuncture study of renal tubular hydrogen ion transport during alterations of acid-base equilibrium in the rat. *American Journal of Physiology*, **222**, 147–158.

Malnic, G. & Giebisch, G. (1972). Mechanism of renal hydrogen ion secretion. *Kidney International*, **1**, 280–296.

Malnic, G. & Steinmetz, P. R. (1976). Transport processes in urinary acidification. *Kidney International*, **9**, 172–188.

Maren, T. A. (1967). Carbonic anhydrase: chemistry, physiology and inhibition. *Physiological Review*, **47**, 595–781.

McKinney, T. D. & Burg, M. B. (1977a). Bicarbonate and fluid reabsorption by renal proximal straight tubules. *Kidney International*, **12**, 1–8.

McKinney, T. D. & Burg, M. B. (1977b). Bicarbonate transport by rabbit cortical collecting tubules. Effect of acid and alkali loads in vivo on transport in vitro. *Journal of Clinical Investigation*, **60**, 766–768.

Morel, F., Chabardes, D. & Imbert, M. (1976). Functional segmentation of the rabbit distal tubular by microdetermination of hormone dependent adenylate cyclase activity. *Kidney International*, **9**, 264–277.

Morris, R. C., Jr., Sebastian, A., & McSherry, E. (1972). Renal acidosis. *Kidney International*, **1**, 322–340.

Murer, H., Hopfer, U. & Kinne, R. (1976). Sodium/proton antiport in brush-border membrane vesicles isolated from rat small intestine and kidney. *Biochemistry Journal*, **154**, 597–604.

Orloff, J., Kennedy, T. J., Jr., & Berliner, R. W. (1953). The effect of potassium in nephrectomized rats with hypokalemic alkalosis. *Journal of Clinical Investigation*, **32**, 538–542.

Pitts, R. F. & Alexander, R. S. (1945). The nature of the renal tubular mechanism for acidifying the urine. *American Journal of Physiology*, **144**, 239–254.

Purkerson, M. L., Lubovitz, H. & White, R. W. (1969). On the influence or extracellular fluid volume expansion on bicarbonate reabsorption in the rat. *Journal of Clinical Investigation*, **48**, 1754–1760.

Puschett, J. B. & Zurbach, P. (1976). Acute effects of parathyroid hormone on proximal bicarbonate transport in the dog. *Kidney International*, **9**, 501–510.

Rau, W. & Frömter, E. (1974). Electrical properties of medullary collecting ducts of the golden hamster. I. The transepithelial potential difference. *Pfluegers Archiv*, **351**, 99–111.

Rector, F. C., Jr., Carter, N. W. & Seldin, D. W. (1965). The mechanism of bicarbonate reabsorption in the proximal and distal tubules of the kidney. *Journal of Clinical Investigation*, **44**, 278–290.

Rector, F. C., Jr. (1973). Acidification of the urine. In *Handbook of Physiology*, Section 8, ed. Orloff, J. & Berliner, R. W. pp. 431–454. Washington, D.C.: American Physiology Society.

Reineck, H. J., Osgood, R. W. Ferris, T. F. & Stein, J. H. (1975). Potassium transport in the distal tubule and collecting duct of the rat. *American Journal of Physiology*, **229**, 1403–1409.

Reineck, H. J., Osgood, R. W. & Stein, J. H. (1977). Distal nephron potassium (K) transport in the rat: effect of amiloride. *Clinical Research*, **25**, 508A.

Rocha, A. S. & Kokko, J. P. (1973a). Sodium chloride and water transport in the medullary thick ascending limb of Henle. Evidence for active chloride transport. *Journal of Clinical Investigation*, **52**, 612–623.

Rocha, A. S. & Kokko, J. P. (1973b). Membrane characteristics regulating potassium transport out of the isolated perfused descending limb of Henle. *Kidney International*, **4**, 326–330.

Schaefer, J., Troutman, S. & Andreoli, T. (1974). Volume reabsorption, transepithelial potential differences and ionic permeability properties in mammalian superficial proximal tubules. *Journal of General Physiology*, **64**, 582–607.

Schmidt, U., Schmid, J., Schmid, H. & Duback, U. C. (1975). Sodium and potassium activated ATPase. A possible target of aldosterone. *Journal of Clinical Investigation*, **55**, 655–660.

Schwartz, J. H., Finn, J. T., Vaughan, G. & Steinmetz, P. R. (1974). Distribution of metabolic CO_2 and the transported ion species in acidification by turtle bladder. *American Journal of Physiology*, **226**, 283–289.

Seldin, D. W., Welt, L. G. & Cort, J. H. (1956). The role of sodium salts and adrenal steroids in the production of hypokalemic alkalosis. *Yale Journal of Biological Medicine,* **29,** 229–247.

Silva, P., Haylsett, J. P. & Epstein, F. H. (1973). Chronic K^+ adaptation: role of Na-K-ATPase. *Journal of Clinical Investigation,* **52,** 2665–2671.

Silva, P., Ross, B. D., Charney, A. N., Besarab, A. & Epstein, F. H. (1975). Potassium transport by the isolated perfused kidney. *Journal of Clinical Investigation,* **56,** 862–869.

Sohtell, M. & Karlmark, B. (1976). In vivo micropuncture pCO_2 measurements. *Pfluegers Archiv,* **363,** 179–180.

Steinmetz, P. R., Omachi, R. S. & Frazier, H. S. (1967). Independence of hydrogen ion secretion and transport of other electrolytes in turtle bladder. *Journal of Clinical Investigation,* **46,** 1541–1548.

Stoner, L. C., Burg, M. B. & Orloff, J. (1974). Ion transport in cortical collecting tubule: effect of amiloride. *American Journal of Physiology,* **227,** 453–459.

Swingle, W. W., Maxwell, R., Ben, M., Baker, C., Lebrie, S. J. & Eisler, M. (1954). A comparative study of aldosterone and other adrenal steroids in adrenalectomized dogs. *Endocrinology,* **55,** 813–821.

Vieira, F. L. & Malnic, G. (1968). Hydrogen ion secretion by rat renal cortical tubule as studied by an autimony microelectrode. *American Journal of Physiology,* **214,** 710–718.

Walser, M. & Mudge, G. H. (1960). *Renal Excretory Mechanisms in Mineral Metabolism,* ed. Comar, C. L. & Bronner, F. Vol. 1A, pp. 287–336. New York: Academic Press.

Wiederholt, M. & Wiederholt, B. (1968). Der einfluss von dexamethason auf die wasser and electrolyte ausscheidung adrenalektomierter ratten. *Pfluegers Archiv,* **302,** 57–78.

Wiederholt, M., Schoormans, W., Fischer, F. & Behn, C. (1973). Mechanism of action of aldosterone on potassium transfer in the rat kidney. *Pfluegers Archiv,* **345,** 159–178.

Wiederholt, M., Agulian, S. K. & Khuri, R. N. (1974). Intracellular potassium in the distal tubule of the adrenalectomized and aldosterone treated rat. *Pfluegers Archiv,* **349,** 295–299.

Williamson, H. E. (1963). Mechanism of the antinatriuretic action of aldosterone. *Biochemical Pharmacology,* **12,** 1449–1450.

Woodhall, P. B. & Tisher, C. C. (1973). Response of the distal tubule and cortical collecting duct to vasopressin in the rat. *Journal of Clinical Investigation* **52,** 3095–3108.

Wright, F. S., Strieder, N., Fowler, N. B. & Giebish, G. (1971). Potassium secretion by distal tubule after potassium adaptation. *American Journal of Physiology,* **221,** 473–448.

Wright, F. S. (1977) Sites and mechanisms of potassium transport along the renal tubule. *Kidney International,* **11,** 415–432.

2 | The renal acidoses

ROBERT G. NARINS

Introductory terms and concepts
Acid secretion versus acid excretion
Acid excretion and bicarbonate added to the body
The anion gap (AG) and the differentiation of the renal acidoses
Uremic acidosis
Pathophysiology of acidosis in chronic renal failure
    Bicarbonate reabsorption
    Failure to acidify urine
    Buffer availability
    Therapy
Renal tubular acidosis (RTA)
Pathophysiology
    Acidosis
    Hyperchloremia
    Potassium metabolism
    Calcium, phosphate, and citrate metabolism
Distal RTA
    Primary distal RTA
    Dysproteinemias
    Disordered calcium metabolism
    Edema-forming states
    Drugs
    Renal transplantation
    Medullary sponge kidney (MSK)
    Hydronephrosis
    Sickle cell anemia (SCA)
    Hypoaldosteronism
Proximal RTA
    Childhood forms
    Adult forms
Low-buffer excretion
    Phosphate depletion
    Diminished ammonia excretion
Diagnosis of RTA
    Incomplete RTA
    Complete RTA
Therapy
    Distal RTA
    Proximal RTA

INTRODUCTORY TERMS AND CONCEPTS

Metabolic acids, produced through intermediary metabolism, gain access to extracellular fluid and consume bicarbonate. The kidney, by coupling sodium reabsorption with hydrogen secretion, excretes these metabolic acids and simultaneously replenishes lost alkali. The two mEq/l/day decrease in serum bicarbonate observed during transient or permanent loss of renal function highlights this vital interplay between metabolic acid production and renal acid excretion (Relman, 1964; van Ypersele de Strihou and Frans, 1970).

In order to carry out its prime acid-base function (i.e., regulation of serum bicarbonate), the kidney must do two things. First, it must reclaim virtually all alkali from the glomerular filtrate, thereby preventing urinary bicarbonate wasting and the systemic acidosis that would otherwise ensue. Second, as noted above, it must excrete the daily load of metabolic acid, some 50–100 mEq/day, and in the process, synthesize new bicarbonate, thereby replacing systemic losses incurred during normal daily metabolism. Both processes (i.e., bicarbonate reclamation and regeneration) are effected by renal tubular epithelial cells and both are mediated by tubular secretion of hydrogen ions. The renal acidoses result from loss of bicarbonate in the urine or failure to adequately excrete acid. Since both processes are effected by tubular epithelium, it follows that renal acidoses are always tubular in origin.

In order to maintain electroneutrality, bicarbonate lost in metabolic acidosis must be replaced by another negatively charged anion (Fig. 2.1). The

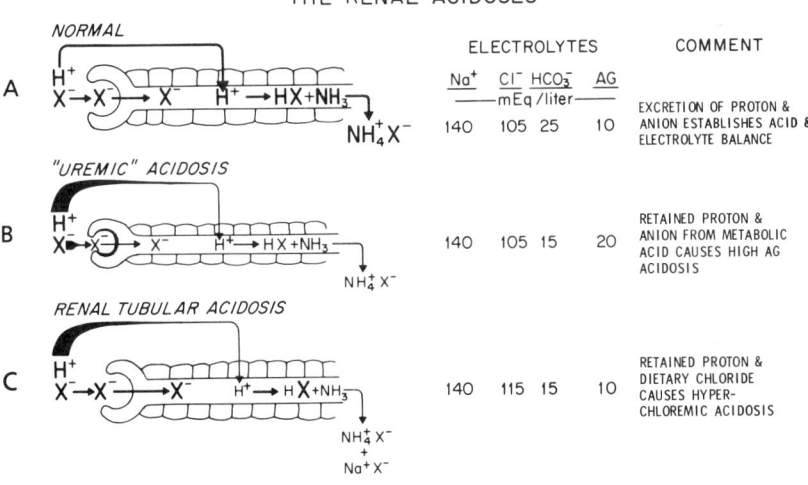

Figure 2.1 Schematic representation of the pathogenesis and effects of the renal acidoses. (*a*) Acid anions (X⁻) are excreted by glomerular filtration, while hydrogen ions (H⁺) must be secreted by tubular epithelium and carried into the urine with buffer (NH₃). (*b*) The serum electrolyte pattern resulting from uremic acidosis, where proportionate reduction of acid and anion excretion occur, and (*c*) in renal tubular acidosis, where H⁺ excretion is disproportionately reduced, are depicted. See text for details.

nature of the substituting anion is greatly influenced by the prevailing level of glomerular filtration. If a proportional degree of glomerular and tubular damage occurs, as seen in generalized renal failure, anions of metabolic acids (e.g., sulfate, phosphate, etc.) are retained. When damage is primarily tubular, however, metabolic anions continue to be filtered and excreted, necessitating retention of dietary chloride to replace the lost bicarbonate anion. In this formulation, glomerular insufficiency influences only the character of the anion replacing bicarbonate and does not otherwise influence the degree of acidosis.

Metabolic acidoses of renal origin are conveniently divided into those developing during the course of generalized renal disease, termed *uremic acidosis*, and those in which acidosis occurs in the absence of azotemia, termed *renal tubular acidosis*. Although these acidoses have many overlapping features, their separation is warranted based on major differences in clinical course, prognosis, and therapy.

Prior to discussion of individual entities, it will be clarifying to stress first certain aspects of acid-base metabolism which become disordered in these diseases.

Acid secretion versus acid excretion (Fig. 2.2a)

Under normal circumstances, 20 percent of renal blood flow is filtered at the glomerulus thereby presenting 5,000 mEq/day of bicarbonate to the tubular lumen. Bicarbonate is not well reabsorbed directly, necessitating its conversion to the more permeable H_2CO_3 by hydrogen ions secreted by proximal and distal tubular cells. The luminal H_2CO_3, in the presence of carbonic anhydrase, rapidly dehydrates to CO_2 and water, which freely enter tubular cells. Intracellular rehydration of CO_2 reforms H_2CO_3 that is in turn, rapidly separated into its component parts, H^+ and HCO_3^-. The latter is returned to blood along with reabsorbed sodium, whereas the former continues to serve the catalytic function of converting poorly permeable, luminal HCO_3^- to CO_2 and H_2O. Several important facts emerge: (1) each mEq of HCO_3^- reabsorbed requires secretion of one mEq of H^+; or, *HCO_3^- reabsorption is synonymous with H^+ secretion;* (2) since H^+ secreted is eventually reclaimed as water (Fig. 2.2a), the 5,000 mEq of H^+ secreted for this daily reclamation of HCO_3^- does not enter the urine. Thus, *acid secretion* is not synonymous with *acid excretion;* (3) certain characteristics of this system should be remembered in order to understand the disease entities described below. Approximately 85 percent of HCO_3^- reabsorption occurs in the proximal tubule and is very responsive to changes in the state of *ECF volume, potassium balance,* and pCO_2 (Ch. 1). The distal tubule reabsorbs the small amount of bicarbonate that normally escapes proximal reclamation but its reabsorptive capacity is easily overwhelmed when significant increments in delivery occur.

Acid excretion and bicarbonate added to the body (Fig. 2.2b)

The ability of tubular cells to synthesize H_2CO_3 rapidly and cheaply from CO_2 and water and then separately metabolize its ionic components enables the kidney to excrete acid and to synthesize alkali simultaneously. Following ionization of H_2CO_3 to H^+ and HCO_3^-, the H^+ ion is secreted into the tubular lumen, and sodium from a nonbicarbonate salt is returned to the cell. In contrast to the situation seen with bicarbonate reabsorption, in which

RENAL HCO₃ REABSORPTION AND ACID EXCRETION

A. HCO₃ REABSORPTION

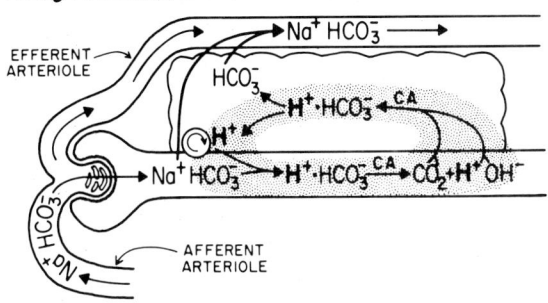

B. ACID EXCRETION

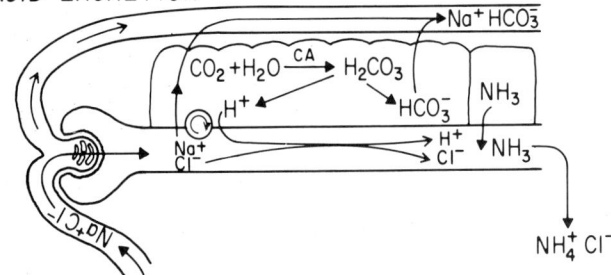

Figure 2.2 A schematic illustration of $NaHCO_3$ reabsorption (details outlined in text). The shaded area (*a*) emphasizes the cyclic nature of carbonic acid ($H \cdot HCO_3$) formation and breakdown in effecting the transfer of HCO_3^- from lumen to peritubular capillary. Hydrogen ions are given in boldface to stress their catalytic role. Their secretion and reabsorption may be clearly followed.

secreted H^+ is ultimately reclaimed, the H^+ in this instance is carried out into the urine as titrated buffer. In the example shown (Fig. 2.2*b*), Na^+, filtered as the chloride salt, is returned to the venous effluent as bicarbonate, while filtered Cl^- appears in the urine with NH_3 and H^+ (i.e., NH_4Cl). The renal requirements for excretion of acid and replenishment of bicarbonate stores are the ability to secrete hydrogen ions, the ability to acidify the urine sufficiently to ensure titration of urinary buffers, and the provision of ample buffer to carry out physiologically important amounts of acid.

The primary urinary buffers are *phosphate* and *ammonia*. The supply of phosphate is directly dependent on dietary intake whereas ammonia is synthesized by tubular epithelium from amino acid precursors. It should be stressed that in the absence of urinary buffers, even maximal acidification of the urine (e.g., to pH 4.0) could *not* result in excretion of enough acid to maintain daily balance. The urinary pH refers only to the free, or active, hydrogen ions in solution and does not imply anything of the quantity of bound, or buffered, acid present. The pH is the negative logarithm of the concentration of free hydrogen ions, and a pH of 4 defines a hydrogen ion concentration of only 0.10 mEq/l. Thus, one liter of urine devoid of buffer would result in excretion of only a minute fraction (0.1–0.2 percent) of the 50–100 mEq of acid produced daily. Provision of buffer to the urine allows for the continued secretion of acid by binding hydrogen ions, thereby limiting development of hydrogen ion gradients that would otherwise restrict further acid secretion. The excretion of acid in amounts that match daily production can be accomplished only by the tubular secretion of hydrogen ions and the simultaneous provision of the buffer needed to excrete it.

The anion gap (AG) and the differentiation of the renal acidoses

The anion gap (AG) is derived from routine measurement of serum electrolytes and serves as an important parameter of acid-base disorders and as a clue to the presence of a variety of other potentially life-threatening but unsuspected diseases (Emmett and Narins, 1977). It is defined as the difference, in mEq/l, between serum Na^+ and the sum of Cl^- and HCO_3^- $[AG = Na^+ - (Cl^- + HCO_3^-)]$. This difference represents those anions, not accounted for by chloride or bicarbonate, that counterbalance sodium's positive charge.

In "uremic acidosis," serum Na^+ and Cl^- remain relatively constant while acid retention caused by tubular disease depresses serum bicarbonate. Coexisting glomerular insufficiency ensures replacement of lost HCO_3^- with retained nonchloride anions of intermediary metabolism. The stable Na^+ and Cl^-, but low HCO_3^-, defines the high anion gap acidosis that is so typical of uremia (Table 2.1).

Table 2.1 Electrolytes in the renal acidoses

	mg/dl		mEq/l			
	BUN	Cr	Na^+	Cl^-	HCO_3^-	AG
Normal	15	1.0	140	105	25	10
Pure Uremic acidosis (UA)	50	5.0	140	105	15	20
Pure Hyperchloremic acidosis (HCA)	15	1.0	140	115	15	10
Mixed UA/HCA	35	3.5	140	110	15	15

Diseases that cause acidosis primarily by altering tubular function, but by leaving glomerular filtration intact, do not alter the anion gap. The electrolytes behave as if hydrochloric acid had been added to blood causing the stoichiometric replacement of HCO_3^- with Cl^-. The normal anion gap is defined by the unchanged serum sodium concentration, and failure of the *sum* of chloride and bicarbonate to change (Table 2.1). The pathogenesis of hyperchloremic metabolic acidosis is discussed more fully in the section on renal tubular acidosis (page 50).

Thus, patients with "uremic acidosis" have marked azotemia and elevated AG's, whereas those with primary tubular disorders have mild to absent azotemia and normal AG. As will be discussed below, some patients demonstrate components of both uremic and hyperchloremic acidosis. The recognition of this *mixed acidosis,* which is not uncommon, is based upon failure of observed changes in the anion gap to explain the entire decrement in bicarbonate. Each mEq/l of nonchloride acid retained, lowers serum bicarbonate concentration by one mEq/l and raises the AG by a like amount. Thus, the AG in a patient with pure "uremic acidosis," whose bicarbonate has fallen from 25 to 15 mEq/l, will increase by 10 mEq/l (a rise from 10 to 20 mEq/l). Many patients pass through a phase of generalized renal disease in which tubular damage is proportionally greater than glomerular. The acidosis in this setting will share characteristics of both high and normal AG acidosis. Some of the bicarbonate lost is replaced by nonchloride anions, thereby elevating the AG, and the remainder is replaced by chloride. In this setting, the increment in the AG accounts for only a portion of the bicarbonate decrement (Table 2.1). The author has seen this form of mixed acidosis more commonly in the interstitial diseases. But mixed acidosis not uncommonly complicates the glomerulopathies as well.

UREMIC ACIDOSIS

Metabolic acidosis is a universal feature of advanced renal disease and accounts for significant morbidity and occasional mortality. Characteristically, it develops late in the course of all generalized nephropathies, usually occurring when glomerular filtration falls below 20–30 ml/minute and when BUN and creatinine exceed 40 and 4 mg/dl, respectively (Bricker, Bourgoignie and Weber, 1976).

Renal acidosis results when the number of milliequivalents of acid excreted falls below that normally produced by metabolism. Renal diseases that attack interstitium and tubules more aggressively (e.g., pyelonephritis) might be expected to cause earlier and more severe acidosis. Studies comparing onset and degree of acidosis in pyelonephritic and glomerulonephritic patients, however, have not convincingly demonstrated such differences (Elkinton, 1962; Elkinton, 1966; Johnson and Morgan, 1965). Nevertheless, many nephrologists have the impression that acidosis does begin earlier and is occasionally quite severe in interstitial renal disease. Since tubular damage tends to exceed glomerular damage, nonchloride anion retention is not a major feature of the acidosis complicating early interstitial renal disease, (i.e., the acidosis tends to be hyperchloremic). In time, as the interstitial nephritis progresses, azotemia develops and the anions of metabolic acids are retained, replacing chloride. A more classical "uremic," or high anion gap, acidosis then evolves. The early tendency to hyperkalemia in this setting probably reflects failure of distal potassium secretion, another manifestation of advanced tubular damage (Lathem, 1958). This syndrome of early hyperchloremic acidosis and hyperkalemia may, in some cases, result from early damage of the juxtaglomerular apparatus with impaired renin secretion. The hypoaldosteronism that results can lead to these chemical findings (Schambelan, Stockigt and Biglieri, 1972). This is discussed more fully in Chapter 8.

Another characteristic of uremic acidosis is its relatively mild and nonprogressive nature. Bicarbonate concentrations of 12–18 mEq/l and arterial pH ≥ 7.30 are the general rule unless sepsis or some other catabolic stress is superimposed. The stability of serum HCO_3^- is even more remarkable when one realizes that it occurs despite slow and inexorable deterioration of all renal excretory functions. How is bicarbonate stabilized during progressive renal failure? Goodman, Lemann and Lennon et al., (1965) have heightened the mystery by showing that patients with chronic stable renal acidosis are actually in *positive* acid balance (i.e., daily renal acid excretion is less than acid production), which causes a retention of up to 10–20 mEq/day. Although the body burden of acid increases daily, serum HCO_3^- somehow escapes titration and systemic acid-base parameters remain unchanged (Table 2.2). The mysterious survival of HCO_3^- and the failure of acidosis

Table 2.2 Acid balance in a patient with stable uremic acidosis

Day	mEq/day			mEq/l
	Acid Produced	Acid Excreted	Net Acid Retained	Serum Bicarbonate
1	65	51	+14	18.2
2	60	48	+12	18.9
3	61	48	+13	19.7
4	70	53	+17	18.3
5	70	52	+18	18.2
6	74	65	+ 9	18.3
		Cumulative acid balance =	+83[a]	

[a] Retention of 83 mEq acid should lower serum bicarbonate by 3 mEq/l.

to progress can only be explained by the participation of some nonextracellular system that absorbs, and buffers the acid being retained. Further studies by Lemann, Litzow and Lennon (1966) have shown that the positive hydrogen ion balance characteristic of chronic acidosis is associated with negative calcium balance, lending support to their suggestion that *bone* serves as this nonextracellular buffer. The calcium carbonates in bone are felt to absorb H^+ ions, thereby stabilizing serum HCO_3^-, but at the price of progressive loss of bone mineral. Goodman et al. have estimated that approximately 35,000 mEq of carbonate are potentially available for buffering. Consistent with this thesis is the observation that bone carbonates are reduced in animals fed acid loads for 5–10 days (Burnell and Teunber, 1971).

More recently, Coe et al. (1975) have reported that acute as well as chronic acidosis can stimulate *parathyroid hormone* release and thereby mobilize even more bone buffer. Parathormone release in acute acidosis results from mild hypocalcemia caused by acidosis-induced phosphate mobilization with resulting hyperphosphatemia. The mechanism by which chronic acidosis caused PTH release is not quite so clear. The hypercalciuria induced by chronic metabolic acidosis, like that in patients with idiopathic hypercalciuria, or subjects treated with furosemide, seems to stimulate PTH release. When the sodium and calcium retaining stimulus of volume depletion is superimposed upon metabolic acidosis, the same systemic acid-base changes become incapable of causing hypercalciuria. In this circumstance, increased PTH release does *not* occur, suggesting that hypercalciuria must be present for sustained acidosis to cause PTH release. Since hypocalciuria (Epstein, 1968) and *not* hypercalciuria accompanies the acidosis of renal failure, it is unlikely that acidosis plays a significant role in the secondary hyperparathyroidism of renal failure.

In the parlance of Bricker et al. (1976), the direct effects of acidosis on bone would represent one more "trade-off" that uremic patients must make in order to maintain acid-base and fluid-electrolyte balance. In this case, the trade-off prevents progressive acidosis, but at the expense of progressive bone destruction.

Although the acidosis of uremia is only one component of the complex problem of renal osteodystrophy, its treatment is simple and effects significant. Tolerance of bone to acidosis is especially poor in children, in whom chronic acidosis can markedly retard growth. The mechanism by which this occurs has not been well worked out, but it is most likely related to the buffering of acid by bone carbonates or to the recently described inhibition of vitamin D activation by acidosis (Lee, Russell and Avioli, 1977). Regardless of mechanism, treatment with alkali has been shown to restore normal growth patterns in children with renal acidosis (Nash et al., 1972; Richards, Chamberlain and Wrong, 1972).

Other clinical signs and symptoms of uremic acidosis (Table 2.3) are dictated by its rapidity of onset as well as its ultimate severity. In general, the acidosis complicating glomerular and interstitial nephritides has an insidious onset and remains relatively asymptomatic. Apart from dyspnea on exer-

Table 2.3 Clinical signs and symptoms of uremic acidosis

Clinical	Laboratory
1. Bone disease	1. Low HCO_3^-
2. Exertional dyspnea	2. Compensatory fall in P_{CO_2}
3. Fatigue	3. Fall in arterial pH
4. Hypotension ("warm shock")	4. Elevated anion gap (occasionally hyperchloremic)
5. Myocardial depression	5. Tendency to hyperkalemia
6. Venoconstriction	6. BUN $\geq$ 40 mg/dl; Cr $\geq$ 4 mg/dl

tion and from fatigue and general malaise, little else is found unless acidosis is acutely worsened by diarrhea, sepsis, etc. Patients with acute renal failure, especially young patients in a surgical setting who tend to be more catabolic, develop acidosis more rapidly and more often are symptomatic (i.e., dyspneic, hyperkalemic, and more susceptible to the hemodynamic effects of acidosis). The latter, also found in end-stage chronic renal failure with severe acidemia (pH < 7.15), are manifested by arteriolar dilatation, depressed myocardial contractility, and constriction of the venous capacitance bed (Mitchell, Wildenthal and Johnson, 1972). The peripheral vasodilation and venous constriction place severe demands on an already compromised myocardium, and may result in arrhythmias and pulmonary edema.

In addition to chemical changes already commented on, a predictable degree of pulmonary compensation (i.e., hypocapnia) also occurs in response to the acidosis of chronic renal failure. Equation 1, derived from the Henderson-Hasselbalch equation (Kassirer and Bleich, 1965), defines the relationship between serum hydrogen ion concentration, P_{CO_2} and bicarbonate:

$$(H^+) = 24 \times \frac{pCO_2}{HCO_3} \quad \text{(EQ. 1)}$$

The retention of metabolic acids in renal failure causes bicarbonate consumption, which elevates the pCO_2/HCO_3 ratio and, thereby, the hydrogen ion concentration of body fluids. Respiration is stimulated by the effect of acidosis on central and peripheral chemoreceptors (Mitchell and Singer, 1965) and lowering of pCO_2 results. Decrease in the numerator causes the pCO_2/HCO_3 ratio to return *toward*—but not *to*—normal. Previous studies of the effects of a variety of metabolic acidoses on respiration have shown that: (1) increased ventilation (i.e., lowering of the pCO_2) is directly proportional to the degree of metabolic acidosis (i.e., extent of HCO_3^- depression); and (2) Albert, Dell and Winters (1967) have shown that the expected pCO_2 for any given degree of sustained, stable metabolic acidosis can be reliably predicted by Equation 2:

$$pCO_2 = 1.5 \, (HCO_3^-) + 8 \pm 2 \quad \text{(EQ. 2)}$$

Others (Elkinton, 1966; Lennon and Lemann, 1966; Poppell et al., 1956) have reported a number of variations of this formula which are quite similar. On balance, all studies seem to indicate that the pCO_2 will fall by 1.0–1.3 mmHg for each mEq/l fall in HCO_3^-. In steady state metabolic acidosis, it tuns out that the last two digits of the measured arterial pH will equal the observed pCO_2. Thus, the azotemic subject whose serum HCO_3^- concentration is 14 mEq/l and whose pH is 7.30, will have a pCO_2 of 30. This interesting mathematical "twist" never ceases to amaze and electrify the uninitiated and adds a certain theatrical flare to the acidophile's rounds. It had been suggested (Moller, 1959) that respiratory compensation for the acidosis of renal failure was less than seen with similar degrees of ketoacidosis. Different degrees of compensation might be anticipated since, as Relman has pointed out (Relman, 1964), chronicity of acidosis may modify the respiratory response, and only the renal acidoses are sustained for more than several days to a week. Furthermore, retained metabolic products in the azotemic patient could conceivably alter the respiratory response to acidosis. Nevertheless, it turns out that respiratory compensation for renal acidosis is not different than that seen in other metabolic acidoses of equal severity and dialysis does not alter the slope of the bicarbonate: pCO_2 relationship (Albert et al., 1967; Hellman, Au and Bartter, 1965). Thus, the above formulae, derived from a variety of metabolic disorders, are equally effective in the renal acidoses.

Pathophysiology of acidosis in chronic renal failure

Metabolic balance studies in subjects with chronic renal failure have shown that the daily rate of acid *production* remains unchanged despite advanced azotemia or substantial degrees of metabolic acidosis or alkalosis (Goodman et al., 1965). Azotemics, like normals, produce approximately 1.0–1.5 mEq/kg/day of acid, and, in order to maintain acid balance, net renal acid excretion must, of course, equal production. Urinary ammonia and titrated phosphate (TA) represent acid excreted by the kidney and, as noted previously, reflect the new bicarbonate synthesized by the kidney and returned to the circulation. Since bicarbonate excretion leaves unneutralized acid behind (i.e., reflects acid retained), net acid excreted (NAE) must equal the difference between the sum of ammonia and TA, on the one hand, and the HCO_3^- excreted, on the other, as shown in Equation 3:

$$\text{NAE (mEq/day)} = (NH_4^+ + TA) = HCO_3^- \qquad (EQ.\ 3)$$

On a normal diet and in acid balance, a 70 kg man produces 70 mEq of metabolic acid and excretes 40 mEq of NH_4^+, 35 mEq TA, and 5 mEq of HCO_3^-.

Bicarbonate reabsorption

Schwartz et al. (1959) first called attention to the reduced bicarbonate reabsorptive capacity of patients with advanced renal disease. Although previous studies in humans (Roberts et al., 1959) and dogs (Morrin et al., 1962) had failed to demonstrate any significant effects of chronic renal failure on the maximum capacity to reabsorb bicarbonate, Schwartz et al. recognized the insensitivity of alkali-loading techniques for detecting small bicarbonate "leaks." These authors normalized the serum bicarbonate of 12 acidotic patients in stable renal failure and, after discontinuing alkali therapy, simply followed bicarbonate excretion while serum levels were allowed to fall spontaneously. When serum bicarbonate is reduced below 24 mEq/l, urine of normal subjects becomes essentially free of bicarbonate and intensely acidic, reflecting the kidney's effort to normalize serum bicarbonate by excreting acid and synthesizing new alkali. Despite serum levels well below 20–24 mEq/l, patients with renal failure had inappropriately alkaline urine and a degree of bicarbonaturia that drained body stores significantly. This leak continued until the serum level reached 10–15 mEq/l and only then could maximal acidification of urine occur. Five of the 12 patients studied excreted more than 10 mEq of HCO_3^- per day despite serum levels below 24 mEq/l. Serum HCO_3^- fell during this period of renal wastage not only because it was physically removed but also because the alkaline urine prevented titration of ammonia and phosphate and thereby impaired renal synthesis of the new HCO_3^- needed to replenish ongoing daily losses.

Recent studies using a modified HCO_3^- titration technique have clearly established that maximal reabsorption is impaired in renal failure (Slatopolsky et al., 1970). In order to define maximal reabsorptive rates, the filtered load of HCO_3^- must be progressively increased by infusing increasingly greater amounts of alkali. Since expansion of the extracellular space inhibits tubular reabsorption of sodium and its attendant anions, and since fluid expansion is an unavoidable concomitant of HCO_3^- infusion, the technique can profoundly modify the very process it attempts to measure. Slatopolsky et al. have standardized the titration technique by using concentrated $NaHCO_3$ solutions in an effort to minimize volume expansion. They have demonstrated decreased capacity to reabsorb HCO_3^- by patients with GFRs below 20 ml/min, and the capacity diminishes further as renal function deteriorates (Fig. 2.3).

Since each mEq of HCO_3^- reabsorbed requires the countersecretion of H^+ (see above), the fall in HCO_3^- reabsorption seen in uremics may be equated with a decrease in total hydrogen secretion. While the presence of this defect in uremics is now well established, its pathogenesis remains controversial.

Although possible, it seems unlikely that *structural damage* alone accounts for this HCO_3^- wastage. Azotemic patients with GFRs $\leq$ 20 ml/min have impaired reabsorption, while those whose GFR is $\geq$ 30 ml/min, demonstrate normal HCO_3^- handling (Slatopolsky et al., 1970). When HCO_3^- reabsorption of damaged kidneys is compared with that of contralateral normal kidneys in nonazotemic dogs with unilateral kidney disease, minimal differences are found (Morrin et al. 1962). This finding suggests that some aspect of the uremic state causes a functional change in the kidney that reduces hydrogen secretory capacity. Of numerous possible explanations for bicarbonaturia, the three that have received most attention include: increased solute load per nephron; increased circulating levels of PTH; and increased fractional sodium excretion.

Solute diuresis: The ability to excrete the salts, nitrogenous and nonnitrogenous wastes that comprise the daily load of solute, is well maintained despite even advanced renal failure. Since excretion of the same solute load

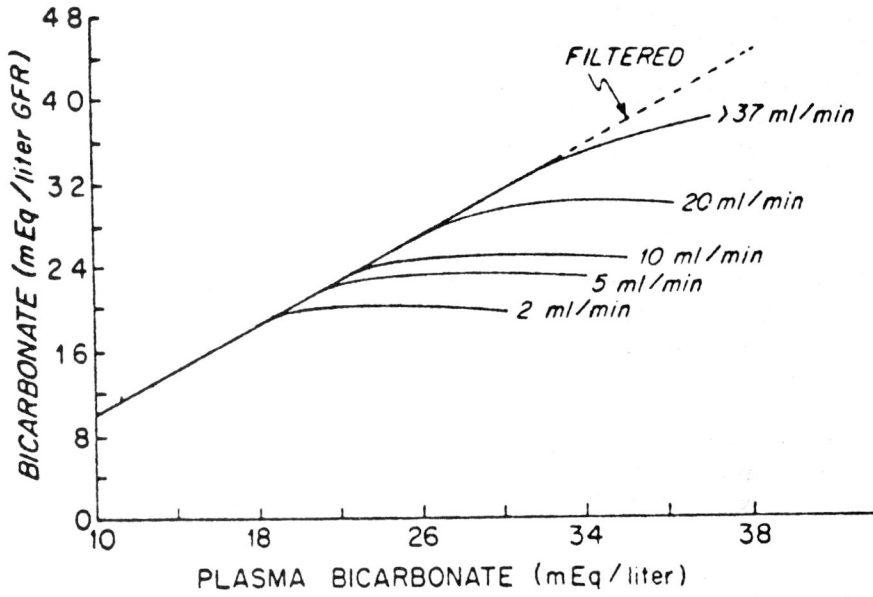

Figure 2.3 Bicarbonate reabsorption is plotted on the ordinate and plasma bicarbonate on the abscissa for representative subjects whose GFR is shown (ml/min) at the end of the line. The line, > 37 ml/min, is a composite of five subjects. (From Slatopolsky, E., Hoffsten, P., Purkerson, M. et al., 1970, *Journal of Clinical Investigation*, **49**, 988. Reproduced with publisher's permission)

must be effected by progressively fewer nephrons, it follows that each of the remaining nephrons is exposed to an ever-increasing amount of solute. In effect, nephrons are undergoing a solute, or osmotic diuresis, which is much like that seen in normals receiving mannitol or urea infusions. The osmotic effect of these poorly reabsorbed solutes impairs reabsorption of fluid from the glomerular filtrate thereby diluting the luminal concentration of sodium. As the concentration gradient for sodium reabsorption becomes less favorable, its reabsorption slows and a fraction escapes into the urine with its attendant anions, chloride and HCO_3^- (Poole-Wilson et al., 1972). Although solute diuresis may play an important role under some circumstances, it is unlikely to contribute significantly to the bicarbonaturia usually seen. In unilateral renal disease, removal of the normal kidney exposes each nephron of the remaining diseased kidney to a sudden increase in solute load. The failure of significant bicarbonaturia to develop minimizes the importance of solute diuresis (Morrin et al., 1962).

Parathyroid hormone (PTH), calcium, and phosphate: A complex interaction exists between bicarbonate reabsorption, on the one hand, and the availability of PTH, calcium, and phosphate, on the other. Each of the three elements of this system, PTH, calcium, and phosphate, has been shown, under various circumstances, to alter renal hydrogen ion secretion, and, therefore, bicarbonate reabsorption.

Acute infusions of *PTH* have been shown to increase bicarbonate excretion in experimental animals (Rodman and Heinemann, 1975) and in man (Hellman et al.). Recent micropuncture studies in dogs (Agus, et al., 1971) revealed that proximal sodium reabsorption is impaired by PTH, causing sodium to be swept into the distal nephron where reclamation of chloride exceeds that of bicarbonate. The resulting bicarbonaturia should eventually lower serum levels causing a metabolic acidosis. Attempts to induce significant degrees of acidosis with acute infusions of PTH have usually been unsuccessful (Bank and Aynedjian, 1976; Karlinsky et al., 1974; Puschett, Zurbach & Sylk, 1976).

In two of these studies (Bank et al.; Puschett et al.), proximal bicarbonate reabsorption was inhibited by PTH, but bicarbonaturia was prevented because of a simultaneous increase in distal tubular reclamation. In the third study, (Karlinsky et al.), bicarbonate wasting induced by PTH was so trivial that serum levels were hardly altered.

Bicarbonate wasting has been shown to be an occasional feature of diseases associated with *chronic* hypersecretion of PTH (Rodman and Heinemann, 1975). The impression has developed, however, that the renal defect is more frequently seen in states of secondary hyperparathyroidism than in the primary disorder (Muldowney, Freaney and McGeeney, 1968) (Muldowney et al., 1971). Thus, Muldowney and associates have demonstrated that reduced bicarbonate reabsorption in chronic renal failure and in intestinal malabsorption, states of *secondary* hyperparathyroidism, improves when PTH secretion is reduced by calcium infusion or following parathyroidectomy. In *primary* hyperparathyroidism, however, HCO_3^- wasting and acidosis are not consistent findings, nor does serum HCO_3^- reproducibly increase following parathyroidectomy (Coe, 1972). Serum calcium serves to distinguish these two states and may function as the linchpin that explains the dissimilar prevalence of bicarbonaturia and acidosis.

Crumb et al. (1974) have shown that elevation of serum calcium stimulates acid secretion, which in turn enhances HCO_3^- reabsorption. Thus, maximal rates of HCO_3^- reabsorption are increased by hypercalcemia and reduced by hypocalcemia. These changes in serum calcium can be expected to modify the response of renal hydrogen secretion to PTH.

The final link in this chain that alters tubular function is *phosphate*. Phosphate depletion reduces renal HCO_3^- reabsorption and leads to acidosis in animals, but this has yet to be described in man (Emmett et al., 1977; Gold et al., 1973). The cause of this defect is uncertain, although the intracellular pH of renal tubular cells may be inappropriately high thereby impairing hydrogen ion secretion and HCO_3^- reabsorption (Gold, et al., 1973).

In primary hyperparathyroidism, the attendant hypercalcemia may offset the bicarbonate-wasting effect of PTH and phosphate depletion. The prevailing balance of these forces, which may be quite variable, dictates whether or not HCO_3^- reabsorption will be altered. It is not surprising therefore that the struggle for dominance between hypercalcemia, on the one hand, and phosphate depletion and PTH excess, on the other, leads to a variable prevalence of hyperchloremic acidosis in primary hyperparathyroidism. The circulating level of PTH in patients with secondary hyperparathyroidism, especially those with chronic renal failure, tends to greatly exceed that of patients with adenomata (Reiss, Canterbury and Egdahl, 1968). Additionally, the driving force that initiates and sustains PTH hypersecretion is the low serum calcium. The protection afforded HCO_3^- reabsorption by hypercalcemia in primary hyperparathyroidism is lost in chronic renal failure thereby unmasking the bicarbonate-wasting effects of the high circulating PTH levels.

Although certain inconsistencies appear in the reported data, it is hard to escape the conclusion that chronic exposure of the kidney to excessive levels of PTH, in the absence of hypercalcemia, contributes to the bicarbonate wasting seen in many patients with renal failure.

Influence of altered sodium metabolism: The relentless decline of renal function seen in most diffuse nephropathies leaves an evershrinking number of nephrons behind to administer and regulate fluid and solute metabolism. This worsening imbalance between unremitting functional demands and decreasing functional mass places an increasing regulatory burden on *each* remaining nephron. The regulatory changes required for maintenance of sodium balance in progressive renal failure are particularly relevant to our discussion of acid-base changes.

Each of the 180 liters of plasma filtered daily delivers 140 mEq of sodium to the nephrons, resulting in cumulative daily filtration of 25,000 mEq of sodium. The ingestion of 125 mEq of sodium in the normal daily diet demands that a like amount be excreted if salt balance is to be maintained. Thus, normal kidneys need reject only 125 of the 25,000 mEq of sodium filtered in order to maintain this balance. The loss of nephrons and fall in GFR, which attend advancing renal disease, cause progressive decline in the filtered load of sodium. Since daily salt intake remains relatively constant, it follows that the fraction of filtered sodium that must be excreted to sustain salt balance must increase progressively. When GFR falls from 180 to 18 l/day, the filtered load of sodium falls to 2,500 mEq/day. Excretion of the same 125 mEq demands that tubular rejection of sodium increase from the normal value of 0.5 to the elevated value of 5.0 percent of the filtered load.

It has been demonstrated that proximal and distal tubular sodium reabsorption share in this regulatory process (Bricker et al., 1976). The inhibition of proximal reabsorption sweeps Na^+ and its accompanying anions distally where further Na^+ reabsorption is somehow attuned to the needs of extracellular fluid and effective circulating arterial volumes. The urinary loss of chloride with sodium serves the best interests of the host in that NaCl excretion rids dietary salt and acts to maintain balance. The loss of HCO_3^- with sodium worsens acidosis and therefore is counterregulatory. In this formulation, acid-base balance is sacrificed in order to stabilize vascular and interstitial fluid volumes. Herein follows a brief discussion of how this adaptive change in renal sodium handling plays a central role in the bicarbonaturia of renal failure.

The low HCO_3^- threshold and diminished maximal rates of reabsorption seen in patients with GFRs below 20 ml/min, is abolished when the nephrotic syndrome complicates similar degrees of azotemia. The hypoalbuminemia increases sodium reabsorption and simultaneously elevates HCO_3^- threshold and maximal reabsorption (Slatopolsky et al., 1970). If the need to excrete sodium in advancing renal disease modifies HCO_3^- reabsorption, then reduction of this excretory burden ought to improve renal HCO_3^- handling. Espinel (1975) has shown that profound reduction in serum HCO_3^- occurs in rats with experimental renal disease when they continue to ingest diets rich in sodium. When salt intake is reduced in proportion to the degree of experimentally induced renal failure, the natriuretic demands on the remaining nephrons remain unchanged. In the example cited above, it was pointed out that salt balance is achieved when 0.5 percent of the 25,000 mEq of filtered sodium is excreted by subjects ingesting 125 mEq of sodium daily. If filtration rate and filtered sodium are reduced by 50 percent (i.e., to 90 l/day and 12,500 mEq/day) and if salt intake is simultaneously reduced by one half (i.e., 62.5 mEq), the fraction of filtered sodium excreted remains at 0.5 percent. Azotemic rats whose dietary sodium is reduced in proportion to their decreased GFR have lesser degrees of acidosis, higher HCO_3^- thresholds, and greater maximal rates of reabsorption. These data strongly support the notion that in renal failure the ECF volume is protected at the expense of body HCO_3^- stores because of a very sensitive interplay between fluid, volume, and hydrogen ion secretion. This resetting of hydrogen secretion is likely to play the major role in the bicarbonaturia of renal failure. A discussion of how the need for sodium excretion is perceived and by what mechanism this need is translated into a natriuretic response is beyond the scope of this paper.

Failure to acidify the urine

By the time systemic acidosis develops, urinary pH is consistently near 5.0. This pH virtually ensures the absence of HCO_3^- and the complete titration of excreted phosphate and ammonia buffers. Schwartz et al. (1959) have shown, however, that as serum HCO_3^- falls spontaneously, the urine remains inappropriately alkaline until a new steady state is reached. At this point, the fall in serum HCO_3^- has reduced the filtered load below threshold, allowing tubular epithelium to reclaim once again the entire filtered load and to acidify the urine maximally. This "HCO_3^- leak" prevents the azotemic patient from maximally acidifying his urine until substantial falls in serum HCO_3^- occur. There is no evidence to show that the common renal disorders impair acidification, once the alkali leak is removed.

Buffer availability

Phosphate and ammonia constitute the two major urinary buffers. Impaired excretion of either one or both would severely restrict acid excretion and thereby contribute to systemic acidosis.

Phosphate: Since phosphate balance is well maintained until renal failure is far advanced, urinary phosphate(i.e., titratable acid) does not decline until dietary intake declines. Protein restriction and aluminum hydroxide gels limit availability of dietary phosphate and, on occasion, will impair acid excretion, thereby worsening systemic acidosis (Franklin et al., 1966). Under normal circumstances, acid retention consequent to a fall in phosphate excretion would stimulate renal ammonia synthesis. The enhanced ammonia excretion would then counter the fall in phosphate, and net acid excretion would not necessarily suffer. In advanced renal disease, however, biosynthesis of ammonia cannot be substantially increased, and a fall in phosphate buffer excretion is poorly tolerated.

Ammonia: Renal ammonia excretion, when expressed per unit of functional mass, is well maintained in advanced renal failure. The progressive shrinkage of functional mass, however, decreases the total quantity of ammonia excreted, thereby limiting total acid excretion. This reduction of ammonia excretion is the primary factor underlying the development of uremic acidosis. While simple reduction in renal mass can account for the observed decline in ammonia excretion, other possibilities exist.

The complex biochemical mechanisms whereby the kidney synthesizes ammonia are subject to perturbation at many different sites (Table 2.4).

Table 2.4 Steps in renal ammoniagenesis

I.	**Substrate synthesis** Extrarenal glutamine production
II.	**Substrate delivery** Renal cell uptake of glutamine
III.	**Intracellular transport** Glutamine entry into mitochondria
IV.	**Catabolism of glutamine** A. Reactions that generate NH_3 by removing glutamine's 2 nitrogen atoms B. Reactions that dispose of glutamine's remaining carbon skeleton
V.	**Excretion of NH_3** Trapped as NH_4^+ in acid urine

Recent studies in azotemic patients by Welbourne, Weber and Bank (1972) have shown that availability of ammonia's precursor, glutamine, does not limit production. The kidney in renal failure, indeed, seems to be producing ammonia at its maximal rate, which does not increase when more glutamine is provided. The acid urine and relatively high flow rates in renal failure undoubtedly ensure trapping and excretion of ammonia. Very little information is available regarding various renal diseases on the intrinsic renal metabolism of glutamine.

Although other functional defects may be demonstrated in the future, it presently appears that reduced ammonia excretion in renal failure simply reflects reduction of functional renal mass.

In summary, the acidosis of renal failure reflects the diminished capacity of the diseased kidney to excrete acid. The consequent retention of metabolic acids produced via intermediary metabolism consumes HCO_3^- and replaces it with anions of the retained acids. The progressive reduction of renal ammoniagenesis is the major factor limiting renal acid excretion. The solute diuresis in remaining nephrons, the excessive PTH secretion, and the reduction of sodium reabsorption all act to mildly reduce renal HCO_3^- reabsorption. The bicarbonate leak, although significant, does not account for a major portion of the acidosis. Finally, chronic renal failure patients appear to be consistently in positive acid balance despite a stable serum HCO_3^- concentration. The latter reflects the titration of bone minerals.

Therapy

The acidosis accompanying chronic renal failure is generally mild and usually without obvious symptoms or signs. The potential risks and benefits of routine use of alkali can therefore reasonably be questioned. Unfortunately, prospective controlled studies bearing on this point do not exist.

If one restricts therapy to the dose of alkali required to maintain plasma HCO_3^- at 20-23 mEq/l, (usually 30-100 mEq of $NaHCO_3$ daily) then, this author believes the risks are minimal. The threat of volume overload always exists, but salt balance is generally well maintained until the development of far advanced renal failure. In fact, a recent study (Husted, Nolph and Maher, 1975) has verified the long-held clinical impression that mEq for mEq, azotemics tolerate $NaHCO_3$ better than NaCl. Weight gain and edema are unusual during $NaHCO_3$ therapy. This study suggests that a chloride leak, perhaps from the thick ascending limb of Henle, allows NaCl to be excreted while $NaHCO_3$ is retained.

Alkalinization of blood reduces the ionized fraction of serum calcium. Patients already hyperphosphatemic and hypocalcemic may demonstrate increased neuromuscular excitability or precipitate frank tetany with alkali therapy. This is an unusual problem if one avoids overzealous use of HCO_3^- and follows routine clinical signs and serum electrolytes.

The benefits derived from alkali therapy are limited to prevention of bone dissolution, therapy of symptomatic acidosis, and prophylaxis against developing severe life-threatening acidosis. The absence of controlled studies make subjective clinical impression rather than hard fact the prime guideline in this area.

The detrimental effects of acidosis on bone are more overtly manifest in young children, whose growth pattern becomes strikingly impaired. Adults, especially those with slowly progressive interstitial nephropathies, which are often complicated by early onset of acidosis, also show osteopenic effects of acidosis. The ability of alkali therapy in renal tubular acidosis and chronic renal failure to decrease calcium wasting, normalize growth, and heal bones is demonstrated in several studies (Lemann et al. 1966; Nash et al. 1972; Richards et al. 1972) that support the need for routine use of alkali.

The serum pH is particularly sensitive to additional decrements in HCO_3^- or increments in pCO_2 in renal failure patients whose HCO_3^- and pCO_2 are already low. Thus, superimposition of diarrhea, sepsis, or even mild CO_2 retention can dramatically acidify the blood of patients whose initial HCO_3^- was 10–15 mEq/l. The use of alkali therapy to maintain a serum HCO_3^- of 20–23 mEq/l affords renal failure patients an extra measure of reserve against other acidifying events to which they are prone.

The various alkalinizing agents and their dosage are discussed at the end of the next section.

RENAL TUBULAR ACIDOSIS (RTA)

Secretion of protons by renal tubular epithelium enables the kidney to regulate serum HCO_3^-. Approximately 5,000 mEq of H^+ are secreted and reabsorbed daily in order to catalyze reabsorption of the 5,000 mEq of HCO_3^- filtered daily. Eighty-five percent of this reclamation occurs proximally and the remainder occurs in the distal nephron. In stark contrast to the 5,000 mEq of H^+ required to mediate HCO_3^- reabsorption, only 50–100 mEq—*secreted* largely in the distal nephron—need be *excreted* with phosphate and ammonia to rid the daily load of fixed acid and to concomitantly replenish daily HCO_3^- losses. Thus, 98–99 percent of secreted H^+ is used, primarily by proximal nephrons, to prevent HCO_3^- wasting, while only 1–2 percent is used by the distal nephron to resynthesize lost HCO_3^-.

Renal tubular acidosis (RTA) results from ineffective secretion of H^+ ions. Depending upon its extent and location, the defect may manifest itself as pure, but self-limited, HCO_3^- wasting, or as failure to excrete the daily load of metabolic acid and thereby fail to replenish systemic HCO_3^- losses. The latter may result from tubular secretory failure or from lack of buffers required to carry acid into the urine.

While it is recognized that overlapping syndromes exist, the great majority of cases are easily separable into proximal (HCO_3^--wasting) or distal (acidification-defect) forms. Although both forms express themselves as hyperchloremic acidosis, their pathophysiology, clinical consequences, and therapy are sufficiently different to warrant such separation. The following discussion attempts to convey to the reader an understanding of the pathophysiologic principles that underlie RTA. From these insights, any physician should be able to recognize and treat all forms of RTA.

Pathophysiology

Acidosis

The defect in H^+ ion secretion is more extensive in the proximal than in the distal form. The H^+ secretory defect of proximal RTA is manifested by *self-limited* bicarbonaturia. The impairment of proximal Na^+/H^+ exchange allows filtered HCO_3^- to escape reabsorption, causing it to be swept into the distal nephron in amounts that overwhelm the latter's reabsorptive capacity. The resulting bicarbonaturia lowers serum HCO_3^-, which in turn progressively diminishes the amount of alkali presented for proximal reabsorption. This repetitive cycle of alkali wasting and serum HCO_3^- lowering continues until the filtered load of alkali once again matches the new, low proximal reabsorptive capacity. At this point, bicarbonaturia ceases, allowing for acidification of the urine and for renewed titration of ammonia and phosphate buffers. Excretion of titrated buffer at this new low serum HCO_3^- concentration allows patients with proximal RTA to excrete their daily load of acid, thereby stabilizing serum HCO_3^- and maintaining acid balance.

Bicarbonate wasting almost always occurs in association with other proximal reabsorptive defects. Thus, some combination of phosphaturia, glycosuria, uricosuria, or amino aciduria accompanies the alkali loss. Only in infancy has pure HCO_3^- wasting been recognized with any significant frequency.

In contrast to patients with proximal RTA, those with the distal form do not acidify their urine despite severe systemic acidosis. Failure to establish an acid urine prevents titration of available buffer, thereby decreasing net acid excretion. The inability of net acid excretion to keep pace with acid production results in positive H^+ ion balance and progressive titration of body buffers.

When H^+ ion generation and secretion by distal epithelial cells are normal, and when the luminal membrane is intact, a 1000:1 blood-to-lumen H^+ ion gradient is established and maintained. The three log unit difference between minimal urine pH—4.4—and normal blood pH 7.4—defines the magnitude of this gradient. Distal RTA—or failure to maximally acidify the urine—could result from one of the following mechanisms (Fig. 2.4). (1) Despite an intact luminal membrane and secretory mechanism, cells may be unable to generate H^+ ions. (2) The H^+ "pump" may be incapable of transporting cellular protons "uphill" against an everincreasing luminal concentration of free H^+. (3) Normal H^+ generation and secretion may be rendered ineffective by a "leaky" luminal membrane that allows backleak of secreted H^+ ions.

If an alkaline distal fluid pH could be maintained, the more favorable H^+ gradient would allow the "weak pump" to secrete more normally (2). This high luminal pH would also diminish any tendency to backleak of secreted H^+ ions (3). If RTA were caused by either of the latter two mechanisms, alkalinization of the urine would therefore increase net acid secretion. If,

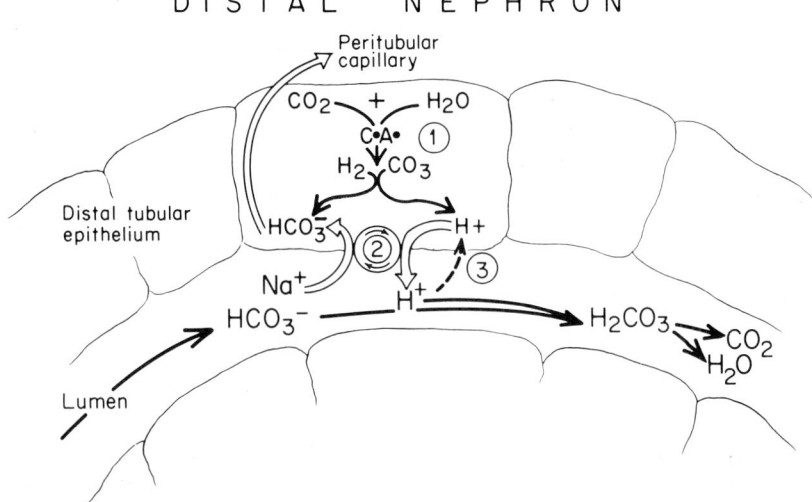

Figure 2.4 Schematic diagram of distal tubular acid secretion. Defects at any of the numbered sites may be important in the pathogenesis of distal RTA. (1) Hydration of CO_2 by carbonic anhydrase reaction (CA), with formation of H^+ and HCO_3^-. (2) Active transport of H^+ into lumen in exchange for luminal Na^+. (3) Potential back diffusion of H^+ into the cell. The generation of CO_2 by H^+ secretion into a bicarbonate-rich lumen is discussed in the text.

however, failure to generate H^+ ions were causative, alkalinization would not be expected to increase distal acid secretion.

During alkali infusion, the distal epithelium continues to secrete H^+ ions, and the urinary pCO_2 reflects this process (Halperin et al., 1974). Secreted H^+ converts luminal HCO_3^- to H_2CO_3. The distal lumen, unlike the proximal, lacks carbonic anhydrase, causing H_2CO_3 to slowly dehydrate to CO_2 and H_2O, thereby raising urinary pCO_2 (Fig. 2.4). When HCO_3^- excretion rates are high, distal H^+ secretion normally causes urine pCO_2 to exceed that of plasma. Recent studies suggest that most patients with distal RTA cannot significantly elevate urine pCO_2. This finding strongly suggests that H^+ ion generation and *not* gradient limitation is etiologic (Halperin et al. 1974).

The importance of buffer availability for normal acid excretion was noted above. Hypophosphaturia, which occurs early in phosphate depletion, limits TA excretion and tends to acidify the host. Since acidosis stimulates renal ammonia synthesis, the fall in TA must be compensated for by the increase in ammonia excretion. Phosphate depletion causes HCO_3^- wasting (Emmett et al. 1977; Gold et al., 1973) in animals but this has yet to be recognized in man. One would expect primary defects in ammonia excretion to be tolerated less well since compensatory increments in phosphate buffer excretion are more restricted.

Hyperchloremia

Volume contraction caused by loss of NaHCO$_3$ with isotonic amounts of fluid, stimulates renal salt retention. Ingested NaCl is retained, effectively replacing NaHCO$_3$ with NaCl and causing a hyperchloremic acidosis.

Invasion of the ECF by strong acid causes NaHCO$_3$ to be replaced by the salt of the acid, as indicated in Equation 4:

$$H^+X^- + NaHCO_3 \longrightarrow NaX + CO_2 + H_2O \qquad (EQ.\ 4)$$

The sodium salt (NaX) is eventually filtered at the glomerulus and Na$^+$ is reabsorbed in exchange for secreted H$^+$ (Fig. 2.2b), thereby returning NaHCO$_3$ to the blood and excreting metabolic acid in association with buffer. Lack of acid secretion or buffer availability allows the acid salt to escape with isotonic amounts of fluid. The failure to regenerate HCO$_3^-$ causes acidosis, while the volume contraction leads to retention of dietary chloride. Hyperchloremic acidosis ensues.

The clinical presentation of RTA is greatly influenced by the changes in potassium, calcium, phosphate, and citrate metabolism that accompany both forms of RTA.

Potassium metabolism

Urinary K$^+$ wasting with depletion of body stores characterizes both forms of RTA. Hyperaldosteronism, stimulated by the salt and water losses that attend proximal and distal forms, and that plays a major role in the ensuing K$^+$ loss.

Potassium wasting in proximal RTA is proportional to the degree of bicarbonaturia. As previously noted, HCO$_3^-$ wasting occurs transiently during the induction of acidosis and disappears during the steady state. Therapy with poorly reabsorbed HCO$_3^-$ leads to more persistent bicarbonaturia. It follows, therefore, that potassium wasting is found only early in the course and when alkali therapy is given.

Neutralization of the negative luminal charge created by active distal Na$^+$ reabsorption is normally effected by countersecretion of H$^+$ and K$^+$. In distal RTA, this process is stimulated by hypersecretion of aldosterone. The H$^+$ secretory defect, however, allows only K$^+$ transport to dissipate the luminal charge. In this setting, alkali therapy repletes volume losses, thereby removing the stimulus to aldosterone secretion. Furthermore, alkalinization of the urine removes any gradient restriction to H$^+$ secretion, allowing H$^+$ to substitute for K$^+$ in the cation exchange for reabsorbed Na$^+$. Thus, NaHCO$_3$ therapy of RTA exacerbates kaliuresis in the proximal form and ameliorates it in the distal form. Despite volume and HCO$_3^-$ repletion, an occasional patient will have sustained hyperaldosteronism. The cause of its persistence is uncertain (Sebastian, McSherry, and Morris, 1971a; Sebastian, McSherry, and Morris, 1971b).

The neuromuscular and cardiac effects of potassium depletion (see Ch. 7) frequently play a prominent role in the clinical presentation of RTA.

Particularly devastating is the occasional hypokalemic paralysis that may involve muscles of respiration. The additive effects of respiratory acidosis on preexisting metabolic acidosis may be fatal.

Calcium, phosphate, and citrate metabolism

The interaction of acidosis with calcium, parathyroid hormone (PTH), and growth has been discussed previously in the section on PTH, calcium, and phosphate. Although systemic acid-base parameters may be equal, bone disease, hypercalciuria, nephrocalcinosis, and nephrolithiasis frequently complicate distal but rarely proximal RTA. Perhaps the positive H^+ ion balance in the former is required to mobilize bone calcium and simultaneously decrease renal calcium reabsorption, thereby causing marked hypercalciuria. The reabsorption of citrate—a major chelator of calcium—from the glomerular filtrate, is greatly enhanced by acidosis. The low concentration of this chelator in the calcium-rich urine of patients with distal RTA may impair solubilization of calcium, thereby predisposing them to nephrocalcinosis and stones. Bicarbonate therapy reverses the effects of acidosis on bone and kidney, thereby preventing the primary cause of morbidity in this disease. Citraturia and amino aciduria persist in proximal RTA as manifestations of the diffuse tubular transport defect—and chelate any increase in calciuria. Thus, bone pain, flank pain, and hematuria are frequent presentations of distal but *not* proximal RTA.

Distal RTA (Table 2.5)

The distal variant may present in its *complete* or *incomplete* form. Patients with incomplete RTA have normal acid-base parameters but a latent inability to maximally acidify their urine in response to an acid load. Complete RTA defines the presence of the same tubular defect, but subjects manifest a persistent hyperchloremic acidosis.

For unexplained reasons, patients with incomplete RTA are able to increase ammonia excretion and thereby excrete their daily load of acid despite a relatively alkaline urine. Hypercalciuria and hypocitraturia may occur despite normal arterial acid-base indices. Some of these patients go on to complete (i.e., acidotic) RTA, usually after developing nephrocalcinosis and consequently losing ammonia-producing renal mass. Thus, incomplete RTA probably represents an early compensated stage of RTA, which often, but not always, evolves into the complete form.

Symptoms of acidosis are relatively nonspecific (i.e., anorexia, fatigue, and dyspnea on exertion). Distal RTA is suspected, therefore, in any patient with unexplained bone disease, nephrocalcinosis or recurrent calcium stones. Muscle weakness, occasionally progressing to flaccid quadriplegia, polyuria, nocturia, and EKG abnormalities are expressions of hypokalemia and require prompt recognition of this disorder. With the advent of automated procedures for serum electrolytes, many cases are being discovered following the serendipitous finding of hyperchloremic metabolic acidosis. Knowledge of the numerous diseases that secondarily cause RTA allows the astute clinician to suspect it from "the company it keeps." Table 2.5 outlines those dis-

Table 2.5 Distal (gradient) renal tubular acidosis[a]

I. Primary
   A. Hereditary
   B. Sporadic
II. Secondary
   A. Dysproteinemias
      1. Hyperglobulinemias
      2. Amyloidosis
   B. Disordered calcium metabolism
   C. Edema-forming states
   D. Drugs
      1. Amphotericin-B
      2. Lithium carbonate
      3. Toluene
   E. Miscellaneous
      1. Renal transplantation
      2. Medullary sponge kidney (MSK)
      3. Hydronephrosis
      4. Wilson's disease
      5. Sickle cell anemia (SCA)
      6. Hypoaldosteronism

[a] Narins and Goldberg, 1977.

orders known to cause distal RTA. The clinical picture is often dominated by the primary disorder, but the development of acidosis and potassium and calcium wasting may have disastrous effects on the primary disease (see below). A brief commentary on some of the causes of the distal form follows. A more exhaustive analysis of these disorders has recently appeared (Narins and Goldberg, 1977).

Primary distal RTA

Distal RTA may develop sporadically without any other associated disorder or as part of an inherited disorder. The genetic form is poorly defined, since some cases have been associated with primary hypercalciuria, others with genetic dysproteinemias, while still others were unassociated with other metabolic disorders (Narins and Goldberg, 1977). Distal RTA may be the renal expression of many genetically transferred metabolic disorders that only secondarily affect the kidney.

Dysproteinemias

The frequency with which distal RTA complicates hyperglobulinemic states is especially striking in the dysglobulinemic "autoimmune disorders." Failure to maximally acidify the urine can be demonstrated in up to half of patients with biliary cirrhosis, Sjögren's syndrome, and hyperglobulinemic purpura. Round cell infiltration of the kidneys is frequently found in these disorders and, although yet unproven, the tubular dysfunction may have an immunologic basis.

Disordered calcium metabolism

The distal acidification defect that complicates the major disorders of calcium metabolism is usually, but not always, associated with nephrocalcinosis. The latter may not yet be demonstrable by x-ray when RTA develops. Thus, hypercalciuria and nephrocalcinosis may cause or result from distal RTA.

Edema-forming states

There is evidence to suggest that avid Na^+ reabsorption by the proximal tubule may so severely limit distal Na^+ delivery that Na^+-H^+ exchange suffers. Urine is inappropriately alkaline, net acid excretion is diminished, and a hyperchloremic acidosis may ensue. This has been most clearly defined in cirrhosis, where hyperglobulinemia may have an additive effect. The acid-base disorder improves with the onset of natriuresis.

Drugs

Amphotericin-B increases permeability of epithelial cell membranes to protons, and the ensuing backleak impairs maximal urinary acidification. The rise in urinary pCO_2 during HCO_3^- infusions shows that backleak and not H^+ ion generation are at fault (see above) (Roscoe et al., 1977). After a cumulative dose of 1 gm, most patients have an incomplete RTA that is reversible with discontinuance of the drug. Complete, but reversible, RTA occurs with greater dosage. *Lithium* impairs distal acidification in therapeutic doses in most patients. Only incomplete RTA has been shown to date. The inability to elevate urine pCO_2 with HCO_3^- infusions suggests lithium impairs distal H^+ generation.

Renal transplantation

Approximately 50 percent of patients receiving cadaver transplants will have permanent incomplete RTA when challenged with an acid load months to years later. Kidneys from living related donors seem to have a lesser incidence. The hyperglobulinemia and round cell infiltration of the transplanted kidney suggests that the functional disorder may be an expression of immunologic attack. The development of overt (i.e., complete) acidosis is prevented by increased ammoniagenesis; therefore, *any renal insult—* including rejection—that impairs ammonia production will unmask the RTA by causing a hyperchloremic acidosis to develop.

Medullary sponge kidney (MSK)

The increased incidence of complete and incomplete distal RTA in patients with MSK suggests that the cystic dilatation of collecting ducts that characterizes this disorder may disrupt acid secretion. The disease is generally benign unless complicated by RTA, stones, or infection.

Hydronephrosis

Reversible, complete, or incomplete RTA often accompanies chronic hydronephrosis thereby making hyperchloremic acidosis a potential clue to the presence of otherwise unsuspected urinary tract obstruction. Distal tubular dysfunction also accompanies relief of acute obstruction (Thirakomen et al., 1976).

Sickle cell anemia (SCA)

Incomplete RTA and nephrogenic diabetes insipidus are functional counterparts of the interstitial nephritis complicating long-standing SCA. Progression to complete (i.e., overt or permanent acidosis) RTA has yet to be documented, but should be watched for since low pH increases sickling of S hemoglobin and should increase the incidence of crises. Interestingly, nephrogenic diabetes insipidus, but *not* RTA, complicates otherwise asymptomatic patients with sickle cell *trait*.

Hypoaldosteronism

The multiplicity of complex biochemical reactions that must take place before renal renin secretion eventuates in Na^+ retention and K^+ and H^+ excretion are all potential targets for disease. Defective hepatic synthesis of renin substrate, primary hyporeninism, failure of converting enzyme to generate angiotensin II from I, impaired adrenal release of aldosterone because of defective biosynthesis or damaged angiotensin II receptors, and, finally, impairment of renal tubular response to mineralocorticoids are all potential or described causes of hyperkalemia and hyperchloremic acidosis. These syndromes are described in greater detail in Chapter 8. It would appear that decreased mineralocorticoid activity causes acidosis by diminishing distal H^+ secretory capacity, which allows the fraction of filtered HCO_3^- normally reabsorbed distally to escape in the urine. Buffer excretion diminishes owing to defective ammonia synthesis. Diminished renal blood flow limits delivery of ammonia's precursor—glutamine—and hyperkalemia impairs renal biosynthesis of ammonia, perhaps by alkalinizing the cell. Lowering of serum potassium with binding resins or treatment with exogenous mineralocorticoids normalizes acid-base balance.

Proximal RTA (Table 2.6)

Unlike the distal variant, incomplete forms of proximal RTA are rarely recognized. The author, however, has seen several children with cystinosis whose early course was characterized by gluco- and aminoaciduria, renal potassium wasting, and normal serum HCO_3^-. It was reasoned that a small defect in proximal HCO_3^- reabsorption forced an increase in distal Na^+/K^+-H^+ exchange, enabling the kidney to reclaim HCO_3^-, at the expense of K^+ wasting. The hypokalemia conceivably increases proximal HCO_3^- reabsorption, thereby establishing a new steady state. When the proximal defect

Table 2.6 Proximal (bicarbonate-wasting) renal tubular acidosis[a]

I. Childhood forms
   A. Primary
   B. Secondary
      1. Amino acid storage diseases
         a. Tyrosinosis
         b. Cystinosis
      2. Altered carbohydrate metabolism
         a. Glycogenosis Type 1
         b. Galactosemia
         c. Hereditary fructose intolerance
      3. Miscellaneous
         a. Lowe's syndrome
         b. Pyruvate carboxylase deficiency
         c. Nephrotic syndrome
II. Adult forms
   A. Primary
   B. Secondary
      1. Heavy metal toxicity
         a. Cadmium   c. Copper
         b. Lead   d. Mercury
      2. Drugs
         a. Carbonic anhydrase inhibitors
         b. Streptozotocin
         c. Tetracycline (outdated)
         d. 6-mercaptopurine
      3. Hormones
         a. Hypervitaminosis D
         b. Hyperparathyroidism
      4. Dysproteinemias and Malignancies
         a. Myeloma
         b. Sjögren's syndrome
         c. Amyloidosis
         d. Nephrotic syndrome
      5. Miscellaneous
         a. Renal transplantation

[a] Narins and Goldberg, 1977.

worsened, HCO_3^- wasting increased, overwhelmed the distal tubule, and an acidosis developed.

Childhood forms

Except for children, HCO_3^- wasting rarely occurs without associated phosphate, glucose, uric acid, and amino acid wasting. Thus proximal HCO_3^- wasting is usually easily diagnosed by "the company it keeps."

Primary bicarbonate wasting: Bicarbonaturia unassociated with other signs of proximal damage occurs as a transient defect, primarily in male infants. Therapy with large doses of $NaHCO_3^-$—often requiring up to 10 mEq/kg/day—reestablishes normal growth and in months to years, a complete remission occurs without loss of renal function. The primary Fanconi syndrome occurs as an inherited or sporadic disorder and has a good prognosis as long as alkali, phosphate, and vitamin D therapy are used appropriately. Growth retardation and bone diseases are its primary presenting signs.

Amino acid storage disorders: The Fanconi syndrome is the functional expression of early cystinosis, which goes on to uremia in 10–20 years. Several children have received renal transplants without recurrence of the disorder.

Disorders of carbohydrate metabolism: Rapidly reversible defects in proximal tubular transport results from ingestion of fructose or galactose in patients with deficiencies of the enzymes fructoaldolase or galactose uridyltransferase. Dietary abstinence is the best therapy. Renal glycogen storage in von Gierke's disease rarely causes functional damage.

Adult forms

Metals: Exposure to certain *heavy metals* may lead to HCO_3^- wasting as part of a Fanconi syndrome. Copper toxicity in Wilson's disease more commonly causes distal RTA. Acidosis is usually very mild and the tubular lesion usually responds to penicillamine therapy.

Drugs: Improved methods of processing and storage have removed tetracycline as a cause of Fanconi syndrome. Carbonic anhydrase inhibitors acetozolamide (Diamox) and sulfamylon cause mild bicarbonaturia with stabilization of serum HCO_3^- at 20-22 mEq/l. The latter drug is an antipseudomonas antibiotic used in burn patients and is systemically absorbed from denuded skin.

Hormones: Interaction of PTH, vitamin D, and calcium with acid-base balance were discussed under uremic acidosis and distal RTA.

Dysproteinemias: In myeloma, Bence Jones proteinuria may disrupt proximal tubular function and the resulting acidosis, phosphaturia, and kaliuresis have disastrous effects on the patient. Myeloma-induced bone disease is exacerbated by acidosis and phosphate depletion, while the defect in urinary concentration induced by potassium depletion may cause dehydration which in turn may predispose to tubular precipitation of protein.

In some nephrotics, continuous hyperfiltration and subsequent reabsorption of protein impairs proximal function. While amino acid wasting is very common, a full blown Fanconi syndrome also may occur. The defect, when present, is usually subclinical.

Hyperglobulinemia and malignancy: Among the causes of hyperglobulinemia associated with RTA, only Sjögren's syndrome and myeloma have caused proximal defects; the rest affect the distal tubule. Patients with myelomonocytic leukemia often develop tubular reabsorptive defects that correlate best with the duration of their lysozymuria. The low molecular weight enzyme is overproduced by white blood cells (not lymphocytes), and its hyperfiltration surpasses normal proximal tubular capacity to reabsorb it, resulting in overflow lysozymuria. Prolonged exposure of proximal cells to excess lysozyme somehow damages proximal transport.

Renal transplantation: Approximately 30 percent of patients receiving a renal transplant will develop a variety of proximal tubular defects—including bicarbonaturia—in the immediate posttransplant period. Bicarbonaturia is a transient defect in the great majority of patients. Slow posttransplant involution of hyperplastic parathyroid glands may play a minor role.

Low-Buffer excretion

Phosphate depletion

The progressive loss of NH_3 excretion with advancing renal disease steadily increases dependence on phosphate as the prime urinary buffer. Limitation of dietary phosphate by restricting protein intake and use of phosphate-binding gels (aluminum hydroxide) diminishes net acid excretion and leads to systemic acidosis. The frequent absence of this complication in many such patients suggests that the fall in PTH levels attending the rise in serum calcium—consequent to phosphate binding—diminishes HCO_3^- wasting. The offsetting effects of diminished buffer excretion and retention of HCO_3^- may leave serum HCO_3^- unchanged. In animals, phosphate depletion mobilizes bone alkali and simultaneously decreases renal HCO_3^- reabsorption (Emmett et al., 1977). The offsetting effect of the alkalinizing process and the acidifying process was to leave serum HCO_3^- unchanged. Whether humans respond similarly is as yet unexplored.

Diminished ammonia excretion

The central role played by impaired NH_3 excretion in "uremic" acidosis has been discussed. Although its pathogenesis and significance are uncertain, gouty patients, despite normal renal function, persistently excrete reduced amounts of NH_3. The reduction of urinary buffer lowers urine pH and allows TA to increase in compensation. The solubility of uric acid decreases at low urine pH thereby adding to the morbidity of gout. The hyperkalemic states (e.g., hypoaldosteronism) impair renal NH_3 synthesis, thereby causing a hyperchloremic acidosis (see above and Ch. 8). The primary defect in NH_3 excretion, shown in a patient with systemic lupus erythematosus (Hadler, Gill and Gardner, 1972), suggests other patients with renal acidosis may have congenital or acquired defects in NH_3 excretion.

Diagnosis of RTA (Table 2.7)

Incomplete RTA

This latent tubular defect is diagnosed by demonstrating failure of appropriate urinary acidification in response to an acid stress. Its presence is suspected in patients with renal and extrarenal disorders known to impair urinary acidification (Table 2.8) and all cases of otherwise unexplained radiopaque renal stones. Early diagnosis and initiation of alkali therapy may prevent the bony and renal consequences that attend the hypercalciuria of RTA. Failure to decrease urinary pH to <5.3 following an NH_4Cl acid load (Wrong and Davies, 1959), defines incomplete RTA. Urine, voided into containers with a few ml of mineral oil (to prevent CO_2 loss), is ana-

Table 2.7 Diagnostic approach to the renal tubular acidoses

I. Clinical suspicion
   A. Bone disease
   B. Muscle weakness
   C. Nephrocalcinosis
   D. Opaque kidney stones
   E. Normal anion gap acidosis
   F. Glycosuria
   G. Associated diseases known to cause RTA (Tables 2.5, 2.6)

II. Laboratory confirmation
   A. Exclude other normal anion gap acidoses (Table 2.8).
   B. Low buffer excretion; persistently low UpH, low PO_4, or ammonia excretion.
   C. Proximal versus distal defect:

	Proximal	Distal
Associated wasting of: PO_4, glucose, amino or uric acids	Usual	Rare
Severe K depletion	Uncommon	Common
Nephrocalcinosis/lithiasis	Rare	Common
Urine pH		
1st AM	< 6.0	> 6.0
Postacid load	< 5.3	> 5.3
Bicarbonate Fractional excretion	May be > 15%	< 5–10%
$UpCO_2$	$UpCO_2 > B\ pCO_2$	$UpCO_2 = B\ pCO_2$
Ease of Replacement	Resistant (Need > 3–5 mEq/kg/day)	Sensitive (Need ≤ 2–3 mEq/kg/day)

lyzed for pH with an electrode before and following acid ingestion. Ammonium chloride, in capsular or liquid form (tablets are poorly absorbed), is ingested—as 0.1 gm/kg—over 30–45 minutes, and hourly urine specimens are collected for the next six hours. Arterial or free-flowing venous blood

Table 2.8 Differential diagnosis of normal anion gap metabolic acidosis[a]

Serum potassium	
Normal–High	Low

A. Hyperalimentation
B. Posthypocapnia
C. Rapid hydration (dilutional acidosis)
D. Hypoaldosteronism
  1. Hyporeninism
  2. Selective or diffuse adrenal damage
  3. Failure of tubular response to aldosterone
E. NH_4Cl, $CaCl_2$ (oral), lysine or arginine hydrochloride therapy
F. Early "uremic" acidosis
	A. Gastrointestinal disorders
	  1. Diarrhea
	  2. Pancreatic fistula
	B. Ureteral diversions
	  1. Ureterosigmoidostomy
	  2. Ileal bladder (obstructed)
	C. Renal tubular acidosis
	  1. Proximal RTA
	  2. Distal RTA
	  3. Lack of buffer

[a] Emmett and Narins, 1977.

should be analyzed for electrolytes prior to and three hours after ingesting the acid. This dose of NH_4Cl should lower serum HCO_3^- by 2–5 mEq/l. If urine pH does not fall to 5.3 or less, incomplete RTA is diagnosed. Urinary citrate and calcium ought to be measured in a 24-hour specimen. Oral $CaCl_2$ acidification may be used in patients with liver failure in whom use of NH_4Cl may be dangerous (Oster, et al., 1975).

Complete RTA

It is usually a simple matter to go through a differential diagnosis of normal anion gap acidosis (Table 2.8) and arrive at RTA.

Tissue buffering and renal acid retention are the compensatory processes that lower serum HCO_3^- in response to the primary decrease in pCO_2 in patients with respiratory alkalosis. Although the low total CO_2 and elevated chloride mimic a normal anion gap acidosis, serum pH is high, not low, and pCO_2 is strikingly depressed.

The clinical history and physical examination allow one to quickly exclude other causes of hyperchloremic acidosis and focus attention on the kidney. Early "uremic" acidosis is excluded by the normal BUN and creatinine. Localization of the defect to the proximal or distal tubule is usually a simple matter since the proximal variant is rarely associated with disordered calcium metabolism and is almost always complicated by hypophosphatemia, hypouricemia, and glycosuria. If doubt remains, a first morning urine pH <6.0 or a fall to ≤5.3 following an acid load, will rule out distal RTA. The response to a *NaHCO3 load* will also differentiate the two. When enough alkali is given to sustain a normal serum HCO_3^- (23–26 mEq/l), patients with a proximal defect may excrete >15 percent of filtered HCO_3^-, while those with distal will rarely excrete >5–10 percent of the filtered load. There is overlap since mild proximal RTA need not reject as much as 15 percent of filtered HCO_3^-. A random, untimed urine specimen and blood specimen are both assayed for creatinine and HCO_3^-, and the fractional excretion (FE) of alkali is calculated, as shown in Equation 5 as:

$$FE_{(HCO_3^-)} = 100 \times \frac{(HCO_3^-)_u \times (Cr)_p}{(HCO_3^-)_p \times (Cr)_u} \quad (EQ. 5)$$

(Subscripts refer to urine (u) and plasma (p) in mEq/l and mg/dl). The $FE_{(HCO_3^-)}$ determines the ease with which alkali may be replaced. Patients with proximal RTA require >3–5 mEq/kg/day of $NaHCO_3$, whereas those with distal are far more sensitive, usually requiring 50–100 mEq/Day (<2–3 mEq/kg/Day).

The assay of urinary phosphate and NH_3 will diagnose the rare patient with renal acidosis on the basis of low buffer excretion. These patients should have persistently acid urine with none of the stigmata of proximal damage. To fully assess NH_3 production, a *three-day acid load* should be given. This persistent and chronic acidosis more fully stimulates renal ammonia synthesis. After collecting a 24-hour control urine, 0.05–0.2 gm/kg of NH_4Cl should be ingested in three divided doses on each of three succeeding days. A second 24-hour urine is collected over the day following the last day of acid load (i.e., on the fourth day). Serum electrolytes should be drawn daily. Ingested NH_4 is converted to urea by the liver and will not significantly influence urinary NH_3. Normal range and mean values (Elkinton et al., 1960) on the fourth day are: urine pH: 4.97 (4.5–5.3); TA: 33 (21–46) mEq/Day; NH_3: 63 (36–99) mEq/Day.

Therapy

The acidosis and associated disorders of potassium, calcium, and phosphate metabolism in patients with various forms of RTA always require chronic therapy and may occasionally present as a crisis or medical emergency.

Distal RTA

Management of "crisis": When potassium depletion accompanying severe normal anion gap acidosis is so profound that clinically apparent muscle weakness is present, patients must be admitted immediately to the intensive care unit. Muscle weakness may cause fatal exacerbation of acidosis and, if therapy is not wisely selected and carefully monitored, survival is unlikely. This "crisis" most commonly complicates distal RTA.

When K^+ depletion causes respiratory muscle paresis, the resulting hypercapnia and hypoxia may be expected to profoundly worsen the acidosis. The rising pCO_2 negates the respiratory compensation that had been stabilizing pH, and adds an element of respiratory acidosis to the metabolic acidosis. Hypoxia, by stimulating anaerobic metabolism and lactic acid production, will worsen the metabolic acidosis.

If initial alkali therapy is not supplemented with K^+, one can expect worsening of preexisting hypokalemia. Furthermore, if alkali were inappropriately given in dextrose-containing solutions, the normal insulin response to hyperglycemia would act to further lower K^+.

Respiratory compromise may be judged clinically and by the appropriateness of pCO_2 vis-à-vis the degree of metabolic acidosis (see above). If pCO_2 is elevated, patients should be intubated, EKG monitoring begun, and parenteral KCl given in large doses (40–60 mEq in the first hour) without HCO_3^-. This dose of KCl is usually sufficient to elevate serum levels ≥ 3.0 mEq/l. Potassium and HCO_3^- may be given together thereafter in non-dextrose-containing fluid. The pCO_2 should be maintained at 20–30 mm/Hg.

The dose of $NaHCO_3$ required for correction may be deceptively high, since K^+ depletion causes a rather striking *intracellular* acidosis. The HCO_3^- "space," in liters, may be calculated as 50 percent of body weight and used for calculating the dose required. If a 70 kg patient had a serum HCO_3^- of 5 mEq/l and it was elected to increase it to 15 mEq/l, then, 10 mEq must be added to each of his 35 liters (0.5×70) of space. Half of these 350 mEq may be given over the next four to six hours and electrolytes, pH, pCO_2 followed. Further therapy is given as needed. Alkali therapy may exacerbate the mild hypocalcemia that frequently complicates this disorder. The Q-T interval on the EKG should be followed and the need for calcium assessed.

Chronic therapy: In cases of complete distal RTA, it must be carefully stressed and clearly understood by all patients that alkali therapy is to RTA as insulin is to diabetes. Both medications must be taken daily for the *remainder of the patient's life.* It is inexcusable to have patients return with nephrocalcinosis and azotemia years after having had RTA diagnosed and therapy prescribed. It is critical to have patients fully understand this need for therapy and the consequences of not adhering to it.

Alkali therapy will stop hypercalciuria and stone formation, heal bones and may lead to resorption of nephrocalcinosis. Sodium bicarbonate in doses of 0.5–3.0 mEq/kg/day in divided doses will suffice. Part of the daily dose may be given as $KHCO_3$. This is one situation in which oral, non-KCl solutions will effectively replenish body K^+ stores. Potassium citrate, gluconate, or acetate may be used. For those patients finding the eructations and abdominal distension often associated with $NaHCO_3$ therapy unacceptable, a sodium citrate solution (Shohl's) is a fine substitute. The citrate is absorbed and converted to HCO_3^- in the liver, thereby avoiding the gastric CO_2 generation caused when oral $NaHCO_3$ combines with HCl in the stomach. Each ml of Shohl's solution represents one mEq of HCO_3^- and may be made up as Na^+ and K^+ salts of citrate, depending on the patient's potassium needs.

If alkali therapy does not stop calcium wasting and ameliorate bone pain, vitamin D and calcium therapy may be initiated. This is usually unnecessary and carries with it added risk of hypercalcemia. Patients should be warned not to markedly increase their activity for the first few months of therapy. The increased exercise following their improved sense of well-being may lead to fractures.

Proximal RTA

Adults with serum HCO_3^- concentrations ≥ 18 mEq/l, without bone disease, probably do not need therapy. Children apparently suffer growth retardation from acidosis and probably should always receive alkali. Doses of 6–10 mEq/kg/day of HCO_3 (half as the potassium salt) are often required. Mild volume depletion induced by diuretics often increases HCO_3^- reabsorptive capacity and allows one to reduce the daily dose of alkali required. Associated defects in phosphate reabsorption will require vitamin D and oral phosphate.

REFERENCES

Agus, Z., Puschett, J. B., Senesky, D. et al. (1971). Mode of action of parathyroid hormone and cyclic adenosine 3'.5'-monophosphate on renal tubular phosphate reabsorption in the dog. *Journal of Clinical Investigation*, 50, 617.

Albert, M. D., Dell, R. B. & Winters, R. W. (1967). Quantitative displacement of acid-base equilibrium in metabolic acidosis. *Annals of Internal Medicine*, 66, 312.

Bank, N. & Aynedjian, H. S. (1976). A micropuncture study of the effect of parathyroid hormone on renal bicarbonate reabsorption. *Journal of Clinical Investigation*, 58, 336.

Bricker, N. S., Bourgoignie, J. J. & Weber, H. (1976). In *The Kidney*, ed. Brenner, B. M., & Rector, F. C., Jr., Ch. 18, pp. 703–736. Philadelphia: W. B. Saunders.

Burnell, J. M. & Teubner, E. (1971). Changes in bone sodium and carbonate in metabolic acidosis and alkalosis in the dog. *Journal of Clinical Investigation*, 50, 327.

Coe, F. L. (1972). Magnitude of metabolic acidosis in primary hyperparathyroidism. *Archives of Internal Medicine*, 134, 262.

Coe, F. L., Firpo, J., Jr., Hollandsworth, D. L. et al. (1975). Effect of acute and chronic metabolic acidosis on serum immunoreactive parathyroid hormone in man. *Kidney International*, 8, 262.

Crumb, C. K., Martinez-Maldonado, M., Eknoyan, G. et al. (1974). Effects of volume expansion, purified parathyroid extract, and calcium on renal bicarbonate absorption in the dog. *Journal of Clinical Investigation*, 54, 1287.

Elkinton, J. R., Huth, E. J., Webster, G. D. et al. (1960). The renal excretion of hydrogen ion in renal tubular acidosis. I. Quantitative assessment of the response to ammonium chloride in acid load. *American Journal of Medicine*, 29, 554.

Elkinton, J. R. (1962). Hydrogen ion turnover in health and in renal disease. *Annals of Internal Medicine*, 57, 660.

Elkinton, J. R. (1966). Clinical disorders of acid-base regulation. *Medical Clinics of North America*, 50, 1325.

Elkinton, J. R., McCurdy, D. K. & Buckalew, S. M., Jr. (1967). In *Renal Disease*, 2/e, ed. Black, D. A. K. Pp. 110–135. Philadelphia: F. A. Davis.

Emmett, M., Goldfarb, S., Agus, Z. S. et al. (1977). The pathophysiology of acid-base changes in chronically phosphate-depleted rats. *Journal of Clinical Investigation*, 59, 291.

Emmett, M. & Narins, R. G. (1977). The clinical use of the anion gap. *Medicine*, 56, 38.

Epstein, F. H. (1968). Calcium and the kidney. *American Journal of Medicine*, 45, 700.

Espinel, C. H. (1975). The influence of salt intake on the metabolic acidosis of chronic renal failure. *Journal of Clinical Investigation*, 56, 286.

Franklin, S. S., Kleeman, C. R., Vilamill, M. et al. (1966). Effect of the Giordano-Giovanetti low-protein diet on the renal response of normal subjects to acute potassium and acid loads. *Clinical Research*, 14, 377.

Gold, L. G., Massry, S. G., Arieff, A. I. et al. (1973). Renal bicarbonate wasting during phosphate depletion: a possible cause of altered acid-base homeòstasis in hyperparathyroidism. *Journal of Clinical Investigation*, 52, 2556.

Goodman, A. D., Lemann, J., Jr., Lennon, E. J. et al. (1965). Production, excretion and net balance of fixed acid in patients with renal acidosis. *Journal of Clinical Investigation*, 44, 495.

Hadler, N. M., Gill, J. R. & Gardner, J. D. (1972). Impaired renal tubular secretion of potassium, elevated sweat sodium chloride concentration and plasma inhibition of erythrocyte sodium outflux as complications of systemic lupus erythematosus. *Arthritis and Rheumatism*, 15, 515.

Halperin, M. L., Goldstein, M. B., Haag, A. et al. (1974). Studies on the pathogenesis of Type I (distal) renal tubular acidosis as revealed by the urinary PCO_2 tensions. *Journal of Clinical Investigation*, 53, 669.

Hellman, D. E., Au, W. Y. & Bartter, F. C. (1965). Evidence for a direct effect of parathyroid hormone on urinary acidification. *American Journal of Physiology*, 209, 643.

Husted, F. C., Nolph, K. D. & Maher, J. F. (1975). $NaHCO_3$ and NaCl tolerance in chronic renal failure. *Journal of Clinical Investigation*, 56, 414.

Johnson, C. W. & Morgan, J. M. (1965). Acidosis: a clue to the etiology of renal failure. *Southern Medical Journal*, 58, 1513.

Karlinsky, M. L., Sager, D. S., Kurtzman, N. A. et al. (1974). Effect of parathyroid hormone and cyclic adenosine monophosphate on renal bicarbonate reabsorption. *American Journal of Physiology*, 227, 1226.

Kassirer, J. P. & Bleich, H. (1965). Rapid estimation of plasma carbon dioxide from pH and total carbon dioxide content. *New England Journal of Medicine*, 272, 1067.

Lathem, W. (1958). Hyperchloremic acidosis in chronic pyelonephritis. *New England Journal of Medicine*, 258, 1031.

Lee, S. W., Russell, J. & Avioli, L. V. (1977). 25-hydroxycholecalciferol to 1,25-dihydroxycholecalciferol: conversion impaired by systemic metabolic acidosis. *Science*, 195, 4282.

Lemann, J., Jr., Litzow, R. & Lennon, E. J. (1966). The effects of chronic acid loads in normal man: further evidence for the participation of bone mineral in the defense against chronic metabolic acidosis. *Journal of Clinical Investigation*, 45, 1608.

Lennon, E. J. & Lemann, J., Jr. (1966). Defense of hydrogen ion concentration in chronic metabolic acidosis. *Annals of Internal Medicine*, 65, 265.

Mitchell, R. A. & Singer, M. M. (1965). Respiration and cerebrospinal fluid pH in metabolic acidosis and alkalosis. *Journal of Applied Physiology*, 20, 905.

Mitchell, J. H., Wildenthal, K. & Johnson, R. L., Jr. (1972). The effects of acid-base disturbances on cardiovascular and pulmonary function. *Kidney International*, 1, 375.

Moller, B. (1959). Hydrogen ion concentration in arterial blood. *Acta Medica Scandinavica*, 165, 348.

Morrin, P. A. F., Gedney, W. B., Newmark, L. M. et al. (1962). Bicarbonate reabsorption in the dog with experimental renal disease. *Journal of Clinical Investigation*, 41, 1303.

Muldowney, F. P., Freaney, R. & McGeeney, D. (1968). Renal tubular acidosis and amino-aciduria in osteomalacia of dietary or intestinal origin. *Quarterly Journal of Medicine*, 38, 517.

Muldowney, F. P., Carroll, D. V., Donohue, J. F. et al. (1971). Correction of renal bicarbonate wastage by parathyroidectomy. *Quarterly Journal of Medicine*, 40, 487.

Narins, R. G. & Goldberg, M. (1977). Renal tubular acidosis: pathogenesis, diagnosis and treatment. *Disease-a-Month*, March.

Nash, M. A., Torrado, A. D., Greifer, I. et al. (1972). Renal tubular acidosis in infants and children. *Journal of Pediatrics*, 80, 738.

Oster, J. R., Hotchkiss, J. L., Carbon, M. et al. (1975). A short duration renal acidification test using calcium chloride. *Nephron*, 14, 281.

Poole-Wilson, P. A., Patrick, J., MacGregor, G. A. et al. (1972). Renal excretion of bicarbonate and hydrogen ions: effects of mannitol diuresis in normal man. *Clinical Science*, 43, 561.

Poppell, J. W., Vanamee, P., Roberts, K. E. et al. (1956). The effect of ventilatory insufficiency on respiratory compensations in metabolic acidosis and alkalosis. *Journal of Laboratory and Clinical Medicine*, 47, 885.

Puschett, J. B., Zurbach, P. & Sylk, D. (1976). Acute effects of parathyroid hormone on proximal bicarbonate transport in the dog. *Kidney International*, 9, 501.

Reiss, E., Canterbury, J. M. & Egdahl, R. H. (1968). Experience with a radioimmunoassay of parathyroid hormone in human sera. *Transactions of the American Academy of Physicians*, 81, 104.

Relman, A. S. (1964). Renal acidosis and renal excretion of acid in health and disease. *Advances in Internal Medicine*, 12, 295.

Richards, P., Chamberlain, M. J. & Wrong, O. (1972). Treatment of osteomalacia of renal tubular acidosis by sodium bicarbonate alone. *Lancet*, 2, 994.

Roberts, K. E., Randall, H. T., Vanamee, P. et al. (1959). Renal mechanisms involved in bicarbonate absorption, *Metabolism*, 5, 404.

Rodman, J. S. & Heinemann, J. O. (1975). Parathyroid hormone and the regulation of acid-base balance. *American Journal of Medical Sciences*, 270, 481.

Roscoe, J. M., Goldstein, M. B., Halperin, M. L. et al. (1977). Effect of amphotericin-B on urine acidification in rats: implications for the pathogenesis of distal renal tubular acidosis. *Journal of Laboratory and Clinical Medicine*, 89, 463.

Sebastian, A., McSherry, E. & Morris, R. C., Jr. (1971a). On the mechanism of renal potassium wasting in renal tubular acidosis associated with the Fanconi syndrome (type 2 RTA). *Journal of Clinical Investigation*, 50, 231.

Sebastian, A., McSherry, E. & Morris, R. C., J. (1971b). Renal potassium wasting in renal tubular acidosis (RTA). *Journal of Clinical Investigation*, **50**, 667.

Schambelan, M., Stockigt, J. R. & Biglieri, E. G. (1972). Isolated hypoaldosteronism in adults. A renin-deficiency syndrome. *New England and Journal of Medicine*, **287**, 573.

Schwartz, W. B., Hall, P. W. III, Hays, R. M. et al. (1959). On the mechanism of acidosis in chronic renal failure. *Journal of Clinical Investigation*, **38**, 39.

Slatopolsky, E., Hoffsten, P., Purkerson, M. et al. (1970). On the influence of extracellular fluid volume expansion and of uremia on bicarbonate reabsorption in man. *Journal of Clinical Investigation*, **49**, 988.

Thirakomen, K., Kozlov, N., Arruda, J. A. L. et al. (1976). Renal hydrogen ion secretion after release of unilateral ureteral obstruction. *American Journal of Physiology*, **231**, 1233.

van Ypersele de Strihou, C. & Frans, A. (1970). The pattern of respiratory compensation in chronic uremic acidosis. *Nephron*, **7**, 37.

Welbourne, T., Weber, M. & Bank, N. (1972). The effect of glutamine administration on urinary ammonium excretion in normal subjects and patients with renal disease. *Journal of Clinical Investigation*, **51**, 1852.

Wrong, O. & Davies, H. E. F. (1959). Excretion of acid in renal disease. *Quarterly Journal of Medicine*, **28**, 259.

3
Lactic acidosis

ARNOLD S. RELMAN

Introduction
Review of lactate metabolism
The biochemistry of lactate
The organ physiology of lactate
Acid-base aspects of lactate metabolism
    Lactate turnover and net acid-base balance
    The feedback control of lactate production by hydrogen ions
Lactate in blood, urine and spinal fluid
Clinical lactic acidosis
Introduction and definitions
The clinical characteristics of lactic acidosis
The clinical causes of lactic acidosis
    Tissue hypoxia
    Drugs

    Fructose and sorbitol
    Epinephrine
    Liver disease
    Neoplastic disease
    Pulmonary embolism
    Sepsis
Diabetes mellitus and lactic acidosis
Idiopathic or spontaneous lactic acidosis
Congenital lactic acidosis
    Deficiency of glucose-6-phosphatase
    Deficiency of fructose-1, 6-diphosphatase
    Pyruvate carboxylase deficiency
    Defects in oxidation of pyruvate
    Defects in oxidative phosphorylation
Treatment

INTRODUCTION

The accumulation of lactate in the blood has long been recognized as a normal response to exercise and as a manifestation of acute tissue hypoxia secondary to shock or asphyxia. However, it was not until 1961 that Huckabee (1961 a,b) first called attention to the fact that hyperlactatemia could also appear under a variety of clinical circumstances in which neither shock nor hypoxemia was evident. Since then a growing profusion of reports in the literature have attested to the frequency and importance of this kind of metabolic disorder. Hyperlactatemia is now emerging as probably the most common cause of acute metabolic acidosis, and is attracting widespread interest as a biochemical sign of paramount clinical significance.

    The purpose of this review is to provide clinicians with a brief summary of what is currently known about the normal metabolism of lactate and how it may become disturbed under many different clinical conditions. Lactic acidosis is defined, and differentiated from other types of metabolic acidosis: its diverse etiologies are described, and its clinical course and management are discussed. No attempt will be made to survey all the literature on this subject; instead, for those readers who wish to explore the field in more detail reference will be made to selected key articles and to reviews that provide background information and more extensive bibliographies. Among the latter is the monograph by Cohen and Woods (1976) and articles by Alberti and Nattrass (1977), Cohen and Simpson (1975) and Oliva (1970).

Emphasis will be placed primarily on those aspects of the subject that have practical clinical significance. However, good clinical practice in this field depends upon a clear understanding of physiological mechanisms. Therefore we shall begin our discussion with a brief review of the biochemistry and physiology of lactate. The relevance of this basic information will become apparent in subsequent sections.

REVIEW OF LACTATE METABOLISM

The biochemistry of lactate

The only biochemical reaction that produces or consumes lactate in the human organism is shown in reaction (1):

$$\text{Reaction 1: } (\text{Pyruvate})^- + \text{NADH} + \text{H}^+ \underset{(\text{LDH})}{\rightleftharpoons} (\text{Lactate})^- + \text{NAD}^+$$

The enzyme lactate dehydrogenase (LDH), present in the cytosol of all tissues, catalyzes a reversible reaction that converts pyruvate to lactate and *vice versa*. The formation of lactate is thus a kind of metabolic "dead-end" in which the lactate, once formed from pyruvate, has no further metabolic path to travel other than to go back again to pyruvate. As shown in Reaction 1 lactate is formed by the reduction of pyruvate with reduced nicotinamide adenine dinucleotide (NADH), and is converted back to pyruvate by oxidation with the oxidized form of the dinucleotide (NAD$^+$). The NADH required for net production of lactate is generated during the conversion of glucose or glycogen to pyruvate in the cytosol, whereas the NAD$^+$ needed for the net transformation of lactate back to pyruvate must be produced by the oxidation of NADH in the mitochondria.

The LDH reaction is usually very close to equilibrium, so it can be rewritten in the form shown in equation (1):

$$\text{Equation 1: } [\text{Lactate}^-] = \text{Keq} \times \frac{[\text{Pyruvate}^-] \times [\text{NADH}] \times [\text{H}^+]}{[\text{NAD}^+]}$$

in which Keq is the equilibrium constant for the LDH reaction and [H$^+$] is the hydrogen ion concentration in the cytosol. In this equation the concentration of lactate in the cytosol is seen to be determined by the product of two variable factors: 1) the ratio [H$^+$] [NADH]/[NAD$^+$], and 2) the concentration of pyruvate. Thus, for a given concentration of pyruvate in tissues the lactate concentration will be a direct function of the [H⁺] [NADH]/[NAD⁺] ratio and, at any given value of this ratio, the lactate concentration will be a direct function of the pyruvate concentration.

Ignoring the term [H⁺] for the moment, and considering only the ratio [NADH]/[NAD⁺] in equation (1) this ratio reflects the state of oxidation or reduction of the cytosolic pyridine nucleotides. In cells containing mitochondria, the ratio is ultimately dependent upon oxidative reactions in the mitochondria. Therefore, any impairment in these reactions, whether due to decreased delivery or availability of oxygen or to some defect in the complex reaction pathway by which NADH is oxidized, will increase the [NADH]/[NAD⁺] ratio and thereby, as predicted by equation (1), increase the lactate concentration. As we shall see below, impairment of mitochondrial oxidative reactions also leads to accumulation of pyruvate, which further contributes to the rise in lactate.

The terms of equation (1) are transposed in equation (2), to give an expression for the ratio of lactate to pyruvate (L/P).

Equation 2: $$\frac{[\text{Lactate}^-]}{[\text{Pyruvate}^-]} = K_{eq} \times \frac{[\text{NADH}]}{[\text{NAD}^+]} \times [\text{H}^+]$$

If we again ignore any variations in [H⁺], we see that the L/P ratio is a direct function of the [NADH]/[NAD⁺] ratio and, hence of the oxidative state of the cell. The normal value of the product of all the terms on the right hand side of equation (2) is approximately 10, so the normal cytosolic ratio of lactate to pyruvate is 10 to 1, which indicates that the LDH equilibrium favors the formation of lactate. Tissue L/P ratios substantially greater than 10 are taken to mean an increase in the ratio: [NADH]/[NAD⁺] in the cytosol and, therefore, to indicate that oxidative reactions are inhibited. In theory, high L/P ratios could also be caused by an increase in the [H⁺] of tissues, but experimental observations have so far shown either no such relationship or even the opposite—probably because the hydrogen ion has multiple and complex effects on metabolic reactions (Relman, 1972).

Turning now to the other primary factor determining lactate concentration, we see from equation (1) that at any given value of [H⁺] × [NADH]/[NAD⁺] lactate will be a direct function of the cytosolic concentration of pyruvate. Unlike lactate, pyruvate is far from being a metabolic "dead-end." It is, in fact, a highly active intermediate situated at the crossroads of several important metabolic pathways, as shown in Figure 3.1. Pyruvate concentration is determined by the balance between the processes producing and consuming pyruvate.

The rate of *de novo* pyruvate production depends mainly upon the rate at which glucose and glycogen are converted to pyruvate via the Embden–Meyerhof glycolytic pathway. One of the essential steps in glycolysis is the oxidation of glyceraldehydephosphate, which utilizes NAD⁺ and converts it to NADH. Hence the maintenance of glycolysis requires a constant supply

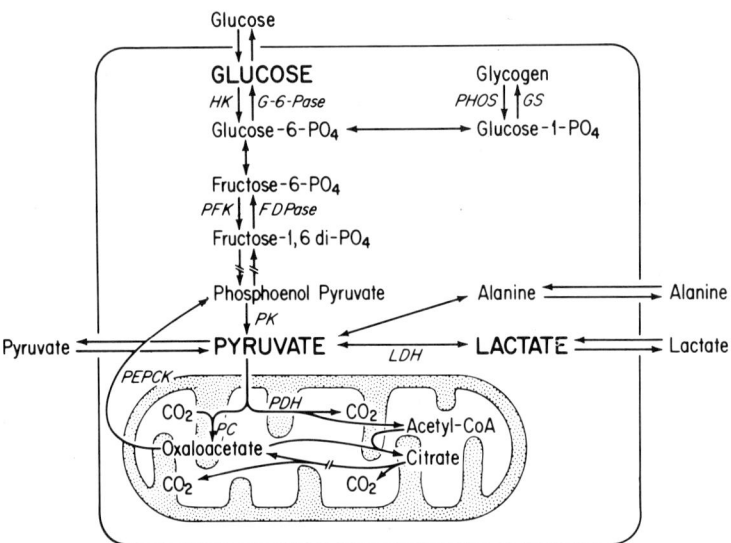

Figure 3.1 Schematic outline of metabolic pathways for interconversion of glucose and lactate in a cell of the liver or the proximal renal tubule. Abbreviations: HK, hexokinase, G-6-Pase, glucose-6 phosphatase; PHOS, phosphorylase; GS, glycogen synthetase; PFK, phosphofructokinase; FD Pase, fructose diphosphatase; PK, pyruvate kinase; LDH, lactate dehydrogenase; PDH, pyruvate dehydrogenase; PC, pyruvate carboxylase; PEPCK, phospho-enol, pyruvate carboxykinase.

of NAD^+, which is generated either through oxidation of NADH in the mitochondrion or through the reduction of pyruvate to lactate (Reaction 1). Glycolysis can therefore proceed with or without oxygen, but it is greatly accelerated under anaerobic conditions and whenever mitochondrial oxidative reactions are impaired. In concert with a shift in the LDH equilibrium and a reduction in pyruvate utilization, stimulation of glycolysis is a major factor in the accumulation of lactate in acute hypoxic states.

In addition to the glycolytic pathway, another significant, although quantitatively less important, source of pyruvate is from the transamination of alanine, which is released into the plasma mainly by skeletal muscle and the gut, and to a lesser extent by the kidney. Alanine is extracted from the plasma and used for pyruvate synthesis almost exclusively by the liver. In the kidney, de novo synthesis of pyruvate is also derived from the carbon skeleton of glutamine after the latter has been extracted from the plasma and has donated its nitrogen groups to form ammonia.

Pyruvate removal is accomplished largely by two aerobic mitochondrial reactions: 1) the oxidative decarboxylation of pyruvate to form acetyl CoA, catalyzed by an enzyme complex called pyruvate dehydrogenase (PDH); and 2) the carboxylation of pyruvate to form oxaloacetate through the action of the enzyme pyruvate carboxylase (PC). The first of these reactions represents the initial step in the oxidation of pyruvate to CO_2 and conversion of pyruvate to fat. The second reaction is the initial step in the conversion of pyruvate to glucose or glycogen, and an example of a process called gluconeogenesis, which occurs exclusively in liver and in kidney cortex. Although gluconeogenesis in effect reverses glycolysis, it is not accomplished simply by a reversal of all the glycolytic reactions between glucose and pyruvate. As shown in Figure 3.1, the conversion of phosphoenolpyruvate (PEP) to pyruvate is irreversible, thereby necessitating a "bypass" through oxaloacetate, the first step of which is the production of oxaloacetate via the PC reaction, and the second step the conversion of oxaloacetate to PEP by the PEP carboxykinase (PEPCK) reaction.

Both pathways of pyruvate utilization are indirectly dependent upon mitochondrial oxidative reactions, the PDH pathway because the initial reaction requires a continuing supply of NAD^+, and the gluconeogenic pathway because of a requirement for high energy phosphate compounds by the PC and PEPCK reactions. As a consequence, any impairment of mitochondrial oxidative function leads to an accumulation of pyruvate which, together with a shift in the LDH equilibrium, contributes to a rise in tissue and plasma lactate.

In the majority of tissues, under most aerobic conditions, conversion to acetyl-CoA (via PDH) and ultimate combustion to CO_2 is the major fate of pyruvate, although in adipose tissue, after feeding, a significant fraction of the acetyl CoA formed by the PDH reaction probably is also converted to fat. During fasting or stress or whenever there is a deficiency of insulin or an increase in epinephrine, cortisol or glucagon, the relative activity of the gluconeogenic pathway in the liver and kidney increases and the activity of the PDH pathway decreases. In the liver net synthesis of glucose is achieved mostly through the utilization of the carbon from lactate and alanine, which are extracted from the blood, converted to pyruvate and then, via the PC and PEPCK reactions, to PEP and ultimately to glucose. The kidney, like the liver, can synthesize glucose from exogenous lactate which it extracts from arterial blood, but a more important substrate for renal gluconeogenesis is glutamine.

To summarize, the concentration of lactate in body fluids is primarily determined by the concentration of pyruvate, with which it is in equilibrium, and by the $[NADH]/[NAD^+]$ ratio in cytosol. The pyruvate concentration is determined by the balance between the reactions producing and consuming pyruvate, while the $[NADH]/[NAD^+]$ ratio is a reflection of the cytosolic oxidation-reduction state. An elevated lactate concentration is therefore due either to an elevated pyruvate concentration, or to an increased $[NADH]/[NAD^+]$ ratio (in which case the L/P ratio is elevated), or to both.

Organ physiology of lactate

All cells of the body form lactate as an intermediate in the metabolism of glucose, but only certain tissues and organs normally add significant quantities of lactate to, or remove it from, the extracellular fluid. The normal plasma level of approximately lmEq per liter therefore reflects the net balance between the various rates of lactate production and consumption characteristic of each tissue in the normal basal state. However, under certain stressful conditions, such as severe ischemia or hypoxia, all tissues are potentially capable of adding lactate to the blood; similarly, under other conditions, such as when adequately perfused with oxygenated blood containing high concentrations of lactate, most tissues (except red blood cells) are potentially capable of extracting lactate from the blood.

Precise information on the total amount of lactate normally added to, and removed from, the blood is lacking. Estimates based on the turnover of isotopically-labeled lactate suggest that about 25–30 milliequivalents per kilogram per day of lactate is produced in a normal resting man (Kreisberg, 1972). However, this technique probably overestimates lactate turnover because the isotope rapidly exchanges with pyruvate and therefore traces the turnover of this compound as well. Estimates derived from direct measurements of A-V differences across organs and from studies of rates of lactate production *in vitro*, both of which are also subject to considerable error, yield figures that are somewhat lower—approximately 15–20 milliequivalents per kilogram per day.

The major organs contributing to the normal turnover of lactate in the blood are shown in Figure 3.2. Red blood cells are obligatory producers of lactate since they have no mitochondria and hence no mechanism for the metabolic removal of pyruvate. They contribute at least one quarter of the total lactate production, according to some authorities possibly more than half. Brain and skin probably also make major contributions to lactate production, but there are few reliable data on the exact amounts. Resting skeletal muscle adds relatively modest quantities of lactate to the blood, but during exercise lactate production from muscle is greatly increased. Maximal exercise may in fact, result in an overwhelming accumulation of lactate, with a rapid rise of blood levels. Not shown in Figure 3.2 are small additions of lactate from leukocytes, platelets and intestinal mucosa which apparently have relatively high glycolytic rates even under aerobic conditions.

The organs normally responsible for lactate uptake are liver and kidney cortex, which can use lactate as a substrate for gluconeogenesis as well as a fuel for oxidation to CO_2. The liver probably accounts for the major share of lactate removal under normal conditions, with the kidney sharing this function mainly when plasma levels are elevated. Renal excretion of lactate, like renal metabolism, is relatively unimportant at normal plasma levels and becomes a significant route of lactate removal only when the plasma

LACTIC ACIDOSIS 71

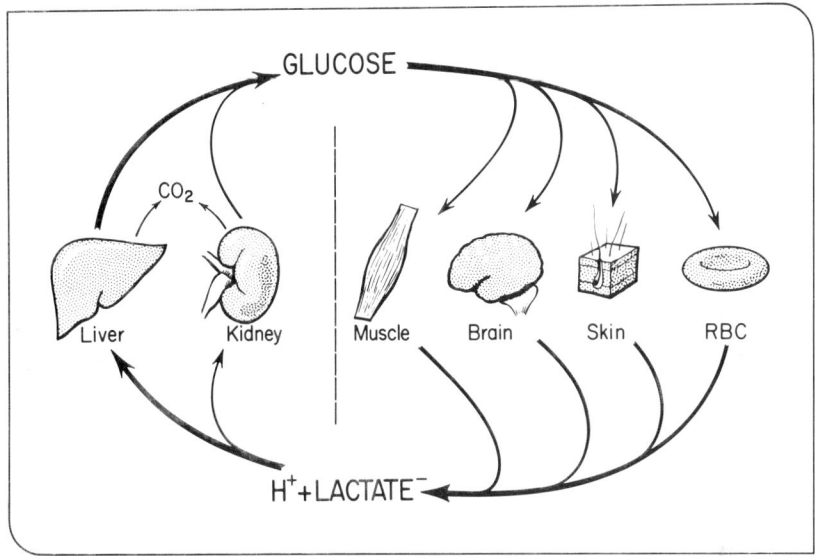

Figure 3.2 Schema of the Cori cycle. Glucose is converted by glycolyzing tissues to lactic acid, which is then reconverted to glucose (or oxidized to CO_2) by the liver and the kidney cortex.

level is very high. Although Figure 3.2 shows skeletal muscle only as a source of lactate, under certain conditions it can also be a consumer. Thus, for example, when plasma lactate is elevated by infusion or by the exercise of some muscle groups, the resting muscles are capable of taking up significant quantities of lactate.

Figure 3.2 emphasizes the cyclical nature of the relationship between glucose and lactate. Glucose is consumed and converted to lactate by those tissues that have obligatory glycolytic activity under normal aerobic conditions (e.g., red blood cells, brain, skin, muscle). The lactate is then taken up by the liver, and to a smaller extent by kidney cortex, and reconverted to glucose. This so-called *Cori cycle* serves to maintain blood glucose levels during fasting and thereby provides a continuing supply of metabolic fuel to organs, such as the brain and red blood cells, which require glucose as an energy source. Another physiological function of the cycle is to maintain normal acid-base equilibrium, which would otherwise be unbalanced by the continuous release of hydrogen ions associated with glycolysis. This aspect of lactate metabolism is discussed in further detail in the next section.

Acid-base aspects of lactate metabolism

Lactate turnover and net acid-base balance:

The net overall reaction for the formation of lactate from glucose through the Embden-Meyerhof glycolytic pathway may be written as:

Reaction 2

$$\text{(at pH 7.4)}$$
$$\text{glucose} + 1.6\ HPO_4^{2-} + 0.4\ H_2PO_4^- + 2\ ADP^{3-} \rightarrow 2\ \text{Lactate}^- + 2ATP^{4-} + 0.4\ H^+$$

Assuming a steady-state the ATP molecules synthesized in the above reaction will be immediately used, as follows:

Reaction 3

$$\text{(at pH 7.4)}$$
$$2\ ATP^{4-} + 2\ H_2O \rightarrow 2\ ADP^{3-} + 1.6\ HPO_4^{2-} + 0.4\ H_2PO_4^- + 1.6\ H^+$$

Adding these two reactions, we get:

Reaction 4

$$\text{glucose} \rightarrow 2\ \text{Lactate}^- + 2H^+$$

Thus the conversion of glucose to lactate results in the release of 2 protons for each molecule of glucose utilized, or one proton for each molecule of lactate produced. Although lactic *acid* is not actually produced at any step in the glycolytic process, the net result is *as if* it were.

The formation of lactate from its other major precursor, alanine, occurs through the pyruvate amino-transferase reaction, followed by the LDH reaction:

Reaction 5

$$\text{alanine} + \alpha\text{-ketoglutarate}^{2-} \rightarrow \text{pyruvate}^- + \text{glutamate}^{2-} + H^+$$
$$\text{pyruvate} + NADH \rightarrow \text{lactate} + NAD$$

Assuming that NADH and NAD are maintained in a steady state, the sum of these two reactions can be written:

Reaction 6

$$\text{alanine} + \alpha\text{-ketoglutarate}^{2-} \rightarrow \text{glutamate}^{2-} + \text{lactate}^- + H^+$$

Here again, as with the glycolytic process, lactate formation results in the production of equivalent amounts of hydrogen ions.

The reactions summarized above show that the normal net production of 15–20 milliequivalents of lactate per kilogram per day releases an equal amount of metabolic acid. This would tend rapidly to use up buffer stores and lower body fluid pH. The size of this acid load can be appreciated when compared to the body's total store of readily available (i.e., non-skeletal) buffers—only 10–15 milliequivalents per kilogram. The load becomes even more impressive when one considers that with vigorous exercise the rate of lactate production in muscles may be increased many fold. The normal rate of excretion of acid in the urine is only of the order of 1 milliequivalent per kilogram per day, so it is clear that renal excretory function ordinarily plays no significant role in the removal of the protons released as a consequence of the normal production of lactate. As we shall see later, renal excretion of acid is of little importance even when the abnormal accumulation of lactate in the blood has resulted in severe acidosis.

It is obvious, therefore, that acid-base balance must be maintained not primarily by the excretion of acid but by the metabolic removal of protons. This is achieved by coupling the utilization of lactate with the consumption of an equivalent number of hydrogen ions. The utilization of lactate, whether through oxidation to CO_2 and water or conversion to glucose, takes up protons, as shown by the following reactions:

Reaction 7

$$Lactate^- + H^+ + 3\ O_2 \rightarrow 3\ CO_2 + H_2O$$

Reaction 8

$$2\ Lactate^- + 2H^+ \rightarrow glucose$$

By these reactions the liver, which normally is the major site of lactate utilization, removes the protons that were released during lactate production and thereby regenerates the body's buffer stores. Like the kidney, therefore the liver must be regarded as a prime regulator of metabolic acid-base balance. Blood leaving the liver must normally be slightly more alkaline than that which is brought to it by the hepatic arterial and portal circulations, the difference representing the net consumption of metabolic acid associated with the hepatic metabolism of lactate. This homeostatic function of the liver has not been generally appreciated, but it is nonetheless vital for survival of the organism. Insofar as the kidney cortex utilizes lactate, it too will share in the metabolic removal of acid. Under many circumstances, when plasma lactate is increased, the kidney probably generates more base through this mechanism than it does through the excretion of acid.

The foregoing discussion should serve to explain not only why acidosis develops when lactate metabolism is disordered but also why it is so often of such sudden onset and extreme severity. The rates of gluconeogenesis and oxidative metabolism in the liver limit its ability to generate bicarbonate. As we shall see below, when the hepatic uptake of lactate is seriously compromised, or when the production of lactate is increased beyond the capacity of the liver to keep up with the load, severe and sometimes intractable acidosis can be expected to develop with great rapidity.

The feedback control of lactate production by hydrogen ions and the role of lactate in the regulation of tissue and blood pH:

As we have seen, the accumulation of lactate is accompanied by the release of hydrogen ions. However, an increase in hydrogen ions inhibits lactate production by directly reducing the activity of PFK (Halperin et al., 1969). Conversely, a reduction in hydrogen ions tends to increase the accumulation of lactate by simultaneously activating PFK and inhibiting pyruvate disposal. These relationships form the basis of an important homeostatic mechanism that helps to stabilize intracellular pH when the latter is threatened by certain types of physiological stress (Relman, 1972). When glycolysis is stimulated by anoxia or ischemia (the Pasteur effect), lactate accumulates in the affected tissues and tissue pH tends to fall. As pH is lowered PFK activity slows down because of inhibition of the enzyme by hydrogen ions. This results in a negative feedback control of lactate production which tends to put a brake on glycolysis and thereby limits the fall in cell pH.

This regulatory mechanism works in the opposite, or positive, direction when severe alkalosis threatens cell pH. A reduction in tissue hydrogen ion concentration causes lactate to accumulate, which in turn tends to moderate the degree of alkalosis and helps to maintain pH within tolerable limits. This phenomenon is of greatest physiological importance in the brain during hyperventilation. The central nervous system is particularly vulnerable to alkalosis, and it is therefore of interest that hyperventilation produces cerebral constriction and tissue hypoxia, which adds a further stimulus to the accumulation of lactate in the brain and spinal fluids.

In contrast to its major role in the defense of certain intracellular fluids against acute alkalosis, lactate plays no significant part in the normal acid-base defenses of the extracellular fluid. Neither respiratory nor metabolic alkalosis elicits more than a slight rise in plasma lactate concentration, probably because any increase in lactate production is balanced by increased removal in the liver and kidney. On the other hand, passive hyperventilation, which may lower cardiac output and hepatic and renal blood flow, has been reported to cause significant hyperlactatemia. Under such circumstances, a combination of tissue hypoxia and alkalosis provides a potent stimulus to glycolysis while at the same time inhibiting lactate uptake.

Lactate in blood, urine and spinal fluid

Lactate and pyruvate readily cross cell membranes, but the mechanism of the transport process is not yet fully understood. Under normal steady state conditions, the concentrations of lactate and pyruvate in the venous blood draining an organ probably are closely related to intracellular concentrations in that organ, but the tissue concentrations are slightly higher. When conditions are not steady, transient variations in local circulation and in the production and consumption of lactate and pyruvate can produce much larger concentration gradients between the venous effluent and the tissue fluids. The interpretation of lactate and pyruvate concentrations in randomly drawn specimens of mixed venous or arterial blood can be further complicated by the heterogeneity of the metabolic and circulatory conditions that may exist in the different organs and tissues of the body. It is therefore not justified to assume that arterial levels bear some fixed relation to the concentrations inside the cells of any particular organ or of the body in general.

Nevertheless, arterial plasma concentrations can be used as a rough clinical guide to intracellular conditions. The normal arterial concentrations of lactate and pyruvate are approximately lmEq per liter, and 0.1 mEq per liter, respectively, so their ratio, even if not their absolute value, is virtually the same as that which exists in cells. A rise in blood lactate necessarily implies a rise in tissue concentrations. A large increase in the blood L/P ratio, to 30 or higher, indicates a probable reduction in the oxidative state of the cytosol of some tissues, although the exact locus of the defect may not be clear.

The concentration of lactate in the urine, even when the latter is concentrated, is normally lower than that in plasma. At normal plasma concentrations virtually all filtered lactate is reabsorbed, although the concentration and total amount of lactate in the urine also tend to rise. The tubular reabsorption mechanism is thought to be nearly saturated at plasma levels of about 6 or 7 mEq per liter and at higher plasma concentrations significant amounts may appear in the urine. However, there are relatively few data on the excretion of lactate by the human kidney when plasma levels are very high. Renal excretion, at any rate, is probably not a major route of lactate removal from the blood under most circumstances.

The concentration of lactate in spinal fluid is normally very slightly higher than that in plasma, presumably because of the glycolytic activity of the brain and the delay in equilibration across the blood-brain barrier. Following hyperventilation, or after brain injury the gradient may increase significantly as lactate rises in the spinal fluid more rapidly than in blood. On the other hand, in some patients with lactic acidosis blood levels may exceed those in spinal fluid, particularly when the mechanism responsible for the accumulation of lactate does not primarily involve the central nervous system.

CLINICAL LACTIC ACIDOSIS

Introduction and definitions

Metabolic acidosis is the most common and ubiquitous of acid-base disorders. It develops whenever the rate of addition of fixed acid to body fluids exceeds the total rate of removal of acid by metabolic and renal excretory processes. It is ordinarily characterized by a reduction in the plasma bicarbonate from whatever its level was prior to the development of the acidosis, and by a tendency for the pH and pCO_2 to fall. However, the absolute levels of plasma bicarbonate, pH and pCO_2 may be normal, low or high, depending on whether there are complicating acid-base disturbances.

We may define *lactic acidosis* simply as a *metabolic acidosis due to the accumulation of lactate in the blood*. Strictly speaking, any increase in lactate concentration, regardless of cause or severity, might fit this description. However, in the discussion that follows it will be convenient to restrict our attention to conditions in which *pathological* mechanisms lead to hyperlactatemia and in which the blood concentration is high enough to be of clinical significance. Different authors have disagreed on the definition of the latter, for there is clearly no one concentration that sharply demarcates "significant" from "insignificant" degrees of hyperlactatemia. However, given the normal physiological variations in blood lactate and the possibilities of error in the sampling and analysis of specimens, concentrations below 2mM per liter obtained at rest should not be regarded as definitely abnormal. Levels between 2 and 4 mM per liter should be considered as abnormal but of uncertain clinical significance. Above that range, rising lactate levels carry increasingly serious implications, because they usually denote a major disturbance of lactate metabolism. Patients with resting blood lactate concentrations much above 4mM per liter are likely to be very ill; those with the highest levels, i.e., 30–50 mM per L, have the gravest prognosis and rarely survive.

It needs to be emphasized that lactic acidosis, often when mild or moderate in degree, and sometimes even when severe, need not cause acidemia, i.e., a reduction in blood pH. As will be noted below, respiratory alkalosis is a frequent concomitant of certain types of lactic acidosis, and under certain other circumstances there may be an associated metabolic alkalosis. Therefore, a reduction neither in blood pH nor in plasma bicarbonate is a necessary accompaniment of lactic acidosis—although both are usually present.

Plasma amino acids: Although they do not contribute significantly to the unmeasured anions, it should be noted here that certain amino acids are often markedly increased in the plasma of patients with lactic acidosis (Marliss et al., 1972). Alanine is increased the most, by far, but proline, valine, lysine, and leucine, are also notably higher than normal. Alanine normally is produced in muscle and then converted to glucose in the liver via transamination to pyruvate. In lactic acidosis it is not known whether alanine release from muscle is increased, but there is much reason to believe that the efficiency with which the liver metabolizes pyruvate (and hence, alanine) is impaired.

Other laboratory findings: The serum *phosphate* is usually elevated (O'Connor, Klein and Bethune, 1977). Whether this is due to release of inorganic phosphate from anaerobic cells or to some abnormality in tubular transport of phosphate is not known. Serum *urate* is also frequently increased, probably because of impaired renal excretion. Serum *potassium* may be elevated, as is to be expected with acidemia; but in some forms of lactic acidosis, for example that following generalized seizures (Orringer et al., 1977) serum potassium is normal despite a very low blood pH. Evidently lactic acid may be released from muscle and taken up by liver without requiring discharge of cellular potassium stores.

The clinical characteristics of lactic acidosis

The diagnosis of lactic acidosis can be made with certainty only by a measurement of blood lactate, but for practical clinical purposes a working diagnosis is justified when the patient is found to have a metabolic acidosis with a large anion gap that cannot be explained by uremia or ketonemia (Emmett and Narins, 1977).

There are no clinical characteristics sufficiently specific to be reliable diagnostic guides. Lactic acidosis often is simply the harbinger of some imminent disaster, or one facet of an underlying disease that produces its own characteristic clinical manifestations. In rare instances the disturbance of lactate metabolism is the only apparent disorder and is not associated with any other recognizable illness. In the vast majority of cases lactic acidosis occurs as a single acute episode, developing rapidly in a matter of minutes or hours or, at most, in a day or two. However, there are an increasing number of reports of cases of lactic acidosis due to congenital or acquired disorders of lactate metabolism, in which the acidosis is chronic and relatively well tolerated for long periods of time, or in which there are recurrent episodes.

The clinical picture is therefore highly varied, depending largely on the nature of the underlying disease and the etiology of the disturbance. Nevertheless, some generalizations about the clinical findings in cases of severe acute lactic acidosis may be useful:

1) The onset of acute lactic acidosis in many ways may resemble that of diabetic ketoacidosis, in that there is the sudden onset of malaise, weakness, anorexia, nausea, blurring of consciousness. Particularly with drug-induced lactic acidosis there may be vomiting and abdominal pain. In contrast to diabetic acidosis, however, the onset of lactic acidosis is not accompanied by polyuria and polydipsia.

2) The first objective sign of acute lactic acidosis is often hyperpnea. Marked hyperventilation is a common feature and, as will be discussed below, respiratory alkalosis is a frequent complication of most types of lactic acidosis, with the notable exception of those cases due to phenformin intoxication. Dehydration is rarely as severe as in diabetic ketoacidosis and the acetone odor characteristic of the latter condition is usually absent.

3) The blood pressure often is low and the pulse rapid, even when there is no evidence of frank vascular collapse. The blood pressure tends to be unstable, and if the acidosis persists or worsens, shock and oliguria often supervene. However, although shock is well known as both a cause and an early consequence of lactic acidosis, there are many patients who initially have perfectly normal blood pressure and renal function. It seems clear that the usual type of low output state is by no means a necessary feature of the onset of the disease.

The *prognosis* is always serious in acute forms of lactic acidosis, depending largely on the nature of the underlying disease, and upon the severity of the acidosis. When the cause is potentially remediable, such as in various types of poisonings and in reversible shock, then the outlook is for recovery in at least half the cases that are appropriately treated. On the other hand, in patients whose underlying condition is basically irreversible, the onset of severe lactic acidosis usually heralds the final phase of the illness, and death is the early outcome in at least 90 percent of the cases.

The clinical causes of lactic acidosis

Tissue hypoxia

By far the most frequent cause of clinical lactic acidosis is tissue hypoxia (Huckabee, 1961a, 1961b). The common factor in all of the clinical conditions listed under this heading is a reduction in tissue oxygen supply, due either to reduced blood supply or arterial hypoxemia, or both. In addition, this type of lactic acidosis is frequently accompanied by respiratory alkalosis (usually the result of hypoxia of the respiratory center) and by an increased release of catecholamines. All three factors, hypoxia, respiratory alkalosis and catecholamine release, would tend to enhance the production and inhibit the consumption of pyruvate. Inasmuch as the [NADH] of severe lactic acidosis usually heralds the final phase of the illness, and [NAD+] ratio is increased, blood lactate always increases proportionately more than pyruvate. The highest recorded levels of blood lactate have occurred in patients suffering from tissue hypoxia.

Lactic acidosis usually develops whenever tissue perfusion is significantly reduced, for whatever reason. Shock, or incipient shock, whether produced by failure of cardiac output, hemorrhage or loss of extracellular fluid, septicemia or any other cause, is often associated with a rise in blood lactate. The first reported instances of clinical lactic acidosis occurred in children who were in circulatory collapse from severe diarrhea (Clausen, 1925). Sudden collapse of the circulation results in a rapid accumulation of lactate in the ischemic tissues, usually promptly followed by an impressive increase in blood lactate. With the more gradual development of shock there may be an early slow rise in blood lactate that precedes any significant drop in blood pressure. This is presumably the result of peripheral vasoconstriction, with resultant ischemia. Selective vasoconstriction often reduces the circulation most markedly in the renal and splanchnic beds, in the skin and in the brain. Lactate production in the latter two areas probably increases, as may the production of lactate by the gut. At the same time the capacity of the kidney and the liver to remove lactate is probably compromised; these organs may even begin to reverse their normal functions and add lactate from the blood.

Accompanying the premonitory rise in blood lactate there is often an increase in ventilation, which is mainly a response to cerebral ischemia and central nervous stimulation. The resulting fall in pCO_2 cannot be described merely as a compensatory response to the metabolic acidosis from lactate, because it is greater than would be normally expected for the degree of reduction in blood pH and bicarbonate. There is, in fact, a second independent disturbance of acid-base balance, i.e. respiratory alkalosis. Thus, insidious development of hyperlactatemia, together with respiratory alkalosis, may be the first objective manifestation of impending shock. Septicemia often begins this way (MacLean et al., 1967), and is particularly likely to produce striking hyperventilation because of the respiratory stimulation from endotoxin. Hyperventilation may also be a prominent feature in patients with lactic acidosis due to traumatic shock, in whom pain and anxiety add to the intensity of the nervous stimulation of respiration. As a result of their respiratory alkalosis, a significant fraction of such patients may have a normal or only slightly reduced blood pH, despite high blood lactate levels. A further consequence is that intracellular pH may be relatively well maintained, thereby allowing continuation of high rates of glycolysis.

Hyperlactatemia clearly has prognostic significance in traumatic shock (Cloutier, Lowery and Carey, 1969), and probably in other forms of tissue hypoxia as well. Mortality rates are higher in patients with significantly elevated blood lactate levels than in those without. Furthermore, the higher the blood level the less likely is the condition to be reversible. The highest concentrations of lactate that have been observed are found in desperately ill patients suffering from some form of tissue hypoxia, usually in circulatory collapse. Under these circumstances the blood lactate may reach levels of 40 to 50 mEq per L or higher before death inevitably ensues.

Tissue hypoxia is less apt to be produced by a reduction in blood oxygen content than by low blood flow, because hypoxemia invokes compensatory responses in cardiac output and in the efficiency of oxygen extraction by tissues. In patients with chronic forms of hypoxemia other than anemia (e.g., pulmonary insufficiency and cyanotic congenital heart disease) an additional compensatory response is the development of secondary polycythemia which, unless so severe as to increase greatly the viscosity of the blood, further improves the delivery of oxygen to tissues. Therefore, lactic acidosis is common in low-blood-flow states of all kinds, but relatively uncommon in clinical hypoxemia. Only the most severe forms of oxygen deprivation and only the most advanced degrees of anemia will lead to significant lactic acidosis. The probability increases when the normal compensatory mechanisms that maintain oxygen transport are impaired. As a general rule, however, arterial oxygen content must be reduced by substantially more than half before there is much likelihood of lactate accumulation on this basis.

An interesting exception to this rule occurs in patients with carbon monoxide poisoning, who have been reported to develop moderately severe lactic acidosis after displacement of much less than half their oxygen carrying capacity by carbon monoxyhemoglobin (Buehler et al., 1975). The explanation is not entirely certain, but two factors have been suggested: 1) the cardiovascular compensatory responses to hypoxemia depend on PO_2

rather than total oxygen content, hence there would be no increase in cardiac output in carbon monoxide poisoning; 2) carbon monoxide increases the affinity of hemoglobin for oxygen, thereby impairing the efficiency of oxygen release in tissues.

In patients with acute left ventricular failure and pulmonary edema who develop lactic acidosis, both factors—reduced blood flow and arterial hypoxemia—probably contribute to the production of tissue hypoxia (Fulop et al., 1973). However, since neither factor is usually of severe degree, lactic acidosis in most cases of pulmonary edema is not often very severe. Of course, if cardiogenic shock should supervene, the acidosis may become overwhelming.

In most patients with tissue hypoxia and lactic acidosis the condition develops rapidly and then either is reversed or progresses inexorably to a fatal termination. Occasionally, however, patients suffering from low-blood-flow conditions or from severe hypoxemia may remain in a precariously balanced state of partial anaerobiosis as evidenced by persistent lactic acidosis lasting over days or even weeks. Included among such instances would be patients with severe low output, congestive heart failure and those with extreme degrees of chronic anemia.

Drugs

Lactic acidosis can occur as a result of the action of a number of drugs and chemical agents, and these are listed in Table 3.1. The two agents of

Table 3.1 Clinical causes of lactic acidosis

TISSUE HYPOXIA
    a) Shock or incipient shock
        cardiogenic
        hemorrhagic or hypovolemic
        septic
    b) Asphyxia
    c) Acute left ventricular failure
    d) Carbon monoxide poisoning
    e) Chronic low flow states
DRUGS AND TOXINS
    a) Ethanol
    b) Phenformin and other biguanides
    c) Methanol
    d) Salicylates
    e) Others: ethylene glycol, dithiazanine, streptozotocin, isoniazid, cyanide, nitroprusside, etc.
    f) Fructose and sorbitol
EPINEPHRINE AND CATECHOLAMINES
LIVER FAILURE
NEOPLASTIC DISEASE
PULMONARY EMBOLISM
SEPSIS
DIABETES MELLITUS
IDIOPATHIC (?)
CONGENITAL

greatest importance in this respect are ethanol and phenformin. These will be discussed below in some detail; the other drugs will be mentioned only in passing.

Ethanol:

It has been known for more than forty years that ethyl alcohol increases blood lactate concentration, due at least in part to a reduction in the efficiency with which lactate is consumed. Alcohol impairs the conversion of lactate to glucose in the liver (Kreisberg, Owen and Siegel, 1971), but there is no evidence that it interferes with the oxidation of lactate or increases its rate of production.

How this effect is to be explained is not yet entirely clear. Alcohol is metabolized in the liver largely through step-wise oxidations, first to acetaldehyde and then to acetate. These reactions are catalyzed by dehydrogenases that require NAD. The conversion of NAD to NADH lowers the redox potential of liver cells (Lieber et al., 1975). This could inhibit the activity of cytosolic PEPCK (Fig. 3.1) and so might contribute to the reduction in gluconeogenic activity.

Whatever the mechanism, it is clear that ingestion of alcohol lowers lactate tolerance and raises lactate blood levels slightly in most patients. In a few who drink excessively, alcohol apparently may cause frank lactic acidosis. It is not known whether this more severe defect is simply an extension of the mechanism that impairs lactate tolerance in normal subjects given test doses of alcohol, or whether it represents a different kind of toxic action on the liver or other tissues resulting from the very high alcohol levels.

Severe acidosis due solely to lactate is relatively uncommon in alcoholics. Much more frequently lactic acidosis in these patients occurs together with non-diabetic ketoacidosis (Miller, Heinig and Waterhouse, 1978). The relative proportions of lactate and ketones in the plasma vary widely; in some lactate predominates while in others ketone acid anions are largely responsible for the increased anion gap. In some patients shock due to dehydration, sepsis or pancreatitis plays a role in the genesis of the lactic acidosis, but in others there is no evidence of circulatory collapse at any time in their clinical course. However, respiratory alkalosis is almost invariably present, sometimes severe enough to cause frank alkalemia, and this factor undoubtedly contributes to the disturbance in lactate metabolism.

The clinical history of alcoholic lactic acidosis is usually that of a recent acute increase in alcohol intake superimposed on a background of chronic alcoholism. Patients usually have not eaten much for several days before their admission to the hospital, because of anorexia, nausea and vomiting. On admission they are acutely ill, hyperpneic and often tremulous. Sometimes delirium tremens supervenes. Analysis of their plasma electrolyte pattern reveals a metabolic acidosis of the anion type, usually complicated by respiratory alkalosis. The BUN is usually low or normal and, if the patient is not also diabetic, so is the blood glucose. There may or may not be evidence of increased ketones in plasma and urine, depending on the concomitant presence of ketoacidosis, but it should be remembered that a reduction in the redox potential of liver cells such as occurs in acute alcoholism will greatly reduce the proportion of aceto acetate to betahydroxybutyrate (BOHB) in the plasma. Since it is only the former that reacts with nitroprusside, patients who have largely BOHB in their blood and urine may not show strongly positive tests for ketones. Blood lactate levels may be only modestly elevated but can be as high as 20 mEq per L, even without any other predisposing factor. L/P ratios are increased, as would be predicted from the high [NADH] [NAD+] ratio in liver cells.

Although excessive alcohol can certainly produce lactic acidosis in non-diabetic patients, it is more likely to do so in diabetics—particularly those being treated with phenformin. Alcohol and phenformin have synergistic effects on lactate metabolism, and the combined use of these two agents is dangerous. Alcohol abuse may suddenly precipitate severe lactic acidosis in a diabetic patient previously well regulated on phenformin.

Unless there are major complications, the prognosis of alcoholic lactic acidosis is usually favorable. Patients may be expected to respond to therapy with intravenous fluids, insulin and glucose within one or two days, with a return of normal lactate and bicarbonate levels in the blood and progressive symptomatic improvement.

Phenformin and other hypoglycemic biguanides:

Shortly after the introduction in 1958 of the oral hypoglycemic agent phenformin (phenethylbiguanide) occasional reports began to appear of acute lactic acidosis occurring in diabetic patients on this medication. For a number of years thereafter as cases continued to be reported the significance of this association was unclear. Many authors suggested that diabetes itself, or possibly some of its major complications, might be the primary cause of the disturbance in lactate metabolism. However, over the past few years evidence had accumulated of a more specific association between the use of phenformin and the occurrence of lactic acidosis, until there no longer could be any reasonable doubt that phenformin itself was an important etiologic factor (Dembo, Marliss and Halperin, 1975; Misbin, 1977).

On July 25th, 1977, the Secretary of Health, Education and Welfare ordered an end to the general marketing of phenformin in the U.S. within 90 days "because of unacceptably high risk of lactic acidosis" (FDA Drug Bulletin, 1977), which was estimated to be between 0.25 and 4 cases per 1000 users per year, with a mortality rate of approximately 50 percent. The relative frequency of phenformin as a cause of lactic acidosis had probably been exaggerated by failure to recognize and report the numerous instances of lactic acidosis that occur under many other circumstances, but the involvement of the drug in the pathogenesis of many cases was clear. What was also becoming clear was that phenformin was likely to cause lactic acidosis in only a very small percentage of cases, and only under particular circumstances. Nevertheless, until its removal from the U.S. market, phenformin was undoubtedly a contributing factor in the genesis of a major fraction of the cases of acute lactic acidosis not clearly due to tissue hypoxia (Conlay and Loewenstein, 1976).

Now that the drug is no longer generally available in the U.S., a detailed discussion of phenformin seems pointless here. However some brief comments are in order not only because of the insight afforded by the drug into the pathogenesis of lactic acidosis, but also because phenformin and metformin (dimethylbiguanide)—a close congener—are still in wide use as oral hypoglycemic agents in other countries.

The therapeutic hypoglycemic effect of low doses of the biguanides apparently results from an insulin-facilitating action in peripheral tissues to enhance the rate of glucose utilization and conversion of glucose to lactate. But unlike insulin, the biguanides also increase the conversion of lactate to glucose and inhibit the oxidation of lactate. Concentration of ingested biguanides in the intestinal mucosa inhibits the active absorption of glucose and amino acids. The net result of all these effects in most patients taking therapeutic doses of drug is a reduction in fasting and post-absorptive blood glucose levels, and a very slight increase in blood lactate (Davidoff, 1973).

Most of the untoward effects of phenformin, however, are associated either with high drug levels (due to overdose or failure of renal excretion) or with complicating conditions that in themselves predispose to lactic acidosis. At high drug levels, phenformin produces not only a greater degree of glucose conversion to lactate by peripheral tissues, but also inhibition of hepatic and renal gluconeogenesis from lactate and amino acids. The mechanism of these effects is not altogether clear but seems to involve a general reduction in mitochondrial oxidative reactions, thereby interfering with the PDH and PC pathways for disposal of pyruvate while simultaneously enhancing the formation of lactate through the glycolytic pathway (Fig. 3.1). The predictable result would be significant accumulation of lactate in the blood, along with increased concentrations of alanine and pyruvate. L/P ratios are usually increased.

The great majority of patients who developed lactic acidosis while on phenformin had *impaired renal function* and where blood levels of phenformin could be measured, drug levels have usually, though not always, been found to be high. Another important predisposing factor is the use of *ethanol*. The latter drug alone, as we have seen, may cause lactic acidosis by inhibiting hepatic gluconeogenesis and reducing the cell redox potential. As might be expected, ethanol and phenformin have synergistic effects on lactate metabolism, and abuse has been responsible in many instances for the sudden appearance of lactic acidosis in a patient previously well compensated on phenformin therapy. *Circulatory failure* appears to have been still another important condition that precipitated the development of clinically severe lactic acidosis in patients taking phenformin. A final predisposing factor is severe liver disease. The liver is the major site of lactate consumption, so patients with impaired liver function are particularly susceptible to the effects of even low doses of phenformin. *Liver disease* may also impair the conversion of phenformin to its inactive hydroxylated metabolite, which would effectively enhance the blood levels of effective drug.

A review of cases of phenformin-associated lactic acidosis suggests that most patients have had in their medical background one or more of the predisposing factors discussed above (Misbin, 1977). They usually have had a moderate degree of renal impairment or circulatory failure or both. Alcoholism and liver disease have also been common complications. In many cases patients had been recently started on therapy or their dosage level had just been increased, suggesting the possibility of high drug levels.

The earliest symptoms are anorexia and nausea, which may appear a few days before the development of severe illness. This may be followed by vomiting and abdominal pain and the rapid progression of prostration and stupor. Polyuria and polydipsia are notably present. On admission to the hospital patients are usually more or less unresponsive and severely hyperpneic. About half, or slightly less, are markedly hypotensive or in frank clinical shock. Although some patients may be moderately dehydrated as the result of vomiting and poor intake, they are not usually as severely dehydrated as are those with typical diabetic ketoacidosis. The pronounced acetone odor so characteristic of ketoacidosis is usually absent.

The laboratory findings are striking. As in ketoacidosis, there is usually a moderate or even marked polymorphonuclear leukocytosis, which is probably a consequence of the acidosis *per se*. Also in common with ketoacidosis, there is a metabolic acidosis due to an increased anion gap. Like ketoacidosis, but in contrast to most other forms of acute lactic acidosis, the phenformin-induced disorder does not usually cause excessive hyperventilation and respiratory alkalosis. The degree of hyperventilation is usually no more than appropriate for the severity of the reduction in plasma bicarbonate and, as a result, the blood pH in most cases is very low.

Lactic acidosis due to phenformin is differentiated from typical ketoacidosis by the absence of a strongly positive qualitative test for ketones in the plasma (nitroprusside test negative at 1:4 dilution or less) and by the finding of a markedly elevated blood lactate concentration. Most patients have blood lactate levels of between 10 and 25 mEq per L. By definition, most of the increased anion is accounted for by lactate, but usually there is some increase in the plasma concentrations of ketone acids as well. Most of the latter is in the form of betahydroxy butyric acid (BOHB), which does not react with nitroprusside, and hence the qualitative ketone test on plasma may be negative despite the presence of significant quantities of ketone anions in the plasma. On the other hand, if renal function is not too seriously impaired, the urine test for ketones is likely to be positive, because the concentration of acetoacetate in urine is ordinarily much higher than in plasma. The hyperlactatemia is accompanied by a significant rise in blood pyruvate, but the latter is never increased in proportion to the lactate, so that the L/P ratio is usually greatly increased. This may reflect an increase in the cellular NADH/NAD ratio due either to circulatory failure and tissue hypoxia or to a specific effect of phenformin on mitochondrial oxidations, as discussed above.

Differentiation from typical diabetic ketoacidosis is often also aided by the blood glucose concentration which, in about half the cases, is less than 100 mgm per 100 ml. By contrast, almost all patients with ketoacidosis have blood glucose levels above 250 mgm per ml. In a few cases, however, phenformin-induced lactic acidosis may be associated with a very high blood sugar, so the presence of severe hyperglycemia does not rule out the diagnosis.

While the above analysis provides a rationale for distinguishing between the typical clinical syndromes of diabetic ketoacidosis and phenformin-induced lactic acidosis in a diabetic, it should be emphasized that they overlap and may occur together in the same patient with varying degrees of relative severity. Many of the circumstances that predispose to lactic acidosis also tend to produce ketosis. As a result, most patients with lactic acidosis have at least some increase in plasma ketone acids. The latter are mainly in the form of betahydroxy butyric acid, so they give a deceptively weak reaction with the clinical test reagent, but they are nevertheless often present in high enough concentrations to contribute significantly to the anion gap. Conversely, as will be discussed later, many of the physiological disturbances accompanying diabetic ketoacidosis favor the development of lactic acidosis. It is therefore not surprising that most patients with typical diabetic ketones have at least some slight increase in blood lactate (> 2 mM per L) and ten to twenty percent have significant lactic acidosis (> 4 mM per L).

Prevention: Lactic acidosis is a relatively infrequent complication of phenformin therapy that usually occurs predictably under circumstances that either increase the drug concentration to toxic levels or aggravate the drug-induced disorder of lactate metabolism. Most cases therefore could be prevented by a more discriminating use of the drug and by careful monitoring and education of patients.

Prognosis: The overall mortality among reported cases of phenformin-associated lactic acidosis was between 40 and 50 percent, but a single figure of this sort is misleading because the prognosis depended heavily on a variety of factors including the presence of major complications (e.g., shock, heart failure, sepsis, or advanced renal or liver disease), the type of therapy and the promptness with which it was begun. In some cases patients were moribund on admission to the hospital and died before treatment could be started, while in others there were serious underlying medical problems which of themselves could have led to a fatal outcome. In the absence of such complications, and when started early enough before the acidosis has caused irreversible changes, appropriate therapy ought to be effective in more than 80 percent of cases. What constitutes "appropriate" therapy is discussed in a separate section.

Other drugs and toxins:

In addition to ethanol and the hypoglycemic biguanide drugs, Table 3.1 lists a number of agents that have been implicated in the production of lactic acidosis. These will be mentioned only briefly.

Methanol is much more toxic than ethanol because its oxidative breakdown products, formaldehyde and formic acid, have deleterious effects on intermediary metabolism. One of these effects is an inhibition of mitochondrial oxidations with consequent stimulation of glycolysis and lactate accumulation. Patients with methanol poisoning develop a very severe metabolic acidosis with an increased anion gap due not only to formate but also to lactate (Emmett and Narins, 1977). Respiratory alkalosis is usually present.

Salicylates in large doses uncouple oxidative phosphorylation in mitochondria and stimulate glycolysis. Blood lactate levels are usually increased. Through a direct action in the brain, salicylates also stimulate respiration. Patients with salicylate poisoning therefore usually have a mixed respiratory alkalosis and metabolic acidosis. The latter is due in part to lactate but, particularly in young children, ketoacidosis often predominates.

Ethylene glycol intoxication results in the accumulation of oxalate in the blood and the characteristic appearance of oxalate crystals in the urine. For reasons that are not entirely clear, lactate simultaneously accumulates; as a consequence, the large anion gap that develops in such cases is a mixture of oxalate and lactate possibly in addition to other organic acids (Emmett and Narins, 1977).

Other drugs that may cause lactic acidosis include streptozotocin, spoiled paraldehyde, isoniazid and dithiazanine.

In all cases of lactic acidosis produced by these various drugs, the clinical picture is that of an acutely developing anion acidosis with prostration, and usually abdominal symptoms. In methanol poisoning and salicylism, hyperventilation is an early and prominent sign. Methanolism also causes optic neuritis and amblyopia. Salicylism is particularly common in young children, in whom ketoacidosis is apt to play a major role. Ethylene glycol poisoning is usually associated with signs of acute neurological, hepatic and renal damage.

The differentiation from other causes of acute metabolic acidosis of the anion type is based on the history of drug ingestion, clinical clues, and the finding of significant amounts of the offending drug or a metabolite in blood or urine. Treatment consists in the management of the acute acid-base disorder, as discussed elsewhere, and in other measures designed to expedite the removal of the drug or its metabolism. Prognosis, as with other types of acute lactic acidosis, depends as much upon particular circumstances of the case as upon treatment. In general, however, if the dose of the drug has not been overwhelming and if irreversible complications have not already developed by the time the patient comes to medical attention, the substantial majority of such cases will respond to appropriate treatment.

Fructose and sorbitol

Fructose and sorbitol (which is converted to fructose) have been used as sources of carbohydrate in parenteral solutions, but lactic acidosis can sometimes result. Fructose has been chosen as a nutrient, rather than glucose, because it does not require insulin and is rapidly taken up by the liver and other tissues. However, the rapid conversion of fructose to fructose 1-phosphate in the liver, catalyzed by fructokinase, can cause sequestration of fructose 1-phosphate and marked depletion of hepatic ATP levels. This loss of ATP, like that caused by hypoxia, inhibits hepatic utilization of lactate, and at the same time enhances the rate of glycolysis in other tissues. A rise in plasma lactate therefore results from increased production as well as impaired utilization.

When infusion rates of fructose exceed 0.5 gm per hour per kg body weight, significant degrees of hyperlactatemia may occur (Hessov, 1974). The risk of lactic acidosis is increased even more if ethanol is included in the infusion as an additional source of calories, or if fructose infusions are given to patients with low blood flow states or serious hepatic dysfunction.

Epinephrine overdosage and the effects of catecholamines

Epinephrine poisoning is a very rare but interesting cause of lactic acidosis (Køllendorf and Møller, 1974). As discussed in the section on regulation of lactate metabolism, epinephrine has important effects on the enzymatic reactions that control the production and consumption of pyruvate (Fig. 3.1). Epinephrine enhances glycogenolysis by activating liver and muscle phosphorylase (PHOS) and by inhibiting glycogen synthetase (GS). It also inhibits PED activity, thereby reducing the flux of pyruvate through the oxidative pathway. At the same time, epinephrine in high dosage has marked peripheral vasoconstrictor activity, causing intense ischemia of skin and the hepatic and splanchnic beds. The net result is probably an increase in lactate production by skin and muscle (and possibly gut), together with reduced uptake by muscle.

Epinephrine regularly produces slight increases in blood lactate when administered in physiological doses to normal subjects, but lactic acidosis due to epinephrine alone is very rare and occurs only with excessive overdosage. Norepinephrine, although a more powerful vasoconstrictor than epinephrine, has less metabolic effect, and in normal subjects does not affect blood lactate level as much. However, both epinephrine and norepinephrine may play a contributory role in the production of lactic acidosis if they are administered frequently or continuously to patients who have low blood flow or severe hepatic dysfunction or are under the influence of ethanol or a hypoglycemic biguanide drug.

Prolonged infusion of norepinephrine or other vasoconstrictor sympathomimetic drugs is often used in the treatment of hypotensive states. While such treatment may be useful in sustaining the perfusing of brain and heart, the price may be an intense prolonged ischemia of the skin and splanchic bed. This effect would contribute to the production of severe lactic acidosis. Acidosis, in turn, impairs the inotropic effects of catecholamines on the heart and may necessitate a further increase in drug dosage. In this manner a vicious circle may develop, characterized by refractory hypotension, intense vasoconstriction, and a rapidly rising blood lactate. Under such circumstances, the circle can be broken only if means can be found to sustain cardiac output without resort to vasoconstrictor amines.

Liver disease:

Since the liver is the major site of lactate consumption, it is predictable that blood lactate would be elevated in severe liver disease. Studies of patients with serious liver functional impairment, particularly those with fulminant forms of hepatic failure due to viral hepatitis or hepatotoxins, support this prediction (Record et al., 1975). Even in the absence of any evidence of circulatory failure or tissue hypoxia, most patients with acute advanced liver failure have some increase in blood lactate concentration, usually in the range of 2–10 mM per L. Pyruvate is increased almost in proportion to lactate, as one would expect if the defect were a non-specific loss of hepatic metabolic function, and therefore L/P ratios are normal or only slightly elevated.

Accumulation of lactate and pyruvate is associated with a marked increase in the blood levels of alanine and other glyconeogenic amino acids, as well as a tendency to hypoglycemia. These are the changes seen in acute alcoholism, in biguanide poisoning and in certain congenital defects in hepatic gluconeogenesis (see below), and they point to a primary hepatic impairment in the utilization of lactate and pyruvate.

Not infrequently, however, the hepatic defect is accompanied by some other complication that increases the peripheral production of lactate. Thus, many patients with liver failure become hypotensive or develop overt shock; others may have sepsis or leukemia or a large intra-abdominal neoplasm or may become hypoxemic. Any of these conditions can increase lactate production significantly (see below) and this can exacerbate the severity of the metabolic acidosis. The most common associated disorder is respiratory alkalosis. By some mechanism not yet understood, severe hepatic damage often results in the stimulation of respiration. Significant degrees of hypocapnea are almost invariable in such patients. Therefore, despite the accumulation of lactate and other organic acid anions, blood pH is often elevated rather than reduced.

Respiratory alkalemia has a particularly potent effect on intracellular pH, and a high pH is known to stimulate glycolysis in muscle, brain and red blood cells (see section on control of lactate metabolism). As a result, lactate production may be increased in these patients and may be partly responsible for the elevated blood lactate levels.

Neoplastic disease

Lactate levels in malignant disease are usually within the normal range, but significant lactic acidosis has occasionally been described. Patients most likely develop hyperlactatemia with acute leukemia or other forms of myeloproliferative disease (Field, et al., 1966) but cases have been described among those "solid" tumors including lymphomas and various types of large intra-abdominal neoplasms.

Leukocytes and erythrocytes have relatively high rates of glycolysis, even under aerobic conditions. Neoplastic cells, in general, also have a high glycolytic rate, so it would be expected that total lactate production might be increased whenever there is a large mass of malignant tissue in the body. Clearly, however, certain factors predispose towards the development of lactic acidosis in only a very small percentage of those with this type of neoplastic disease. In some cases there are obvious causes, such as sepsis, liver failure or shock, but in others the reason for the development of lactic acidosis remains obscure and must be related in some way to the neoplastic process itself. Lactic acidosis may be an acute pre-terminal episode in these cases, but sometimes it manifests itself as a more chronic condition.

Pulmonary embolism

Lactic acidosis occasionally develops in patients who have suffered one or more pulmonary embolic episodes. The latter are often heralded by the onset of hyperventilation. Sometimes there is shock or hypotension and a very significant degree of hypoxemia, which would help to explain the development of lactic acidosis, but in other cases there are no other obvious etiologic factors beyond the pulmonary embolism itself and the hyperventilation.

The appearance of lactic acidosis in these patients is a grave prognostic sign, often portending imminent cardiovascular collapse.

Sepsis

Like pulmonary embolism, septicemia frequently begins with sudden hyperventilation and respiratory alkalosis, followed by the development of lactic acidosis (MacLean et al., 1967). Shock, or impending shock, is often the explanation but many times the blood pressure, skin circulation and arterial oxygen content seem to be well maintained at first, even as the blood lactate is rising. Whether the disturbance in lactate metabolism represents an occult type of peripheral vascular failure, or is the result of some direct metabolic effect of endotoxin is not certain.

Respiratory alkalosis probably reflects a stimulation of the respiratory center by endotoxin. As in so many other conditions associated with lactic acidosis, hyperventilation may play a role in stimulating lactate production, but this cannot be the whole explanation.

Diabetes mellitus and lactic acidosis

The relationship between diabetes mellitus and lactic acidosis is still not entirely clear. The role of phenformin and other hypoglycemic biguanides has already been discussed. But even if one excludes the phenformin-related cases, diabetes still remains as a factor in the background of a significant fraction, perhaps as much as a quarter of all cases (Oliva, 1970) (Cohen and Woods, 1976).

Diabetes is a common disease, of course, and a common cause of many of the major medical complications that lead to lactic acidosis. It is to be expected therefore that many patients with lactic acidosis secondary to circulatory insufficiency, ischemia or sepsis would be diabetic. Many of the reported instances of so-called "idiopathic" or "spontaneous" lactic acidosis (see below) in patients with diabetes mellitus were in fact cases of this sort. Beyond this, however, there is some reason to believe that the diabetic defect itself might predispose to lactic acidosis. In diabetic ketoacidosis slightly increased blood levels of lactate are common, even in the absence of any complications, although most patients with well-controlled diabetes have normal or only borderline lactate concentrations. Insulin facilitates the disposal of pyruvate through the oxidative (PDH) pathway (Fig. 3.1); it also inhibits the release from muscle of alanine, which is a precursor of pyruvate. Therefore in the absence of insulin, or when the action of insulin is less effective, pyruvate and lactate might tend to accumulate. Whether, in fact, these considerations have any practical consequences for the pathogenesis of lactic acidosis is not known. The use of insulin in the treatment of this disorder will be discussed elsewhere.

"Idiopathic" or "spontaneous" acute lactic acidosis

In his original reports in 1961, Huckabee suggested that there was a small group of patients in whom acute lactic acidosis developed without any apparent cause. They had high L/P ratios, which he interpreted to indicate some sort of tissue hypoxia, but there was no clinical evidence of circulatory failure or arterial hypoxemia at the time lactic acidosis first appeared. He termed these cases "idiopathic" or "spontaneous" lactic acidosis, and this nosological category has been generally accepted ever since.

However, critical examination of the cases in the literature that have been described as examples of this entity raises serious doubts about the validity of the concept. In many of these cases, lactic acidosis apparently was simply a premonitory sign of impending vascular collapse, often associated with severe heart failure, sepsis or liver disease. In others the data are simply not sufficient to rule out such complications. A putative clinical entity like "idiopathic" acute lactic acidosis might of course lead to vascular collapse without the need of any other pathological process. The fact remains, however, that virtually all of the adequately described patients in the literature went into frank vascular collapse and died shortly after the onset of lactic acidosis, and had clinical or autopsy evidence of some major underlying disease that could have caused their death.

I therefore remain unconvinced of the significance of this classification. Premonitory shock-like states can exist for hours or days before frank collapse occurs, and a disturbance in lactate metabolism may be one of the earliest harbingers of the condition. Until the role of circulatory failure in the genesis of so-called "idiopathic" lactic acidosis is clarified, it would seem prudent to withhold final judgment about the issue.

Congenital lactic acidosis

Lactic acidosis may result from various kinds of congenital defects which cause increased production or decreased consumption of lactate and pyruvate. These disorders are associated with chronic or recurrent elevation of blood lactate, which usually is discovered in infancy or early childhood. They often produce serious neurological, neuromuscular or hepatic dysfunction that may be incompatible with prolonged survival, but in some instances patients survive into adult life. Occasionally the disease does not even become apparent until adulthood. Congenital lactic acidosis is rare, but it has been recognized with increasing frequency over the past few years and may turn out to be more common than at first supposed. Often the degree of hyperlactatemia is relatively mild and will cause no symptoms of itself, but sometimes the disturbance becomes severe enough to produce severe acidosis and patients may become acutely and dangerously ill as a result.

In most instances congenital lactic acidosis probably represents an inherited defect in one of the enzymes that regulate pyruvate metabolism (Fig. 3.1). However, clear-cut identification of such defects has been achieved in only a fraction of the cases reported so far and there remain at present many cases in which no specific biochemical mechanism has yet been established. Classification of this group of disorders is therefore incomplete, and must be regarded as purely tentative. Nevertheless, it may be useful to attempt a provisional classification because this will help to clarify some important points about the pathogenesis of lactic acidosis.

Table 3.2 lists the different types of congenital lactic acidosis according to what is presently known about the nature of the biochemical defect in each. The disorders in the first group represent deficiencies in three of the

Table 3.2 Congenital lactic acidosis

DEFECTS IN GLUCONEOGENESIS
1. Deficiency of glucose-6-phosphatase
   (Type I glycogen-storage disease)
2. Deficiency of fructose-1, 6-diphosphatase
3. Deficiency of pyruvate carboxylase

DEFECTS IN OXIDATION OF PYRUVATE
1. Deficiency of pyruvate dehydrogenase
2. Defects in oxidative phosphorylation

four enzymes unique to gluconeogenesis in liver and kidney. Lactic acidosis results from impairment in the capacity of the liver and kidney cortex to convert pyruvate (and hence lactate) into glucose.

Deficiency of glucose-6-phosphatase

(Type 1 glycogen-storage disease, or von Gierke's disease) has been recognized for many years as a cause of fasting hypoglycemia and of glycogen accumulation in liver and kidney. It also results in chronic lactic acidosis (Zuppinger and Rossi, 1969). As shown in Figure 3.1, G-6-Pase is the final enzyme in the pathway leading to glucose from all precursors. In its absence neither the liver nor the kidney is able to synthesize or release glucose when gluconeogenesis and glycogenolysis are stimulated by fasting or by hormonal stress (epinephrine, glucagon, glucocorticoids). Uptake of lactate by these organs is therefore inhibited; there may be, in fact, net release of lactate from the splanchnic bed. The metabolic disturbance in many ways resembles that which occurs in fulminant hepatic failure. The blood concentrations of glucose precursors, like lactate and alanine, are increased, there is a tendency to hypoglycemia and there is increased fat mobilization and ketone body formation to provide an energy source other than carbohydrate. A deficiency of G-6-Pase also occurs in the intestinal mucosa, which is the only other tissue normally endowed with this enzyme. What consequences this deficiency may have for the absorption and metabolism of glucose in the intestines is not known.

In the fed state blood lactate is only modestly elevated (2–5 mM per L). Administration of glucose will lower the lactate concentration, as will ethanol. Both of these substances, particularly ethanol, usually increase blood lactate in normal subjects, but in patients with G-6-Pase deficiency glucose appears to facilitate lactate utilization by the liver and alcohol inhibits lactate production. Fasting exacerbates the disturbance in lactate metabolism because it impairs the oxidation of pyruvate through the PDH pathway (Fig. 3.1), while increasing the delivery of alanine to the liver. Since the gluconeogenic pathway is blocked, lactate must accumulate, the blood level rises above 5 mEq per L, and acidosis increases. The acidosis is due largely to lactate but ketone acids also contribute. The L/P ratio in the fed state is nearly normal, as would be expected, but with fasting the ratio rises more than it does in normal subjects.

Lactic acidosis in G-6-Pase deficiency is rarely of great severity, but it tends to be worst during episodes of intercurrent illness in infants unable to eat normally. Hypoglycemia is usually a greater threat than acidosis. Except for the deleterious effects of hypoglycemia, brain metabolism is normal and hence there is no central stimulation to respiration and no excessive hyperventilation. Respiratory alkalosis is therefore usually not a feature of this disorder. Treatment consists in maintaining blood glucose levels by frequent carbohydrate feedings, which usually keep the blood lactate within reasonable limits. However, galactose (derived from milk sugar) and fructose should be avoided because they cannot be used by the liver to make glucose and will tend to aggravate the acidosis.

Deficiency of fructose-1, 6-diphosphatase has recently been identified as another cause of lactic acidosis and hypoglycemia in infancy (Baker and Winegrad, 1970) (Pagliara et al., 1972). The defect involves the liver and probably also the kidney cortex, but not muscle, which has a normal amount of enzyme. Like type I glycogen-storage disease, this disorder produces a chronic lactic acidosis of mild to moderate degree, which is exacerbated by fasting and by any intercurrent illness that prevents adequate nutrition. In the early stages of fasting blood glucose levels are sustained by glycogenolysis and the administration of epinephrine or glucagon will cause an hyperglycemic response (unlike type I glycogen-storage disease). However, after hepatic glycogen is exhausted, blood glucose levels fall, fat stores are mobilized and ketogenesis increases. The pathogenesis of the lactic acidosis that ensues is similar to that described in type I glycogen-storage disease.

The liver is enlarged with fat, rather than glycogen, but otherwise the two conditions are much alike. The L/P ratio in the blood is at first nearly normal but then tends to rise moderately under the influence of fasting. Acidosis can be very severe and, because there is no metabolic disturbance in the brain, there is no respiratory alkalosis.

As with type I glycogen-storage disease, treatment is directed mainly towards relief of the hypoglycemia, by administration of glucose. This results in prompt improvement of the acidosis, presumably because production of alanine, and possibly also lactate, is reduced in the periphery, while at the same time the hepatic redox state is improved. Neither galactose nor fructose can substitute for glucose in therapy nor can fat or protein.

Pyruvate carboxylase deficiency

Like G-6-Pase and FDPase, pyruvate carboxylase (PC) is an enzyme that catalyzes one of the unique steps in the gluconeogenic sequence in liver and kidney (Fig. 3.1). However, it also plays an important role in the synthesis of fat and is very active in adipose tissue. There is virtually no PC activity in muscle and relatively little in brain, although it is possible that PC is involved in the synthesis of structural lipids in the central nervous system.

The syndrome of PC deficiency, like FDPase deficiency, has been reported only a few times. It is characterized by recurrent episodes of lactic acidosis and signs of diffuse brain damage. The L/P ratio is normal or only slightly elevated and the plasma alanine level is increased. In at least some cases, respiratory alkalosis accompanies the metabolic acidosis. Hypoglycemia has been a variable feature of the syndrome, possibly because the enzyme block does not impair gluconeogenesis from glycerol and fructose. There is no evidence of any abnormality in the peripheral metabolism of lactate in these cases (Grover, Auerbach, Patel, 1972).

Neurological damage is severe. The pathological lesion was described by Leigh, as *"subacute necrotizing encephalomyelitis."* It causes mental retardation and various types of movement disorders. The nature of the connection (if any) between the PC defect and the brain lesions is not yet clear. It is not known whether PC plays an important role in human brain tissue, but an obvious possibility is that the defect in hepatic PC also involves the brain, which as a consequence is unable to synthesize structural lipids properly.

Defects in oxidation of pyruvate

This group of cases is characterized by defects in the oxidation of pyruvate to form acetyl-CoA (Fig. 3.1). The series of reactions by which pyruvate is decarboxylated to form acetyl-CoA is catalyzed by a complex of three enzymes collectively referred to as pyruvate dehydrogenase (PDH). The PDH pathway is one of the vital steps in the extraction of chemical energy from food, therefore serious defects in it are not likely to be compatible with life. However, in recent years a few young children have been identified with partial deficiencies in PDH activity, and in one instance an infant survived until the age of six months with a total deficiency in the first component of the PDH complex (Blass et al., 1972) (Farrell et al., 1975) (Stromme, Borud and Moe, 1975).

Since the PDH system is common to all oxidative cells, the enzyme defect is generalized and impairs the utilization of pyruvate in all tissues. In the few cases described so far, the clinical manifestations were primarily those of diffuse neurological damage, with mental deficiency and movement disorders. In the patients with partial PDH defects there was a moderate increase in blood pyruvate level, with an accompanying rise in lactate and a normal or only slightly increased L/P ratio. The enzymes of gluconeogenesis were normal, so there was no disturbance in glucose metabolism. In the one infant described with a total block in the PDH pathway, the neurological disease and the degree of hyperlactatemia and hyperpyruvatemia were more severe. Metabolic acidosis was a major problem, which could be prevented only by drastic restriction of carbohydrate intake. The defect in this patient was shown to be due to a total absence of pyruvate decarboxylase, which is the first of the three enzymes in the PDH complex.

Defects in oxidative phosphorylation

Under this heading can be included a heterogeneous variety of cases that seem to be characterized by a diffuse mitochondrial defect in the processes by which production of high energy phosphate bonds is coupled to oxygen consumption. In many instances there have been morphological changes in mitochondria of skeletal muscle. Clinical signs are those of a generalized myopathy, sometimes associated with neurological defects. Lactic acidosis is produced intermittently when exercising muscles increase their rate of glycolysis excessively in an attempt to compensate for inefficient ATP production. There is no hypoglycemia and glucose metabolism is entirely normal. L/P ratios are usually greatly elevated (Shapira et al., 1975) (Sengers et al., 1975) (Hackett et al., 1973).

The exact nature of the defect or defects in this group of cases has not yet been defined but they appear to represent some type of mitochondrial disease that in effect mimics the biochemical consequences of hypoxia. Minimal exercise that normally would increase blood lactate only slightly and transiently produces in these patients serious and prolonged lactic acidosis. Whether or not the mitochondrial defect involves the liver is not known. If ATP synthesis were impaired in the liver as well as in muscle, the capacity to dispose of lactate would be more seriously affected.

Future developments in this field will undoubtedly necessitate considerable modification of the above classification. However, the evidence to date would seem to justify the conclusion that chronic lactic acidosis may occur as a result of congenital defects in any of the enzymes that are involved in the metabolism of pyruvate or in the production of high energy phosphate that is normally coupled to the oxidation of pyruvate. The best studied of these disorders are those that involve the enzymes of gluconeogenesis in liver and kidney, since they are compatible with longer survival. Defects in the enzymes that facilitate the oxidation of pyruvate or the formation of high-energy phosphates are likely to be more generalized and more lethal.

Treatment

Since lactic acidosis is a biochemical sign produced by a great variety of pathological processes, it should be obvious that there can be no one standard form of treatment. The first and most important step in management is to identify if possible the basic cause of the problem. The outcome will depend primarily on whether or not the cause can be recognized and treated effectively.

The first principle is therefore to identify the cause of the lactic acidosis and treat it as vigorously as possible. Shock, sepsis, low output states, and the like must be promptly dealt with if the patient is to survive. Treatment aimed at the underlying cause is likely to be more effective than attempts to treat the acidosis per se although the latter are usually an essential part of management. In treating low output states associated with lactic acidosis, it is best to avoid vasoconstrictor catecholamines since they will increase ischemia and thereby aggravate lactic acidosis. If central venous pressure is adequate, consideration should be given to a trial of nitroprusside which in one interesting recent case report apparently was useful in reducing peripheral ischemia and thereby dramatically improved the lactic acidosis. When the central venous pressure and other clinical signs point to a reduced circulating volume, it is of course important to expand the circulation with the whole blood or plasma as needed. When lactic acidosis and respiratory alkalosis develop suddenly without other apparent cause, one should always suspect sepsis or pulmonary emboli.

The indication for therapy with *alkali* is severe acidosis with significant acidemia. When indicated, alkali should be given as intravenous sodium bicarbonate solution, at a concentration appropriate for the particular situation at hand. Isotonic solutions should be used when there is no disorder of osmolality and when extra circulating volume is needed and can be tolerated. When there is hyponatremia or when circulatory overload is likely, hypertonic solutions should be used. In theory bicarbonate distributes itself throughout body water and calculations of the amount needed to produce a given rise in plasma bicarbonate are made simply by multiplying the desired rise in plasma bicarbonate concentration by 50 percent of the body weight in kilograms. Thus, for example, if an 80 kilogram patient has a plasma bicarbonate of 10 mEq per L and we wish to raise it to 20, he should be given 10 times 40 or 400 mEq of sodium bicarbonate.

Intravenous alkali should always be given with due precautions to avoid circulatory overload and with regard to possible complications if plasma calcium is low. The calculation of the amount needed to achieve a given plasma bicarbonate level usually underestimates the need because lactate production usually continues for at least a period of time after therapy has begun. In severe cases it is not unusual for the plasma bicarbonate to fail to rise despite the administration of large amounts of alkali over a relatively short period of time. If very large quantities of alkali are needed, it is advisable to assure renal excretion of the excess sodium by the simultaneous administration of a powerful diuretic, preferably furosemide. In this way it may be possible to give hundreds or even thousands of milliequivalents of alkali without greatly expanding the extracellular volume and thereby risking congestive heart failure. When renal failure renders diuretics impotent, peritoneal dialysis may provide a means of exchanging bicarbonate for chloride in the plasma while simultaneously removing unwanted sodium and water. In cases of phenformin intoxication dialysis may also constitute a means of lowering drug levels rapidly.

It is important to remember that over-enthusiastic use of intravenous alkali may lead to severe alkalosis and alkalemia. When there is hyperventilation, restoration of plasma bicarbonate levels to normal will not usually cause the respiration to slow. Respiration may not slow down for many hours after the blood pH has been brought up to normal or has even exceeded normal. Therefore, patients often become alkalemic after treatment because their CO_2 tension remains low even after their plasma bicarbonate has been restored to normal. There may also be an overshoot alkalosis as the result of the metabolism of lactate during the recovery period. It must be remembered that all the lactate measured in the blood is potential alkali. When hepatic consumption of lactate converts it to bicarbonate, the net effect may be a very marked rise in blood bicarbonate and the development of severe alkalemia. The risk from alkalemia is primarily that of convulsions and tetany.

In diabetic patients with lactic acidosis, whether or not phenformin or other biguanide drugs are involved, insulin may be a useful form of therapy. (Dembo, Marliss, and Halperin, 1975). There is no general agreement on this point but there seems to be good theoretical grounds for believing that insulin would be helpful in stimulating the oxidation of pyruvate through the PDH pathway and in reducing the flow of gluconeogenic amino acids to the liver. Evidence that insulin therapy makes a big difference in survival or promptness of response is not available at present. Nevertheless, the rationale seems sound and I believe that it is advisable to include insulin in the regimen whenever one is treating a diabetic patient. The amount to be used depends on circumstances, and no general rules can be made. Usually, however, insulin requirements are relatively modest.

In a few cases intravenous *methylene blue* has been administered in an effort to increase the redox potential of cells and thereby limit the production of lactate. Methylene blue is a hydrogen acceptor and will tend to reduce the amount of NADH while increasing NAD. Although this should work in principle, there is no supporting clinical evidence that it does. For this reason, methylene blue has not been used very widely in recent years and probably ought to be abandoned.

Thiamin and *lipoic acids* act as coenzymes for the PDH enzyme complex. For this reason, these agents have occasionally been administered in massive doses in patients with lactic acidosis. Only when specific defects in PDH are involved, is there likely to be any benefit. Even then, benefits are unproven and the value of such therapy remains uncertain. A recent report (Majoor, 1978) suggests that thiamin injections were useful in treating lactic acidosis in an alcoholic patient with cardiac beri-beri, but more experience is needed to settle this question.

Another agent known to enhance PDH activity is *dichloracetate*, which has been shown to lower lactate levels in non-acidotic diabetic patients (Stacpoole, Moore and Kornhauser, 1978) and in experimental lactic acidosis produced by functional hepatectomy (Blackshear, Holloway and Alberti, 1974) or by phenformin poisoning (Loubatieres, Rebes and Valette, 1977). This compound appears to be a promising agent for the treatment of many types of lactic acidosis *not* associated with acute circulatory failure and hypoxia, but it has not yet had a clinical trial.

Lactic acidosis associated with severe hyperventilation can probably be improved, at least transiently, by an increase in the CO_2 tension of body fluids, which will tend to lower intracellular pH and thereby reduce the glycolytic rate. Obviously, when the patient is significantly acidotic from lactic acidosis, it does not make much sense to increase his acidemia by inhalation of 5 percent CO_2 but this form of therapy might occasionally be useful when respiratory alkalemia is severe.

Some authors have suggested that hyperbaric oxygen might be helpful in cases of lactic acidosis due to tissue hypoxia. Such cases would be identified by very high L/P ratios. To the best of my knowledge there is no experience with this form of therapy. It sounds rational and is probably worth trying when the specialized facilities are available.

Sodium nitroprusside has been reported to have been successfully used in a patient with lactic acidosis of unknown cause who had severe peripheral vasoconstriction but no other evidence of shock (Taradash and Jacobson, 1975). The vasodilator effect of the drug might have been useful in relieving tissue ischemia in this case, but no similar cases have been reported since then, and the value of vasodilator therapy in lactic acidosis remains uncertain.

Summary: In the last analysis the patient's survival will depend upon the nature of the underlying disease that has precipitated the disorder in lactate metabolism. If that can be accurately identified and effectively treated, then the prognosis is good. Recovery is also more likely if the patient is suffering from an acute transient intoxication such as that caused by alcohol or phenformin. In such situations one simply needs to tide the patient over a period of severe acidosis while waiting for the intoxication to subside. Major improvements in therapy must await better understanding of the basic mechanisms involved in this disorder. Dichloracetate may be a promising agent for the treatment of those types of lactic acidosis not due to hypoxia, but it has not yet had a clinical trial.

REFERENCES

Alberti, K. G. M. M., and Nattrass, M. (1977) Lactic acidosis. *Lancet,* **2,** 25–29.

Baker, L., and Winegard, A. I. (1970) Fasting hypoglycemia and metabolic acidosis associated with deficiency of hepatic fructose-1, 6-diphosphatase. *Lancet,* **1,** 13–16.

Blackshear, P. J., Holloway, P. A. H., and Alberti, K. G. M. M. (1974) The metabolic effects of sodium dichloracetate in the starved rat. *Biochemical Journal,* **142,** 279–286.

Blass, J. P., Schulman, J. D., Young, D. S., and Hom, E. (1972) An inherited defect affecting the tricarboxylic acid cycle in a patient with congenital lactic acidosis. *Journal of Clinical Investigation,* **51,** 1845–1851.

Buehler, J. H., Berns, A. S., Webster, J. R. Jr., Addington, W. W., and Cugell, D. W. (1975) Lactic acidosis from carboxyhemoglobinemia after smoke inhalation. *Annals of Internal Medicine,* **82,** 803–805.

Clausen, S. W. (1925) Anhydremic acidosis due to lactic acid. *American Journal of Diseases of Children,* **29,** 761–766.

Cloutier, C. T., Lowery, B. D., and Carey, L. C. (1969) Acid-base disturbances in hemorrhagic shock. *Archives of Surgery,* **98,** 551–557.

Cohen, R. D., and Simpson, R. (1975) Lactate metabolism. *Anesthesiology,* **43,** 661–673.

Cohen, R. D., and Woods, H. F. (1976) *Clinical and Biochemical Aspects of Lactic Acidosis.* Oxford: Blackwell.
Conlay, L. A., and Loewenstein, J. E. (1976) Phenformin and lactic acidosis. *Journal of the American Medical Association,* **235,** 1575–1578.
Davidoff, F. (1973) Guanidine derivatives in medicine. *New England Journal of Medicine,* **289,** 141–146.
Dembo, A. J., Marliss, E. B., and Halperin, M. L. (1975) Insulin therapy in phenformin-associated lactic acidosis. *Diabetes,* **24,** 28–35.
Emmett, M., and Narins, R. G. (1977) Clinical use of the anion gap. *Medicine,* **56,** 38–54.
Farrell, D. F., Clark, A. F., Scott, C. R., and Wennberg, R. P. (1975) Absence of pyruvate decarboxylase activity in man: A cause of congenital lactic acidosis. *Science,* **187,** 1082–1084.
Field, M., Block, J. B., Levin, R., and Rall, D. P. (1966) Significance of blood lactate elevations among patients with acute leukemia and other neoplastic proliferative disorders. *American Journal of Medicine,* **40,** 528–548.
FDA Drug Bulletin Vol 7, No. 3, August 1977.
Fulop, M., Horowitz, M., Aberman, A., and Jaffee, E. R. (1973) Lactic acidosis in pulmonary edema due to left ventricular failure. *Annals of Internal Medicine,* **79,** 180–184.
Graham, D. L., Laman, D., Theodore, J., and Robin, E. D. (1977) Acute cyanide poisoning complicated by lactic acidosis and pulmonary edema. *Archives of Internal Medicine,* **137,** 1051–1055.
Grover, W. D., Auerbach, V. H., and Patel, M. S. (1972) Biochemical studies and therapy in subacute necrotizing encephalomyelopathy (Leigh's syndrome). *Journal of Pediatrics,* **81,** 39–44.
Hackett, T. N. Jr., Bray, P. F., Ziter, F. A., Nyhan, W. L. and Creer, K. M. (1973) A metabolic myopathy associated with chronic lactic acidemia, growth failure, and nerve deafness. *Journal of Pediatrics,* **83,** 426–431.
Halperin, M. L., Connors, H. P., Relman, A. S., and Karnovsky, M. L. (1969) Factors that control the effect of pH on glycolysis in leukocytes. *Journal of Biological Chemistry,* **244,** 384–390.
Hessov, I. (1974) Effects of fructose and glucose infusions on blood acid-base equilibrium in the postoperative period. *Acta Chirurgica Scandinavica,* **140,** 347–351.
Huckabee, W. E. (1961 a) Abnormal resting blood lactate. I. The significance of hyperlactatemia in hospitalized patients. *American Journal of Medicine,* **30,** 833–839.
Huckabee, W. E. (1961 b) Abnormal resting blood lactate. II. Lactic acidosis. *American Journal of Medicine,* **30,** 840–848.
Iles, R. A., Cohen, R. D., Rist, A. H., and Baron, P. G. (1977) The mechanism of inhibition by acidosis of gluconeogenesis from lactate in rat liver. *Biochemical Journal,* **164,** 185–191.
Køllendorf, K., and Møller, B. B. (1974) Lactic acidosis and epinephrine poisoning. *Acta Medica Scandinavica,* **196,** 465–466.
Kreisberg, R. A. (1972) Glucose-lactate inter-relations in man. *New England Journal of Medicine,* **287,** 132–137.
Kreisberg, R. A., Owen, W. C., and Siegal, A. M. (1971) Ethanol-induced hyperlacticacidemia: Inhibition of lactate utilization. *Journal of Clinical Investigation,* **50,** 166–174.
Lieber, C. S., Teschke, R., Hasumura, Y., and Decarli, L. M. (1975) Differences in hepatic and metabolic changes after acute and chronic alcohol consumption. *Federation Proceedings,* **34,** 2060–2074.
Loubatieres, A. L., Rebes, G., and Valette, G. (1977) Reduction par le dichloroacetate de sodium et par l'insuline des hyperlactatemies graves consecutives chez le chien anestheie ou eveille a l'administration de phenformine. *Compte Rendu Academie Science de Paris,* **282,** 325–327.
MacLean, L. D., Mulligan, W. G., McLean, A. P. H., and Duff, J. H. (1967) Patterns of septic shock in man. *Annals of Surgery,* **166,** 543–562.
Majoor, C. L. H. (1978) Alcoholism as a cause of beri-beri heart disease. *Journal of the Royal College of Physicians of London,* **12,** 143–152.

Marliss, E. B., Aoki, T. T., Toews, C. J., Felig, P., Connon, J. J., Kyner, J., Huckabee, W. E., and Cahill, G. F., Jr. (1972) Amino acid metabolism in lactic acidosis. *American Journal of Medicine* **52**, 474–481.

Miller, P. D., Heinig, R. E., and Waterhouse, C. (1978) Treatment of alcoholic ketosis. *Archives of Internal Medicine*, **138**, 67–72.

Misbin, R. I. (1977) Phenformin-associated lactic acidosis: pathogenesis and treatment. *Annals of Internal Medicine*, **87**, 591–595.

O'Connor, L. R., Klein, K. L., and Bethune, J. E. (1977) Hyperphosphatemia in lactic acidosis. *New England Journal of Medicine*, **297**, 707–709.

Oliva, P. (1970) Lactic acidosis. *American Journal of Medicine*, **48**, 209–225.

Orringer, C. E., Eustace, J. C., Wunsch, C. D., and Gardner, L. B. (1977) Natural history of lactic acidosis after grand mal seizures. A model for the study of an anion-gap acidosis not associated with hyperkalemia. *New England Journal of Medicine*, **297**, 796–799.

Pagliara, A. S., Karl, I. E., Keating, J. P., Brown, B., and Kipnis, D. M. (1972) Hepatic fructose-1, 6-diphosphatase deficiency. A cause of lactic acidosis and hypoglycemia in infancy. *Journal of Clinical Investigation*, **51**, 2115–2123.

Record, C. O., Iles, R. A., Cohen, R. D., and Williams, R. (1975) Acid-base and metabolic disturbances in fulminant hepatic failure. *Gut*, **16**, 144–149.

Relman, A. S. (1972) Metabolic consequences of acid-base disorders. *Kidney International*, **1**, 347–359.

Sengers, R. C. A., ter Haar, B. G. A., Trijbels, J. M. F., Willems, J. L., Daniels, O., and Stadhouders, A. M. (1975) Congenital cataract and mitochondrial myopathy of skeletal and heart muscle associated with lactic acidosis after exercise. *Journal of Pediatrics*, **86**, 873–880.

Shapira, Y., Cederbaum, S. D., Cancilla, P. A., Nielsen, D., and Lippe, B. M. (1975) Familial poliodystrophy, mitochondrial myopathy, and lactate acidemia. *Neurology*, **25**, 614–621.

Stacpoole, P. W., Moore, G. W., and Kornhauser, D. M. (1978) Metabolic effects of dichloracetate in patients with diabetes mellitus and hyperlipoproteinemia. *New England Journal of Medicine*, **298**, 526–530.

Stromme, J. H., Borud, O., and Moe, P. J. (1976) Fatal lactic acidosis in a newborn attributable to a congenital defect of pyruvate dehydrogenase. *Pediatric Research*, **10**, 60–66.

Taradash, M. R., and Jacobson, L. B. (1975) Vasodilator therapy of idiopathic lactic acidosis. *New England Journal of Medicine*, **293**, 468–471.

Zuppinger, K., and Rossi, E. (1969) Metabolic studies in liver glycogen disease with special reference to lactate metabolism. *Helvetica Medica Acta*, **35**, 406–422.

4

Metabolic alkalosis

ANTHONY SEBASTIAN
HENRY N. HULTER
FLOYD C. RECTOR, JR.

Respiratory compensation in metabolic alkalosis
Coexistence of metabolic alkalosis with a primary respiratory disturbance of acid-base equilibrium
Pathogenesis of metabolic alkalosis: General mechanisms
Renal regulation of acid-base equilibrium
Renal response to metabolic alkalosis of extrarenal origin
Influence of effective arterial blood volume
Possible role of bicarbonate secretion by the renal tubule
Administration of chloride prerequisite to correction of volume contraction and repair of alkalosis: Effect of "isometric" ECF volume expansion
Perpetuation of metabolic alkalosis in the absence of ECF volume contraction
Metabolic alkalosis of renal origin
Role of potassium deficiency and hypermineralocorticoidism
    Acid-base response to deprivation of dietary potassium uncomplicated by other potential alkalosis-producing conditions
    Potassium depletion as a necessary cofactor in the pathogenesis of alkalosis due to hypermineralocorticoidism
    Mechanism whereby potassium depletion potentiates the alkalosis-producing effect of mineralocorticoid hormone
        Exaggerated urine pH lowering response to mineralocorticoid hormone and enhanced renal ammonia production
        Enhanced proximal bicarbonate reabsorptive capacity
        Impaired chloride reabsorption by the renal tubule
        Facilitated hormone-stimulated Na^+-nondependent H^+ secretion
Induction of metabolic alkalosis with inhibitors of active chloride reabsorption
Hyperglucocorticoidism
Hypercalcemia, hypoparathyroidism, hypervitaminosis D and hyperphosphatemia

The primary disturbance of acid-base equilibrium in metabolic alkalosis is an increase in the concentration of base (chiefly bicarbonate) in the extracellular fluid compartment of the body. The absolute amount of base need not be increased if extracellular volume is greatly reduced, as in so-called contraction alkalosis. The increase in bicarbonate concentration disturbs the normal equilibrium of the bicarbonate-carbonic acid-carbon dioxide buffer system and decreases hydrogen ion concentration, (i.e., increases pH). This effect on acid-base equilibrium is ameliorated by a secondary increase in carbon dioxide tension mediated by a reduction in the rate of alveolar ventilation. Accordingly, when metabolic alkalosis is the predominant disturbance of acid-base equilibrium, it is characterized by alkalemia and hypercapnia as well as by a supernormal plasma bicarbonate concentration. Typically, the increase in plasma bicarbonate concentration is accompanied by an approximately commensurate decrease in plasma chloride concentration, unless there exists a primary disturbance in the regulation of plasma osmolality, (i.e., plasma sodium concentration) or an increase in the concentration of unmeasured anions (lactate, acetoacetate, etc.). Consideration of the pathogenesis of metabolic alkalosis, therefore, compels consideration of the mechanisms of chloride loss as well as bicarbonate gain.

RESPIRATORY COMPENSATION IN METABOLIC ALKALOSIS

The reduced rate of alveolar ventilation in metabolic alkalosis is mediated by diminished activity of neural chemoreceptors resulting from the reduction of extracellular hydrogen ion concentration (Fencl, Miller and Pappenheimer, 1966; Berger, Mitchell and Severinghaus, 1977). Studies in two separate series of patients indicate that the degree of hypercapnia is proportional to the severity of the alkalosis as reflected by the plasma bicarbonate concentration (Fig. 4.1a) (van Ypersele de Strihou and Frans, 1973; Fulop, 1976). Earlier studies in patients with metabolic alkalosis of mild degree and short duration had not revealed a significant degree of respiratory compensation (Roberts et al., 1956). In subsequent case reports of patients with more severe metabolic alkalosis, the occurrence of hypercapnia was considered to be exceptional (Tuller and Mehdi, 1971; Jarboe, Penman and Luke, 1972; Lifschitz et al., 1972; Oliva, 1972; Shear and Brandman, 1973; and Saunders et al., 1974), but in retrospect the degree of hypercapnia in these patients appears to be appropriate for the degree of metabolic alkalosis as judged from the extrapolated relationship between pCO_2 and plasma bicarbonate concentration observed in the larger series of patients (Fig. 4.1a).

Evidence has been presented that a lesser degree of respiratory compensation, or none, occurs when metabolic alkalosis is associated with potassium depletion (Goldring et al., 1968). It was postulated that potassium depletion reduces intracellular pH in the respiratory chemoreceptors, which thereby limits the suppressive effect of alkalemia on alveolar ventilation. This phenomenon might explain in part the "normal variation" in ventilatory response observed at any given plasma bicarbonate concentration (Fig. 4.1a). Experimentally induced potassium depletion in dogs and rats, however, does not influence the magnitude of the ventilatory response to metabolic alkalosis (Penman, Luke and Jarboe, 1972; Aquino and Luke, 1973). The influence of potassium depletion on the degree of respiratory compensation to metabolic alkalosis in man requires further study.

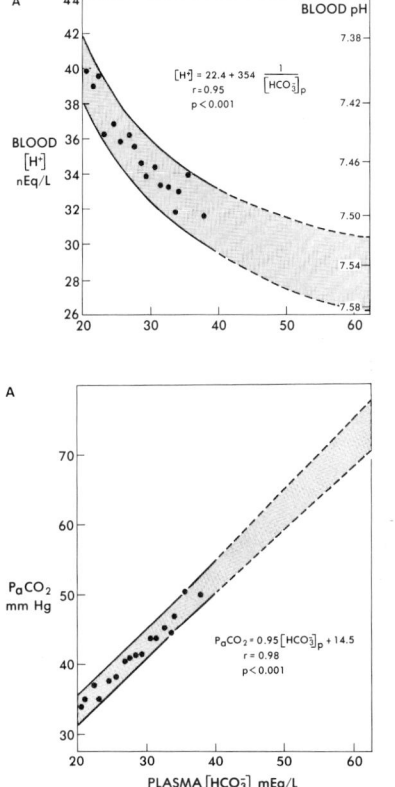

Figure 4.1a The respiratory response to chronic, uncomplicated metabolic alkalosis in man. The upper panel depicts the response of blood hydrogen ion concentration (and pH) to a sustained increase in plasma bicarbonate concentration. The lower panel depicts the corresponding respiratory response (P_aCO_2). For each relationship, the associated 95 percent confidence band is indicated by the shaded area. Extrapolation of the confidence band to values of plasma bicarbonate concentration greater than 40 mEq/l is indicated by the dashed lines. (Drawn from data of van Ypersele de Strihou and Frans, 1973)

COEXISTENCE OF METABOLIC ALKALOSIS WITH A PRIMARY RESPIRATORY DISTURBANCE OF ACID-BASE EQUILIBRIUM

Elucidation of the quantitative limits of the ventilatory response to metabolic alkalosis permits the recognition of coexisting primary disturbances in respiratory regulation of acid-base equilibrium. In patients with metabolic alkalosis, values of carbon dioxide tension in blood lower than or greater than those predicted for a given plasma bicarbonate concentration indicate the coexistence of respiratory alkalosis or respiratory acidosis, respectively. The coexistence of respiratory and metabolic alkalosis can be confirmed (Fig. 4.1b) by comparison of the observed plasma bicarbonate concentration or hydrogen ion concentration with the value expected to result solely from the physiological response to a primary increase in carbon dioxide tension (van Ypersele de Strihou, Brasseur and DeConinck 1966) Fig. 4.1*b*: The physiological increase in plasma bicarbonate concentration that occurs in chronic respiratory acidosis reflects in part titration of carbonic acid by body buffers and in part de novo generation of bicarbonate by the kidney. In patients with respiratory acidosis, values of plasma bicarbonate concentration greater than those expected for the observed carbon dioxide tension indicate the coexistence of metabolic alkalosis.

Recognition of the coexistence of metabolic alkalosis and respiratory acidosis has several clinical implications. In patients with chronic obstruc-

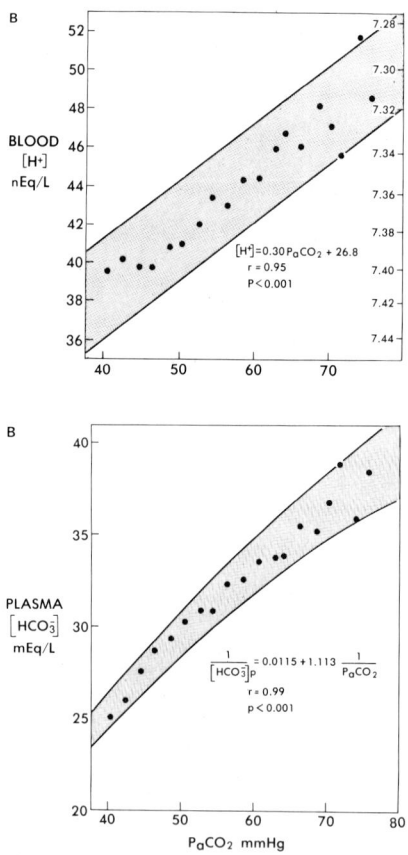

Figure 4.1*b* The metabolic response to chronic, uncomplicated respiratory acidosis in man. The upper panel depicts the response of blood hydrogen ion concentration (and pH) to a sustained increase in P_aCO_2. The lower panel depicts the corresponding alteration in plasma bicarbonate concentration. For each relationship the associated 95 percent confidence band is indicated by the shaded area. (Drawn from data of van Ypersele de Strihou et al., 1966)

tive pulmonary disease, the degree of hypercapnia and attendant hypoxemia is greater when metabolic alkalosis is superimposed; correction of the metabolic alkalosis leads to a significant reduction in carbon dioxide tension and amelioration of hypoxemia (Bear et al., 1977). It should be emphasized that the reduction in plasma bicarbonate concentration sufficient to correct metabolic alkalosis in patients with coexisting chronic respiratory acidosis does not return the concentration to normal (22–26 mEq/l). Normalization of the plasma bicarbonate concentration in such patients would offset the physiological compensation for the persisting hypercapnia and would therefore subject the patients to a greater degree of acidemia than would otherwise attend uncomplicated chronic hypercapnia (Fig. 4.1b). Normalizing the plasma bicarbonate concentration in this circumstance thus superimposes acute metabolic acidosis on chronic respiratory acidosis. This consideration is important in clinical circumstances in which correction of metabolic alkalosis requires titration of extracellular bicarbonate by administration of exogenous acid (Petzel et al., 1976).

PATHOGENESIS OF METABOLIC ALKALOSIS: GENERAL MECHANISMS

The increased concentration of bicarbonate in the extracellular fluid in most common types of clinical and experimental metabolic alkalosis can be explained by at least one of three general physiological disturbances:

(a) Loss of hydrogen ion from the body (which occurs characteristically in association with chloride loss in approximately equivalent amount); for each mEq of H^+ lost one mEq of base is generated in the body from the shift in equilibrium of body buffers: $CO_2 + H_2O \rightleftharpoons H_2CO_3 \rightleftharpoons HCO_3^- + H^+$.

(b) Gain of exogenous base (bicarbonate or metabolic precursors of bicarbonate).*

(c) Reduction in volume of the extracellular fluid compartment without proportional reduction of extracellular bicarbonate content (contraction alkalosis); for each liter reduction in ECF volume (as reflected by reduction in body weight) the plasma bicarbonate concentration increases by approximately 1.4 mEq/L (Cannon et al., 1965) (Fig. 4.2)).

Numerous congenital, acquired, iatrogenic, and experimental conditions have been identified as capable of initiating these disturbances. (Table 4.1) When the initiating conditions persist or recur at sufficiently frequent intervals, the resultant alkalosis tends to persist. If the initiating disturbance

---

* Gain of base in the extracellular fluid might also occur because of decreased base excretion in the stool. This mechanism might contribute to the pathogenesis of metabolic alkalosis in patients with congenital chloride diarrhea (Bieberdorf, Gorden and Fordtran, 1972). The amount of base normally excreted in the stool is too small, however, to cause appreciable metabolic alkalosis in the absence of other alkalosis producing factors.

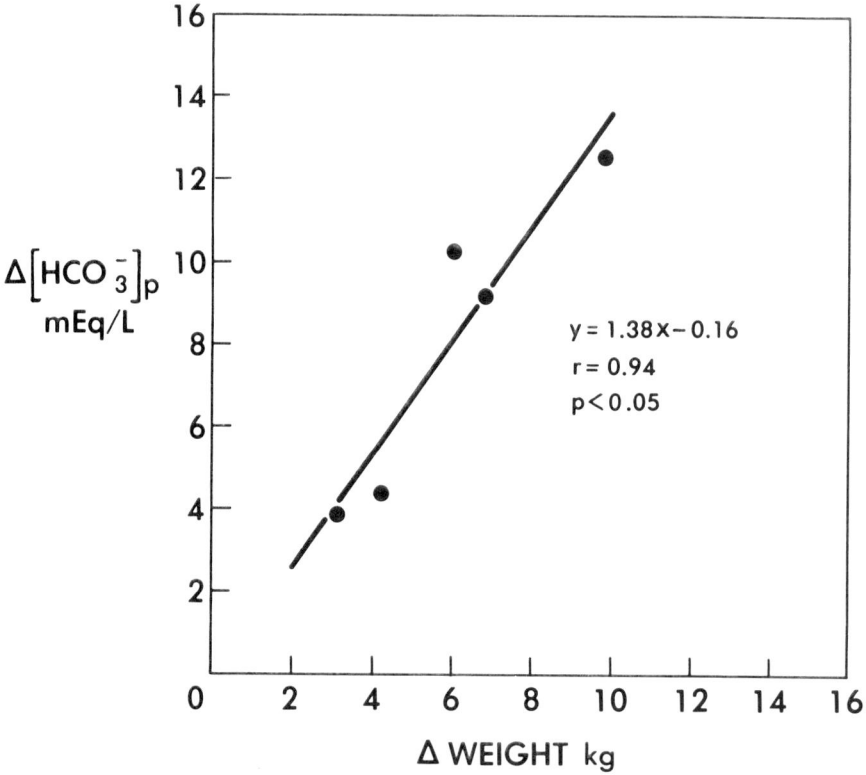

Figure 4.2 The plasma bicarbonate response to rapid diuresis in congestive heart failure in man. The prompt reduction of ECF volume associated with the acute excretion of sodium and chloride results in "contraction alkalosis" even though external hydrogen ion balance is not significantly altered. For each liter of ECF lost, plasma bicarbonate concentration increases by approximately 1.4 mEq/l. (Drawn from data of Cannon et al., 1965)

is only transient, the alkalosis is correctable spontaneously by action of the kidneys provided that pathophysiological constraints have not been imposed on the homeostatic function of the kidneys to maintain systemic acid-base equilibrium.

RENAL REGULATION OF ACID-BASE EQUILIBRIUM (FIG. 4.3)*

Under normal physiological conditions the kidneys maintain the concentration of extracellular bicarbonate constant (at approximately 22–26 mEq/L) by delivering to the extracellular fluid compartment (via renal

---

* This subject is considered in detail in Chapter 1.

Table 4.1 Pathogenesis of metabolic alkalosis: mechanisms and syndromes

I. External loss of H^+ and Cl^-
   A. In gastric fluid, as unbuffered H^+
      1. Gastric drainage
      2. Vomiting
   B. In urine, as buffered H^+ (NH_4^+, $H_2PO_4^-$)[a]
      1. Combined potassium depletion and hypermineralocorticoidism
         a) Aldosterone
            (1) Primary aldosteronism
            (2) Primary reninism
            (3) Secondary aldosteronism without diminished distal sodium delivery (diuretics, Bartter's syndrome)
         b) Desoxycorticosterone
            (1) 11-hydroxylase deficiency
            (2) 17-hydroxylase deficiency
            (3) Adrenal carcinoma
         c) (?) Liddle syndrome (unidentified mineralocorticoid)
         d) Licorice
         e) Hydrocortisone (mineralocorticoid effect?)
         f) Carbenoxolone
      2. Inhibition of renal chloride reabsorption
         a) Chloruretic drugs
         b) (?) Bartter's syndrome
      3. Hypercalemia and/or hypoparathyroidism
      4. (?) hyperglucocorticoidism
   C. In stool, buffer state of H^+ unknown[a]
      Congenital chloride diarrhea
II. Excessive load of HCO_3^- or HCO_3^--precursor
   A. Intake of alkalinizing salts
   B. Metabolism of endogenously produced organic anion (recovery from ketoacidosis or lactic acidosis)
   C. Disproportionate loss of carbonic acid relative to HCO_3^- during recovery from chronic hypercapnia
III. Contraction of ECF volume without proportional reduction of ECF HCO_3^- content.

[a] Indicates chloride or saline resistant types of metabolic alkalosis in the chronic state. Chloride resistant metabolic alkalosis can be distinguished by the finding of substantial chloride excretion in urine or stool.

vein) more bicarbonate than that delivered to the kidneys (via renal artery) by an amount just sufficient to neutralize the net amount of hydrogen ion added to the extracellular fluid as a result of gastrointestinal and metabolic processing of foodstuffs (Fig. 4.3). In healthy adult subjects ingesting typical American diets, the metabolic production of noncarbonic acids (sulfuric acid, phosphoric acid, various organic acids), in combination with the excretion of base in the stool in excess of the amount of base ingested in the diet, provides a net "endogenous" production of H^+ slightly less than 1 mEq/kg body weight per day (approximately 50 mEq/day) (Lennon, Lemann and Litzow, 1966). These noncarbonic acids, represented by HA, titrate body buffers represented by $NaHCO_3$ and thereby tend to reduce extracellular bicarbonate concentration and increase the concentration of A^-, the associated anions: $HA + NaHCO_3 \rightleftharpoons NaA + H_2CO_3$. The resultant H_2CO_3 is excreted as CO_2 by the lungs. By providing a net input of bicarbonate into ECF equal to the net endogenous acid production,

108 ACID BASE AND POTASSIUM HOMEOSTASIS

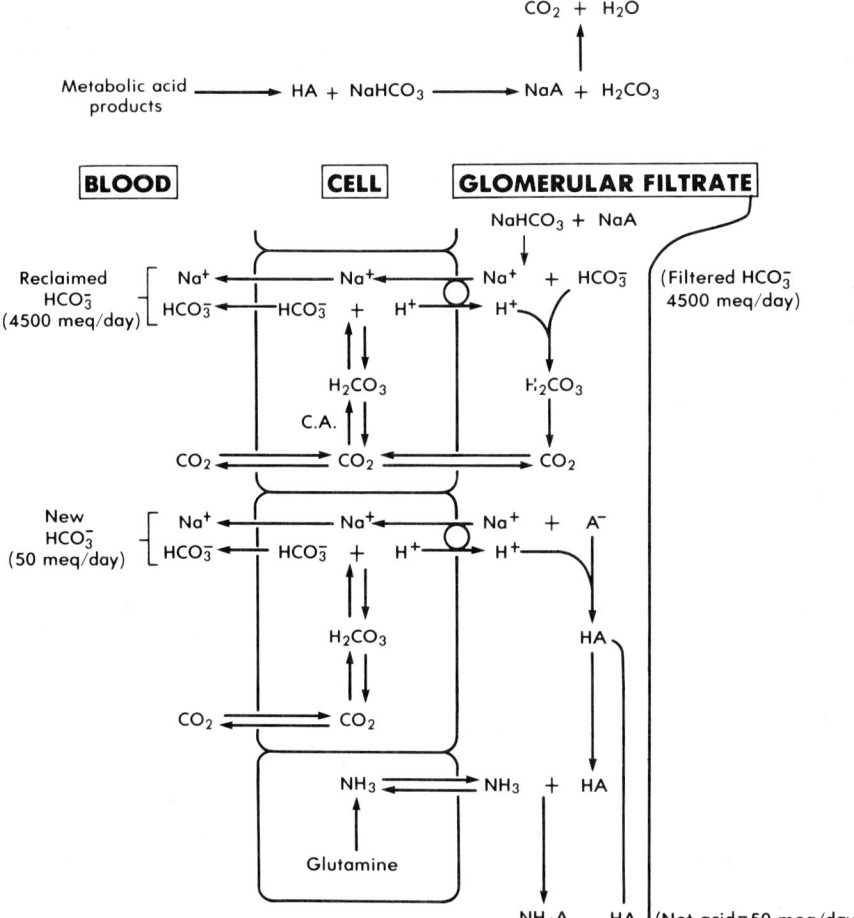

Figure 4.3 Normal renal acidification: Distribution of secreted hydrogen ion between bicarbonate reabsorption and net acid excretion.

and by excreting A^- in the urine at a rate equal to the rate of A^- accession to the ECF, the kidneys prevent an inexorable decline in extracellular HCO_3^- concentration and rise in A^- concentration.

The kidneys, therefore, maintain systemic acid-base equilibrium by recycling the ECF HCO_3^- delivered to them and regenerating the HCO_3^- and other body buffers exhausted in the process of neutralizing H^+ produced endogenously. Recycling delivered bicarbonate requires reclamation by the renal tubule of bicarbonate filtered by the glomerulus (approximately 4,500 mEq/day); regeneration of body buffers is accomplished by de novo generation ("synthesis") of bicarbonate by the renal tubules at a rate equal to the net rate of endogenous acid production (approximately 50 mEq/day). The renal tubules accomplish these two related processes by secreting H^+

into the tubular fluid (Fig. 4.3) (Pitts and Lotspeich, 1946; Rector, Carter and Seldin, 1965; Rector, 1973; Rector, 1976). For each mEq of H^+ secreted, one mEq of HCO_3^- is generated in renal cellular fluid and delivered to the blood and body fluids. Secreted H^+ titrates (1) all of the filtered HCO_3^-, which is converted to H_2CO_3 and CO_2 reabsorbed, and (2) an amount of nonbicarbonate buffer ($HPO_4^=$, NH_3) sufficient to provide for the elimination by the body of a net amount of buffered H^+ (as $H_2PO_4^-$ and NH_4^+), which is equal to the net endogenous acid production. The amount of secreted H^+ excreted in the urine as $H_2PO_4^-$ is measured as "titratable acid," and that excreted bound to NH_3 (ammonia) is measured as "ammonium" ion, NH_4^+. The sum of titratable acid and ammonium excretion minus the normally negligible bicarbonate excretion is referred to as "net acid excretion." Thus net acid excretion is a quantitative measure of the net renal input of bicarbonate into body fluid and normally equals the net endogenous acid production (Relman, Lennon and Lemann, 1961).

Reclamation of filtered HCO_3^- is accompanied by reabsorption of sodium in equivalent amounts. The net input of bicarbonate generated de novo, and the coupled excretion of H^+, is also accompanied by reabsorption of an equivalent amount of sodium, and with the excretion of an equivalent amount of A^- (of which $H_2PO_4^-$ is a part). Thus, maintenance of the constancy of ECF HCO_3^- concentration, reabsorption of filtered sodium, and excretion of the anion load associated with the production of endogenous noncarbonic acids are closely interrelated processes. (Schwartz and Cohen, 1978).

H^+ secretion and HCO_3^- reabsorption occur in both proximal and distal segments of the nephron. As the concentration of HCO_3^- in the lumen decreases, the concentration of H^+ increases and, as a result, a limitation is imposed on the net rate of H^+ secretion. In the proximal convoluted tubule, the gradient limitation on H^+ secretion is mitigated by the action of the enzyme carbonic anhydrase located in the brush-border membrane. By accelerating the conversion of luminal H_2CO_3 to CO_2 and H_2O, carbonic anhydrase serves to accelerate the titration of secreted H^+ by filtered HCO_3^-, thereby maintaining a cell-to-lumen H^+ concentration gradient smaller than it would be otherwise and as a result, reducing the limitation on net H^+ secretion ordinarily imposed by the increasing H^+ concentration gradient. This facilitatory effect of carbonic anhydrase permits the reabsorption of 85 percent or more of filtered HCO_3^- in the proximal convoluted tubule (Rector, Carter and Seldin, 1965).

In the distal nephron, H^+ is secreted in exchange for Na^+ at a rate sufficient to reclaim the small fraction of filtered HCO_3^- that normally escapes reabsorption proximally and that titrates $HPO_4^=$ and NH_3 in amounts sufficient to provide the normal rate of net acid excretion and coupled de novo HCO_3^- generation. An adequate supply of Na^+, as NaA, is available for "exchange" with secreted H^+ even when the amount of Na^+ delivered distally as NaCl is greatly reduced, because A^- is produced (as HA) and excreted at a rate essentially equal to the rate of net acid excretion. A^- is poorly reabsorbable as compared with chloride, hence conditions that enhance NaCl reabsorption in proximal segments of the nephron (e.g., hepatic cirrhosis, congestive heart failure) do not limit the amount of NaA delivered to the distal nephron and therefore do not characteristically impair the ability of the kidney to maintain normal plasma bicarbonate concentrations.

The hydrogen ion secretory capacity of the distal nephron is smaller than that of the proximal nephron, but the intraluminal bicarbonate load is also smaller. Hydrogen ion secretion in the distal nephron is not facilitated by catalyzed intraluminal dehydration of carbonic acid; carbonic anhydrase is not present on the luminal membrane. At the normal rate of distal HCO_3^- delivery, and given the amounts of $HPO_4^=$ and NH_3 available for titration and the physicochemical characteristics of the $HCO_3^-/H_2CO_3/CO_2$, $HPO_4^=/H_2PO_4^-$, and NH_3/NH_4^+ buffer systems, the secretion of H^+ at a normal net rate by the distal nephron is dependent on the ability of the distal nephron to increase and maintain the concentration of H^+ in the lumen tenfold greater or more than it is in the blood:

1. The reduction of urinary bicarbonate to negligibly small concentrations necessitates a reduction of urinary pH to less than 6.4, a tenfold hydrogen ion concentration gradient between urine and blood.

2. The pK of the $HPO_4^=/H_2PO_4^-$ buffer system is 6.8, hence more than 90 percent of the available $HPO_4^=$ is titrated at urinary pH of 5.8. Except for phosphate, the anion components of the endogenously produced acids are relatively weak H^+ acceptors and do not provide substantial buffer capacity for excreted H^+. With typical rates of inorganic phosphate excretion, even maximal rates of titratable acid excretion are insufficient to provide a normal net acid excretion.

3. The remainder is excreted as NH_4^+. NH_3 is generated within the renal tubular cells and freely diffuses into the lumen and peritubular capillaries, whereas NH_4^+ is poorly diffusable across cell membranes. Because the concentrations of NH_3 in lumen and capillary are similar (and reflect the rate of cellular NH_3 production), but the concentration of H^+ in the lumen is typically greater than in the blood, and because NH_3, H^+, and NH_4^+ are in equilibrium instantaneously both in lumen and in capillary, a substantial fraction of NH_3 produced by the kidney is "trapped" in the lumen as NH_4^+ and excreted in the urine rather than delivered out of the kidney in renal venous blood.

RENAL RESPONSE TO METABOLIC ALKALOSIS OF EXTRARENAL ORIGIN

The renal response to an acute input of base of extrarenal origin is characterized by an increase in urine pH and bicarbonate excretion and a decrease in the excretion rates of titratable acid and ammonium. Net acid excretion and the net renal input of bicarbonate into the ECF therefore decrease. The reduction in net renal input of bicarbonate serves to limit the magnitude of the increase in plasma bicarbonate concentration. Since the reduced net renal input of bicarbonate is less than the normal rate of endogenous acid production, the plasma bicarbonate concentration decreases towards normal when the pathophysiological events initiating the alkalosis cease to occur.

The pH of the urine increases in acute metabolic alkalosis of extrarenal origin because of the greater buffering of secreted hydrogen ion by the increased filtered bicarbonate. The associated increase in bicarbonate excretion indicates that the filtered load of bicarbonate exceeds the prevailing hydrogen ion secretory rate in both proximal and distal nephron segments. Bicarbonate, $HPO_4^=$, and NH_3 compete for secreted H^+ in the distal nephron: As the amount of bicarbonate delivered to the distal nephron increases, the rates of excretion of titratable acid and ammonium decrease. The excretion rates of titratable acid and ammonium vary inversely with urinary pH and bicarbonate excretion. If the increase in urinary bicarbonate excretion greatly exceeds the combined reduction in titratable acid and ammonium, net acid excretion becomes negative. Hence, net base is excreted in the urine. In this circumstance the kidneys serve to mitigate the severity of metabolic alkalosis by returning less bicarbonate to the ECF in the renal veins than is delivered in the renal arteries. The net renal input of bicarbonate becomes negative.

Influence of effective arterial blood volume

During the genesis of acute extrarenal alkalosis, the rapidity and the magnitude of the increase in bicarbonate excretion (and the reduction in net acid excretion) appears to be dependent on the extent to which homeostatic mechanisms regulating effective arterial blood volume permit urinary sodium excretion to increase. The balance of forces that determines the fraction of the filtered load of sodium reabsorbed by the renal tubule appears to be a major determinant of the extent to which a supernormal filtered load of bicarbonate will exceed the prevailing hydrogen ion secretory rate (bicarbonate reabsorptive rate) of the kidney. If the primary increase in plasma bicarbonate concentration is associated with a gain of sodium (e.g., administration of $NaHCO_3$), effective arterial blood volume increases, fractional sodium reabsorption decreases and, as a result, excretion of the increased filtered load of bicarbonate is unconstrained by a limitation on sodium excretion: Excretion of $NaHCO_3$ will not reduce ECF volume below normal. By contrast, if the increase in plasma bicarbonate concentration occurs predominantly with a commensurate loss of chloride and without gain of sodium or other cation (e.g., loss of gastric fluid), effective arterial blood volume does not increase, and, as a result, excretion of the increased filtered load of bicarbonate is constrained by volume regulatory forces limiting an increase in urinary sodium excretion: Excretion of $NaHCO_3$ would result in hypovolemia.

These considerations are exemplified by the studies carried out by Toussaint and coworkers (Toussaint, Telerman and Vereerstraeten, 1958) in dogs in which plasma bicarbonate concentration was experimentally increased without ECF volume expansion by hemodialysis with an HCO_3^--rich, Cl^--poor dialysate, and the results compared with those obtained when plasma bicarbonate concentration and ECF volume were increased concurrently by the administration of sodium bicarbonate (Fig. 4.4): The rate of HCO_3^- excretion for any given plasma concentration was greater in the dogs that were volume expanded. Subsequently, Purkerson et al. (1969) demonstrated in rats that the bicarbonaturic response to $NaHCO_3$ administration could be amplified by preexpansion of ECF volume by saline administration. Kurtzman (1970) and Slatopolsky et al. (1970) observed that

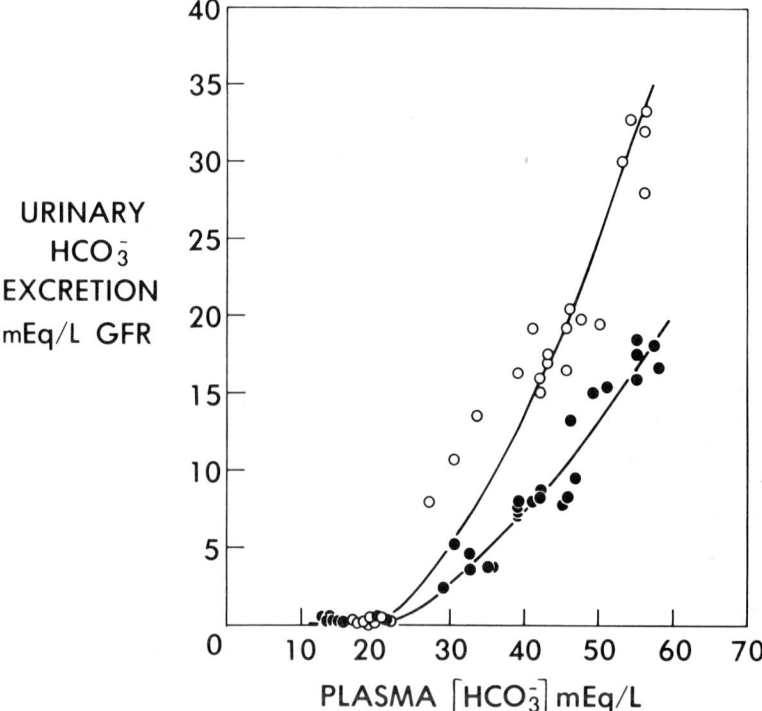

Figure 4.4 The bicarbonaturic response to an acute increase in plasma bicarbonate concentration in dogs. The renal response in animals infused with sodium bicarbonate solution (open circles) is compared with that in animals in which plasma bicarbonate concentration was increased without expansion of ECF volume by hemodialysis with an HCO_3^--rich, Cl^--poor dialysate. For any plasma HCO_3^- concentration, the HCO_3^- excretory response was greater in the volume-expanded dogs. (Drawn from data of Toussaint et al., 1958).

experimentally induced contraction of ECF volume (hemorrhage or furosemide administration) in dog and man greatly limited the bicarbonaturic response to $NaHCO_3$ administration. Reduction of effective arterial blood volume despite expansion of ECF volume, as in the nephrotic syndrome or experimentally induced thoracic vena cava obstruction, also greatly limited the bicarbonaturic response to $NaHCO_3$ administration. These observations indicate that the rate of renal bicarbonate reabsorption relative to the filtered load of bicarbonate is modulated by the volume of ECF as it reflects effective arterial blood volume.

The mechanism whereby changes in extracellular fluid volume modulate the reabsorption of filtered bicarbonate has not been established. The results of micropuncture studies in rats indicate that volume expansion suppresses the reabsorption of sodium, bicarbonate, and chloride in the proximal convoluted tubule and that bicarbonate reabsorption is suppressed to a greater extent than chloride over a wide range of subnormal to supernormal plasma bicarbonate concentrations (Kunau et al., 1968; Galla, Beaumont and Luke, 1977). Reduction in proximal sodium reabsorption during volume expansion has been attributed to an alteration in the Starling forces (hydraulic and colloid osmotic pressures) that govern the rate of capillary uptake of fluid from the renal interstitial space. It has been postulated that the alteration in Starling forces leads to an increase in "backleak" of proximal reabsorbate into the tubular lumen through the "tight junctions" between cells (Seldin, 1976). Because the ratio of bicarbonate to chloride in the tubular fluid normally decreases along the length of proximal tubule, backleak of the presumably bicarbonate-enriched reabsorbate would be expected to reduce net reabsorption of bicarbonate to a greater extent than that of chloride (Rector, 1973). Direct evidence for the backleak hypothesis, however, has not been adduced.

Possible role of bicarbonate secretion by the renal tubule

The bicarbonaturic response to acute extrarenal alkalosis might be mediated in part by secretion of bicarbonate by the renal tubule. In studies of bicarbonate transport in isolated perfused rabbit cortical collecting tubules, McKinney and Burg (1977b) recently reported that net bicarbonate secretion (bath-to-lumen) occurred in tubules from animals that had been pretreated with sodium bicarbonate (by gavage) prior to study. By contrast, net bicarbonate reabsorption occurred in tubules obtained from animals pretreated by acid loading. The directional differences in net bicarbonate transport were not dependent on differences in composition of bathing and perfusing solutions, which were identical. Induction of the bicarbonate secretory state was therefore not due to the establishment of a transepithelial bicarbonate concentration gradient favoring luminal entry of bicarbonate, but rather appeared to be the result of a persisting alteration in the transport characteristics of the epithelium. A preliminary report suggests that the bicarbonate secretory mechanism in rabbit collecting tubule is sodium-linked but not chloride-linked, and is electroneutral (McKinney and Burg, 1977a).

A bicarbonate secretory mechanism has been previously described in the isolated turtle bladder, a urinary epithelium with anatomic and functional characteristics similar to mammalian collecting tubule epithelia (Leslie, Schwartz and Steinmetz, 1973; Steinmetz, 1974). In this preparation, bicarbonate transport from serosal to mucosal fluid compartments occurs via an energy-dependent, electroneutral process that operates in opposition to the acidifying effect of active hydrogen ion secretion. In the presence of a serosal to mucosal bicarbonate concentration gradient, net secretion of bicarbonate occurs. Luminal entry of bicarbonate is accompanied by countertransport of chloride in approximately equivalent amounts, and reduction of chloride concentration in the mucosal solution greatly reduces the net rate of bicarbonate secretion. Whether such a HCO_3^-/Cl^- exchange mechanism operates in the mammalian distal nephron is not known. If bicarbonate secretion via such a mechanism contributes to the bicarbonaturic response to acute extrarenal alkalosis, the modulating effect of luminal chloride concentration on the rate of bicarbonate secretion might account for the difference in magnitude of the bicarbonaturic response in alkalosis induced by $NaHCO_3$ administration and HCO_3^-/Cl^- exchange hemodialysis (Fig. 4.4): The suppression of proximal fluid reabsorption that accompanies $NaHCO_3$ administration would be expected to increase distal delivery of chloride (as well as bicarbonate) and thereby possibly enhance bicarbonate secretion.

Administration of chloride prerequisite to correction of volume contraction and repair of alkalosis: effect of "isometric" ECF volume expansion

The limitation on bicarbonate excretion imposed by volume contraction can be sufficiently severe to prevent correction of metabolic alkalosis by the kidney. This constraint is illustrated in studies reported by Kassirer and Schwartz (1966a) in which metabolic alkalosis was induced experimentally in normal subjects by aspiration of gastric fluid (Fig. 4.5). ECF volume contraction was induced by prolonged restriction of dietary intake of sodium chloride prior to induction of alkalosis. In response to the acute increase in plasma bicarbonate concentration, urine pH and bicarbonate excretion increased and net acid excretion decreased. The magnitude of changes in bicarbonate and net acid excretion, however, was presumably less than would have occurred had there been no volume constraints limiting an increase in sodium excretion, as is suggested by the minimal increase in urinary sodium excretion that did occur. The reduction in net acid excretion existed transiently following cessation of gastric drainage, and the plasma bicarbonate concentration, therefore, tended to decrease toward normal.

But before the plasma bicarbonate concentration could return to normal, urine pH and bicarbonate concentration decreased and net acid excretion increased to the steady state predrainage control levels, and as a result the alkalosis persisted. When ECF volume contraction was subsequently corrected by returning NaCl to the diet, thereby removing the hypovolemic constraint on sodium excretion, urine pH and bicarbonate excretion increased, and net acid excretion decreased as urinary sodium excretion increased. As a result, correction of the alkalosis occurred.

The changes in plasma and urinary acid-base composition that occurred on reinstitution of dietary NaCl in these studies were associated with retention of administered chloride and correction of hypochloremia. Provision of chloride is prerequisite to correction of both hypochloremia and metabolic alkalosis in volume contracted patients. Provision of sodium as salts of anions such as sulfate or phosphate does not result in sufficient retention of sodium by the kidney to normalize effective arterial blood volume. Sulfate

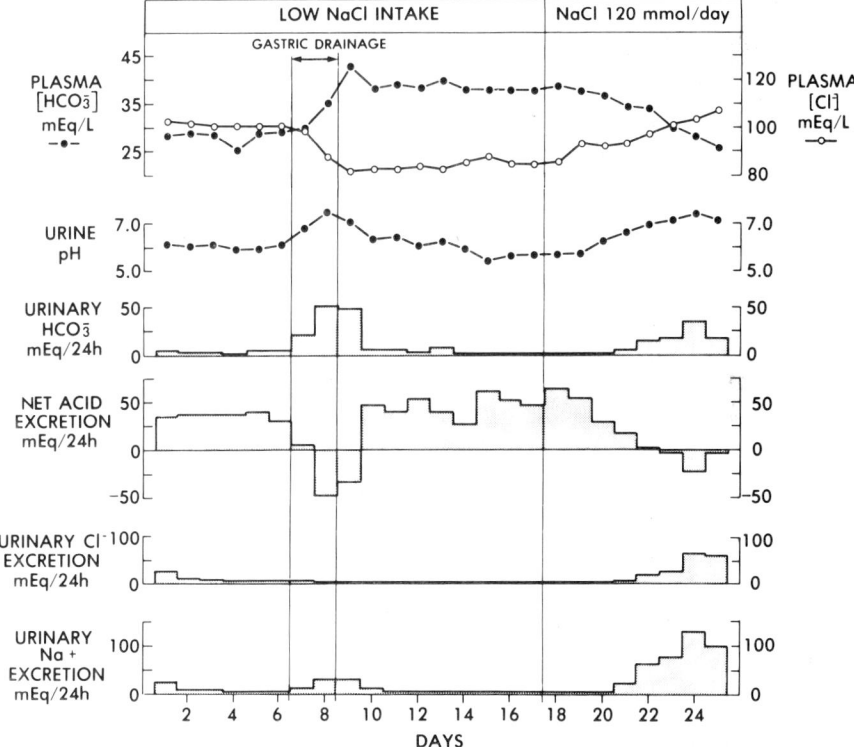

Figure 4.5 Plasma and urinary acid-base and electrolyte response to metabolic alkalosis induced by gastric drainage in a normal subject ingesting a low salt diet. In response to the acute increase in plasma bicarbonate concentration, urinary bicarbonate excretion increased and net acid excretion returned to normal before plasma bicarbonate concentration was reduced to normal, resulting in chronic metabolic alkalosis. When ECF volume contraction was corrected by the administration of NaCl, a prompt increase in urine pH and bicarbonate excretion occurred, resulting in full correction of the alkalosis. (Drawn from data of Kassirer and Schwartz, 1966)

and phosphate are considerably less reabsorbable by the renal tubule than chloride, and as a result, permit less sodium reabsorption for the same reabsorptive stimulus (Schwartz, van Ypersele de Strihou and Kassirer, 1968). The persisting volume constraint on sodium excretion prevents a sufficient increase in bicarbonate excretion and reduction in net acid excretion to correct the alkalosis. Provision of sodium with the readily reabsorbable anion, chloride, permits sufficient retention of sodium to normalize effective arterial blood volume and thereby initiate the renal mechanism that corrects metabolic alkalosis.

In patients with gastric alkalosis, the increase in plasma chloride concentration following reinstitution of dietary NaCl was originally inferred to be the primary event initiating the corrective process rather than the attendant increase in ECF volume (Schwartz et al., 1968). The subnormal concentration of plasma chloride was considered to be directly responsible for perpetuating the metabolic alkalosis. It was reasoned that because chloride is unique among filtered anions by virtue of its ready reabsorbability by the renal tubule, conservation of sodium filtered without the availability of adequate amounts of readily reabsorbable anion would lead to an increased rate of sodium-cation "exchange," of which Na^+-H^+ exchange is part. This inference was based on the observation that acute administration of sodium salts of poorly reabsorbable anions (e.g., sulfate) results in a prompt reduction in urine pH and increase in net acid excretion in normal subjects in whom reabsorption of sodium by the kidney was stimulated (by mineralocorticoid hormone administration) and availability of chloride was reduced (by dietary NaCl restriction) (Schwartz, Jenson and Relman, 1955).

To investigate whether an increase in plasma chloride concentration is necessary to initiate the corrective process in metabolic alkalosis associated with volume contraction, Cohen (1968, 1970) expanded ECF volume with a solution containing chloride, bicarbonate, and sodium in concentrations that duplicated those of the plasma (so-called isometric expansion). These studies were carried out in volume-contracted alkalotic dogs maintained on dietary NaCl restriction. A substantially greater fraction of administered chloride was retained than bicarbonate, which was sufficient to permit correction of the metabolic alkalosis. Similar results were obtained when glomerular filtration rate was prevented from increasing (by aortic constriction) as ECF volume increased. These findings establish that neither an increase in plasma chloride concentration nor an increase in filtered chloride load are required to initiate the renal corrective process. It was inferred that the corrective process was initiated by the increase in ECF volume.

Taken together with the results of the gastric drainage studies of Kassirer and Schwartz (1966a and b), the results of the isometric infusion studies of Cohen (1968, 1970) establish that correction of metabolic alkalosis and of hypochloremia by the kidney in hypovolemic man and animals occurs via a process in which "the repair of volume deficits and the provision of chloride are inseparable and interdependent features." This corrective process is associated with the urinary excretion of sodium and chloride in lesser amounts than those administered (i.e., the external balance of sodium and chloride becomes positive, and ECF volume thereby increases). A larger fraction of the administered chloride is retained than administered sodium. The increase in sodium excretion is accompanied by an increase in the rate of bicarbonate excretion and a reduction in net acid excretion (Fig. 4.5).

The changes in plasma and urinary acid-base and electrolyte composition that occur following ECF expansion indicate that the corrective process entails a reduction in the ratio of the concentrations of bicarbonate to chloride in the renal tubular reabsorbate from supernormal toward normal. It was postulated by Cohen and coworkers that this compositional alteration occurred in the reabsorbate generated by the distal nephron, and was dependent upon an increased supply of tubular fluid delivered to the distal nephron as a result of the suppression of proximal fluid reabsorption that attends ECF volume expansion. It is known that the absolute rates of sodium, chloride, and bicarbonate reabsorption in the distal nephron increases when distal delivery of these ions increases during ECF volume expansion (Kunau, Webb and Borman, 1974). For a given increase in rate of sodium reabsorption in the distal nephron during volume expansion, it would be predicted from the formulation of Cohen (1968, 1970) that in patients with hypochloremic metabolic alkalosis, a larger fraction of the delivered chloride will be reabsorbed than bicarbonate. As a result, the ratio of the concentrations of bicarbonate to chloride in the distal reabsorbate would decrease. The continual recycling of extracellular fluid (via glomerular filtrate and proximal tubular fluid) through such an anion selective reabsorptive process in the distal nephron eventually normalizes the HCO_3^-/Cl^- concentration ratio in ECF.

The mechanism that mediates the selectivity of anion reabsorption during ECF expansion in metabolic alkalosis has not been elucidated. If, as postulated by Cohen (1968, 1970), the reabsorptive selectivity occurs in the distal nephron and requires a suppression of fluid reabsorption in the proximal tubule, the phenomenon of selectivity may reflect the relationship between the anion composition of the fluid delivered out of the proximal nephron and the separate anion reabsorptive capacities in the distal nephron. It seems reasonable to assume that during ECF volume expansion in patients with metabolic alkalosis that the concentration of bicarbonate in the fluid rejected from the proximal tubule is greater than in normal subjects similarly treated, and that conversely, the chloride concentration is less (Kunau et al., 1968; Galla et al., 1977). If the distal nephron selectively reabsorbs the delivered chloride and rejects the delivered bicarbonate, presumably the distal reabsorptive capacity for chloride is farther from saturation relative to the amount of chloride delivered than is the distal capacity for bicarbonate reabsorption relative to the amount of bicarbonate delivered. Accordingly, relative to the ratio of bicarbonate to chloride in alkalotic plasma, the ratio in the distal reabsorbate decreases and the ratio in the urine increases. As this anion selective processing of alkalotic glomerular filtrate continues, plasma concentrations of bicarbonate and chloride normalize.

To further investigate the phenomenon of anion reabsorptive selectivity during ECF volume expansion, Hulter et al. (1976a) examined the effect of isometric expansion in dogs with HCl-induced chronic metabolic acidosis and hyperchloremia. In these studies the kidney selectively rejected administered chloride and retained administered bicarbonate; and as a result, the plasma concentrations of bicarbonate and chloride returned to normal. During expansion, as the expected reduction occurred in the fraction of filtered sodium reabsorbed, a greater reduction occurred in the fraction of filtered chloride reabsorbed, and a lesser reduction occurred in the fraction of filtered bicarbonate reabsorbed. Hence, during the period of expansion, the ratio of the concentrations of bicarbonate to chloride in the tubular reabsorbate gradually increase until the plasma concentrations of bicarbonate and chloride become normal. Taken together with the findings of Cohen (1968, 1970) in metabolic alkalosis, these findings suggest that under conditions of ECF volume expansion the ratio of the reabsorptive capacities of bicarbonate to chloride by the renal tubule approximates the ratio of the plasma concentrations of these anions in normal (nonacidotic, nonalkalotic) animals. The studies of isometric expansion in metabolic acid-base disorders provide a clear demonstration of the capacity of the kidney to regulate plasma bicarbonate and chloride concentrations when the constraints of ECF volume contraction on sodium excretion are removed.

Perpetuation of metabolic alkalosis in the absence of ECF volume contraction

The observation that in a variety of clinical conditions (Table 4.1) metabolic alkalosis persists despite provision of NaCl suggests that factors other than ECF volume contraction can be responsible for maintaining an abnormally high HCO_3^-/Cl^- concentration ratio in the renal reabsorbate. The major pathogenetic factors that have been implicated in such "chloride resistant" metabolic alkalosis are depletion of body potassium stores, hypermineralocorticoidism, hypercalcemia, hypoparathyroidism, and the administration of pharmacologic agents that inhibit renal tubular reabsorption of chloride. The renal alkalosis-sustaining effect of these factors does not appear to be dependent on ECF volume contraction. Hypovolemia may be demonstrably absent, not clinically evident, or if present, not prerequisite to the persistence of alkalosis. In primary aldosteronism, metabolic alkalosis persists despite a sustained increase in ECF volume.

It is generally assumed that the same renal mechanisms that perpetuate chloride resistant metabolic alkalosis can also account for its genesis in the clinical conditions in which it occurs. This has been demonstrated to be the case in metabolic alkalosis induced by furosemide administration (Bosch et al., 1977) and in that resulting from combined potassium depletion and mineralocorticoid excess (Hulter, Sigala and Sebastian, 1977). The pathophysiology of metabolic alkalosis of renal origin is discussed in the next section.

METABOLIC ALKALOSIS OF RENAL ORIGIN

Prototypically, metabolic alkalosis of renal origin results from a primary increase in the net rate of hydrogen ion secretion by the renal tubules. Coupled with the resultant increase in urinary net acid excretion, a commensurate increase occurs in the net renal input of bicarbonate into the extracellular fluid compartment, and the plasma bicarbonate concentration thereby increases. When the resultant increase in filtered bicarbonate load balances the increased rate of hydrogen ion secretion, the excretion of net acid returns to normal, but the metabolic alkalosis persists as long as the pathogenetic stimulation of hydrogen ion secretion persists. The persistent stimulation of hydrogen ion secretion implies that the ratio of the concentrations of HCO_3^- to Cl^- in the renal reabsorbate is maintained at a supernormal level.

Role of potassium depletion and hypermineralocorticoidism

The pathophysiology of metabolic alkalosis has been extensively delineated by investigations carried out over the past thirty years (Darrow et al., 1948; Relman and Schwartz, 1952; Seldin et al., 1954; Womersley and Darragh, 1955; Seldin, Welt and Cort, 1956; Grollman and Gamble, 1959; Huth, Squires and Elkinton, 1959; Atkins and Schwartz, 1962; Roth and Gamble, 1965; Cohen, 1968; Kassirer et al., 1970; Kassirer, Lowance and Schwartz, 1971; Seldin and Rector, 1972; Kurtzman, White and Rogers, 1973a; Kurtzman, White and Rogers, 1973b). A number of important issues however remain unresolved. Central among these is the role of body potassium depletion in the pathogenesis of metabolic alkalosis (Womersley and Darragh, 1955; Seldin et al., 1956; Grollman and Gamble, 1959; Huth et al., 1959; Atkins and Schwartz, 1962; Kassirer et al., 1971; Kurtzman et al., 1973a, b; Elkinton, Squires and Crosley, 1951; Black and Milne, 1952; Struyvenberg, DeGraeff and Lameijer, 1965; Tannen, 1970; Wright, et al., 1971; Levine, Walker and Nash, 1973; Tannen and Kunin, 1976).

Acid-base response to deprivation of dietary potassium uncomplicated by other potential alkalosis-producing conditions

Although it is widely believed that potassium depletion per se can cause metabolic alkalosis, it has not been demonstrated experimentally that induction of an uncomplicated state of body potassium depletion can result in the genesis of metabolic alkalosis, either in man or dog. Excluding studies complicated by the potential alkalosis-producing effects of hypermineralocorticoidism (Grollman and Gamble, 1959; Atkins and Schwartz, 1962; Roth and Gamble, 1965; Kassirer et al., 1970; Kassirer et al., 1971; Kurtzman et al., 1973a; Black and Milne, 1952) or administered alkali or cation-binding agents (Kurtzman et al., 1973a; Black and Milne, 1952; Tannen, 1970), potassium depletion produced simply by deprivation of dietary potassium has been reported to result in no change in plasma bicarbonate concentration (Womersley and Darragh, 1955; Grollman and Gamble, 1959; Moore, et al., 1955; Hulter et al., 1976b), "modest" but statistically insignificant increases (Huth et al., 1959), or slight decreases (Tannen, 1970; Hulter et al., 1976b; Burnell and Dawborn, 1970; Burnell, Teubner and Simpson, 1974). Potassium deprivation in rats has been observed to cause frank metabolic alkalosis (Seldin, Welt and Cort, 1956; Struyvenberg et al., 1965; Tannen and Kunin, 1976), but not by all investigators (Wright et al., 1971; Levine et al., 1973; Welbourne and Francoeur, 1977), and the reason for the conflicting observations has not been determined. Whether the alkalosis is in part of renal origin, when it occurs, has also not been determined.

It might be predicted that potassium deprivation would lead to an increase in urinary net acid excretion since potassium depletion has been found to enhance the reabsorptive capacity of the proximal tubule for filtered bicarbonate (Kunau et al., 1968; Bank and Aynedjian, 1965) and to increase the renal production rate of ammonia (Tannen, 1970; Tannen and Kunin, 1976; Gabuzda and Hall, 1966; Yablon and Relman, 1977). A primary increase in the rate of bicarbonate reabsorption in the proximal tubule would be expected to decrease the buffering of secreted hydrogen ion by bicarbonate in the distal nephron, and as a result, a reduction in urine pH and an increase in the excretion rates of titratable acid and ammonium would occur. To the extent that renal production of ammonia is increased concurrently, the increase in ammonium excretion for a given reduction in urine pH should even be magnified. Yet potassium deprivation has not been demonstrated to result in significant increases in net acid excretion in any species studied, including man. The urinary excretion rate of ammonium increases commonly during potassium deprivation, but because urine pH also increases, titratable acid excretion decreases (Tannen, 1970; Burnell et al., 1974)†. Indeed, a significant reduction in net acid excretion occurs in

---

† The reduction in net acid excretion that occurs during dietary potassium restriction may be due in part to a reduction in aldosterone secretion secondary to hypokalemia. The secretion rate of aldosterone is known to decrease during potassium restriction (Cannon, Ames and Laragh, 1966). Net acid excretion remains unchanged during potassium deprivation in adrenalectomized dogs maintained on constant replacement doses of mineralocorticoid (and glucocorticoid) hormone (Hulter et al., 1976b). Failure of net acid excretion to increase in these studies raises the possibility that a primary increase in bicarbonate reabsorptive capacity per se does not greatly alter distal bicarbonate delivery at normal filtered loads of bicarbonate. Alternatively, potassium depletion might induce a mild impairment of distal hydrogen ion secretion just sufficient to offset any reduction in distal bicarbonate delivery.

the dog during potassium deprivation, and metabolic acidosis occurs as a consequence (Hulter et al., 1976b; Burnell et al., 1974). A similar sequence may occur in man. In a study of potassium depletion in normal subjects, a significant reduction of plasma bicarbonate concentration occurred in subjects studied in whom potassium deficits greater than 2 mEq/kg body weight were produced solely by restricting potassium intake (Tannen, 1970).

Substantial deficits of body potassium (300–500 mEq) attending chronic metabolic alkalosis associated with hypovolemia do not prevent complete correction of the alkalosis following administration of sodium chloride (Kassirer and Schwartz, 1966b). Indirect evidence suggests that very large deficits of potassium (greater than 500 mEq), larger deficits perhaps than generally occur in hypovolemic states or that can be produced solely by dietary restriction, can result in chronic "chloride resistant" metabolic alkalosis. Garella, Chazen and Cohen (1970) reported the occurrence of chloride-resistant metabolic alkalosis (plasma bicarbonate concentrations 39–47 mEq/l) in four patients who presented with severe potassium depletion (serum potassium concentrations less than 2.0 mEq/l). In the one patient in whom detailed balance observations were made, the deficit of body potassium estimated as the cumulative amount of potassium retained during replacement therapy exceeded 1000 mEq. None of the patients exhibited evidence of ECF volume contraction, and none had overt hyperadrenocorticoidism. Metabolic alkalosis persisted despite adequate chloride intake (30–80 mEq/day). Administration of potassium supplements resulted in correction of the metabolic alkalosis. Evidence that the severe degree of potassium depletion had been responsible for the perpetuation of the alkalosis was the finding that the corrective effect of potassium administration was not observed until approximately 500 mEq of potassium had been retained (Fig. 4.6). The corrective response was characterized by bicarbonaturia and reduction in urinary chloride excretion, suggesting that potassium retention reduced the ratio of bicarbonate to chloride in the renal reabsorbate toward normal. Garella et al. (1970) postulated that severe potassium depletion results in metabolic alkalosis by diminishing the permeability of the renal tubule to chloride: "Any change that would decrease the ease with which chloride may accompany reabsorbed sodium would necessitate an acceleration of sodium-cation exchange if sodium stores are to be conserved. In the presence of severe potassium depletion such an acceleration of sodium-cation exchange would almost certainly increase sodium-hydrogen exchange and thereby enhance bicarbonate reabsorption."

Potassium depletion as a necessary cofactor in the pathogenesis of alkalosis due to hypermineralocorticoidism

Whereas questions remain concerning the alkalosis-producing effect of potassium depletion per se, there is no question that potassium depletion is a necessary cofactor in the pathogenesis of metabolic alkalosis induced by excess adrenal mineralocorticoid hormones (Grollman and Gamble, 1959; Roth and Gamble, 1965; Kurtzman et al., 1973a). Although administration of pharmacological amounts of adrenal mineralocorticoid hormones results

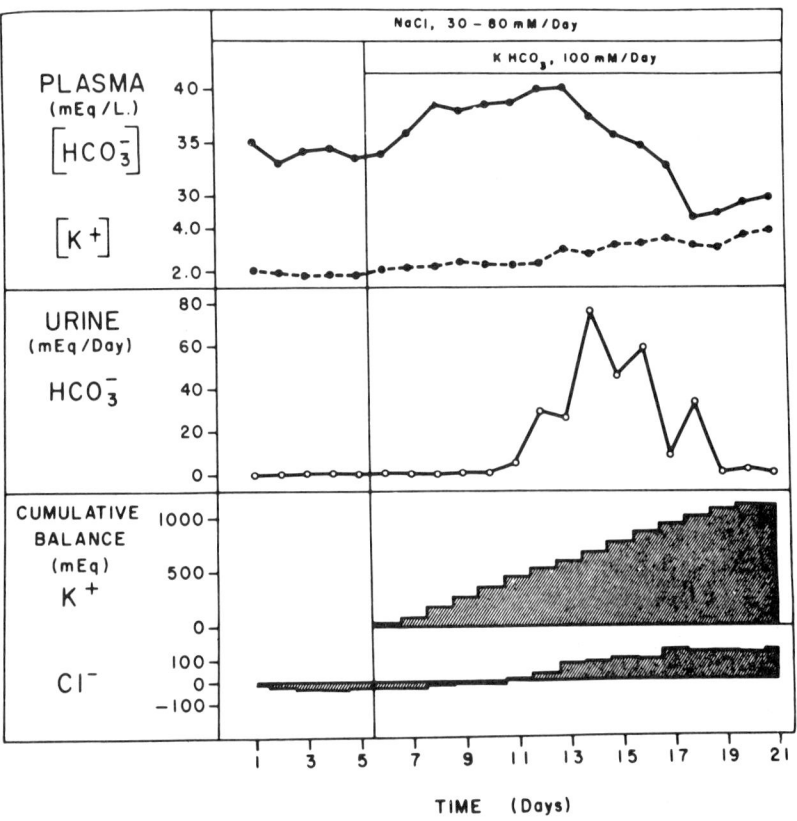

Figure 4.6 Plasma and urine acid-base and electrolyte response to NaCl and $KHCO_3$ in a patient with severe potassium depletion and "saline-resistant" metabolic alkalosis. Plasma cortisol concentration was normal and aldosterone excretion was low-normal. Chloride retention and bicarbonaturia occurred only after potassium supplements were provided and 500 mEq of potassium had been retained. (From Garella et al., 1970 with permission of the publisher and the author)

in a prompt reduction in urine pH and increase in net acid excretion (Lemann, Piering and Lennon, 1970; Mills, Thomas and Williamson, 1960; Mills, Thomas and Williamson, 1961; Relman and Schwartz, 1952; Lifschitz, Schrier, and Edelman, 1973), the magnitude of this effect appears to be insufficient to cause frank metabolic alkalosis. Chronic administration of mineralocorticoid hormones results in little or no increase in plasma bicarbonate concentration in man (Relman and Schwartz, 1952; Kassirer et al.,

1970) or dog (Grollman and Gamble, 1959; Roth and Gamble, 1965; Kurtzman et al., 1973a) when dietary intake of potassium is normal. Even four times daily administration of 10 times normal amounts of aldosterone for 90 days fails to increase plasma bicarbonate concentration by more than 15 percent in normal subjects (Kassirer et al., 1970). Preexisting or coexisting restriction of potassium intake, however, conditions the prompt development of frank metabolic alkalosis that persists until potassium is resupplied or hormone administration is discontinued (Grollman and Gamble, 1959; Roth and Gamble, 1965; Kurtzman et al., 1973a). Discontinuation of mineralocorticoid administration or reduction in the dose to physiological replacement levels in adrenalectomized dogs results in correction of metabolic alkalosis even though potassium retention is precluded by continued restriction of potassium intake (Kurtzman et al., 1973a). Administration of potassium (even as the bicarbonate salt) corrects the metabolic alkalosis despite continued administration of excess mineralocorticoid hormone. These observations establish in the dog that potassium depletion is a necessary cofactor in the pathogenesis of metabolic alkalosis induced by minetralocorticoid excess.

Analysis of data from published cases of primary aldosteronism suggests that the severity of metabolic alkalosis is proportional to the prevailing degree of potassium depletion (as reflected in the serum potassium concentration) (Kassirer et al., 1970) (Fig. 4.7). Differences in degree of potassium depletion among patients perhaps reflect in part differences in dietary potassium intake. Preliminary studies indicate that the degree of potassium depletion is a major determinant of the severity of metabolic alkalosis in experimental hyperaldosteronism in man (Kassirer et al., 1971).

Mechanism whereby potassium depletion potentiates the alkalosis-producing effect of mineralocorticoid hormone

Exaggerated urine pH-lowering response to mineralocorticoid hormone and enhanced renal ammonia production

The mechanism whereby potassium depletion potentiates the alkalosis-producing effect of mineralocorticoid hormone has been investigated in dogs (Hulter et al., 1977). Pharmacological amounts of desoxycorticosterone administered daily in dogs restricted of dietary potassium results in a significantly greater increase in net acid excretion than in dogs maintained on a normal potassium intake. The urine pH-lowering effect of the hormone at any given buffer excretion rate was exaggerated by potassium depletion suggesting that the stimulatory effect of the hormone on hydrogen ion secretion might be exaggerated. The excretion rate of ammonium was greater than predicted for any urine pH in potassium-restricted dogs both before and during hormone administration, suggesting that the availability of

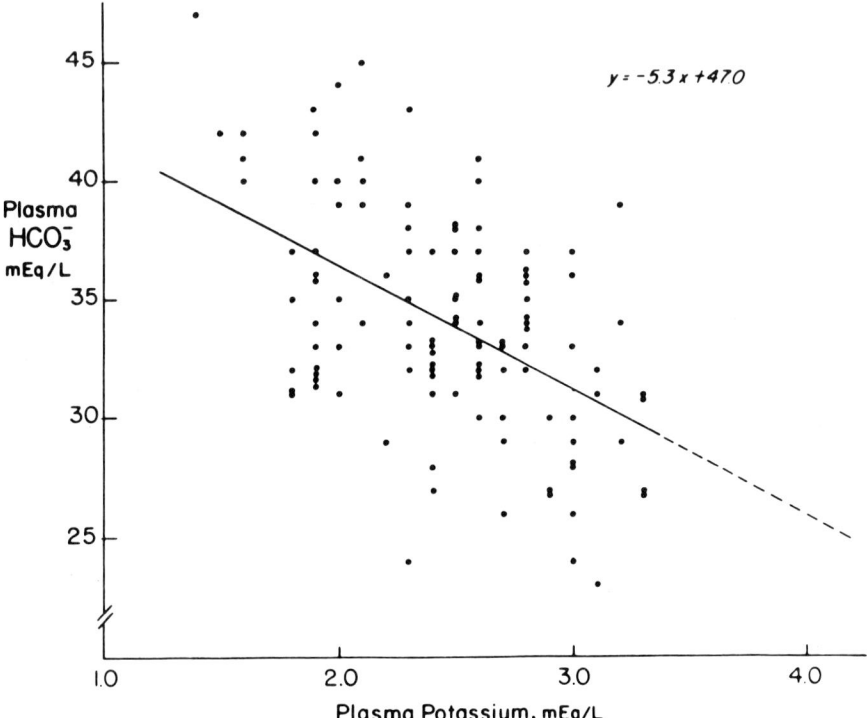

Figure 4.7 The relationship of plasma bicarbonate concentration with plasma potassium concentration in 111 published patients with primary aldosteronism without evidence of a complicating acid-base or electrolyte disturbance. (From Kassirer et al., 1970, with permission of the publisher and the author)

ammonia for trapping in the tubular lumen at any luminal pH was enhanced by potassium depletion. This finding is consistent with the known effect of potassium depletion to enhance renal ammonia production. The combined effects of potassium depletion on ammonia production and on the urine pH-lowering response to mineralocorticoid hormone resulted in a significantly greater increase in net acid excretion following hormone administration in the potassium-restricted dogs. The exaggerated increase in net acid excretion was not accompanied by an exaggerated reduction in sodium excretion, but rather was associated with a greatly attenuated reduction in chloride excretion. Potassium depletion potentiated the acid excretory effect of mineralocorticoid hormone and attenuated its chloride reabsorptive effect without altering its sodium reabsorptive effect. Thus, when sodium reabsorption is stimulated by mineralocorticoid hormone administration, the ratio of HCO_3^-/Cl^- in the renal reabsorbate increases to a much greater extent in potassium-depleted dogs, and as a result, frank metabolic alkalosis ensues.

Enhanced proximal bicarbonate reabsorptive capacity

The effect of potassium depletion to increase the bicarbonate reabsorptive capacity of the proximal convoluted tubule may contribute to the exaggerated acid excretory response to mineralocorticoid hormone. In response to the stimulation of distal hydrogen ion secretion by the hormone, net acid excretion would remain increased until the amount of bicarbonate escaping reabsorption in the proximal tubule increased sufficiently to balance the increased rate of distal hydrogen ion secretion, whereupon net acid excretion would return to normal and initiate a steady state of chronic metabolic alkalosis. In potassium-depleted animals, the cumulative increment in net acid excretion would be greater because the increased proximal bicarbonate reabsorptive capacity would compel a greater increase in plasma bicarbonate concentration to provide the requisite increase in distal bicarbonate delivery that permits achievement of the steady state. The effect of potassium depletion to enhance proximal bicarbonate reabsorptive capacity is opposed, however, by the expansion of extracellular fluid volume that attends the hormone-induced increase in sodium reabsorption. Expansion of ECF volume depresses renal bicarbonate reabsorptive capacity both in potassium-depleted and potassium-replete animals. The effect of potassium depletion to limit the chloride reabsorptive response to mineralocorticoid hormone possibly limits the hormone-induced expansion of extracellular fluid volume and thereby contributes to the genesis of metabolic alkalosis.

Impaired chloride reabsorption by the renal tubule

The mechanism whereby potassium depletion attenuates the chloride reabsorptive response to mineralocorticoid hormone has not been elucidated. It has been postulated that the permeability of the distal nephron to chloride is reduced during potassium depletion. Renal conservation of chloride during restriction of chloride intake is impaired in potassium depleted rats (Luke and Levitan, 1967), and recovery in the urine of radioisotopic chloride microinjected into the distal tubules of potassium-depleted rats is supernormal (Luke et al., 1976). An impairment in distal chloride permeability might attenuate the passive increase in chloride reabsorption that ordinarily occurs when sodium reabsorption is stimulated by mineralocorticoid hormone. Since the passive increase in chloride reabsorption partially shunts the increase in transtubular electrical potential caused by increased sodium reabsorption, a limitation on chloride permeability would tend to exaggerate the increase in potential difference and thereby to augment the acid excretory response to mineralocorticoid hormone. It has been demonstrated in toad bladder epithelia that the rate of hydrogen ion secretion is directly related to the magnitude of the transepithelial electrical potential difference (Ziegler, Fanestil and Ludens, 1976). Based on the combined evidence from clearance (Schwartz et al., 1955; Bank and Schwartz, 1960) and micropuncture (Clapp, Rector and Seldin, 1962) studies in the rat, voltage-driven hydrogen ion secretion appears to be present in the mammalian nephron.

Facilitated hormone-stimulated Na^+- nondependent H^+ secretion

Even if chloride permeability were not impaired, urinary chloride excretion might fail to decrease following mineralocorticoid hormone administration if the hormone-induced stimulation of sodium reabsorption failed to increase the lumen negative transtubular potential. Ordinarily mineralocorticoid hormone stimulates both transtubular potential and chloride reabsorption in collecting tubules (Gross, Imai and Kokko, 1975; Hanley and Kokko, 1977). The normal increase in transtubular potential and resultant increase in chloride reabsorption might be attenuated if mineralocorticoid hormone stimulated an electrogenic hydrogen ion secretory mechanism to a greater extent than the concomitant stimulation of sodium reabsorption. In turtle bladder epithelia aldosterone can stimulate hydrogen ion secretion under experimental conditions in which no transepithelial electrical potential difference is present or is permitted to develop (Al-Awqati et al., 1976). In this epithelium, at least a component of the hydrogen ion secretory mechanism is electrogenic in nature (i.e., capable of transporting positive charge into the lumen) and is not dependent on sodium reabsorption. It has been postulated that aldosterone alters the luminal membrane of the epithelial cell in such a way as to increase the "conductance" of hydrogen ion through the "active transport pathway" (Al-Awqati et al., 1976) (See Fig. 8.2.). The possibility might be considered that potassium depletion in some way so potentiates the electrogenic hydrogen ion secretory response to mineralocorticoid hormone that the characteristic hormone-induced increase in transtubular potential is greatly blunted, thereby accounting both for the diminished chloride reabsorptive response to the hormone and for the augmented acid excretory response. Perhaps potassium depletion in some way facilitates the postulated hormone-induced increase in hydrogen ion conductance.

Whatever the mechanism whereby potassium depletion alters the acid excretory and chloride reabsorptive response to mineralocorticoid hormone, it is conceivable that the degree of potassium depletion influences the magnitude of the effect. At severe degrees of potassium depletion, even subnormal concentrations of circulating aldosterone may have a hyperaldosteronelike effect and result in metabolic alkalosis (Garella et al., 1970).

Figure 4.8 summarizes the interrelated effects of potassium, mineralocorticoid hormone and extracellular fluid volume on net acid excretion.

Effect of inhibitors of active chloride reabsorption

Net acid excretion increases and metabolic alkalosis characteristically occurs during chronic administration of diuretic drugs that specifically in-

METABOLIC ALKALOSIS 127

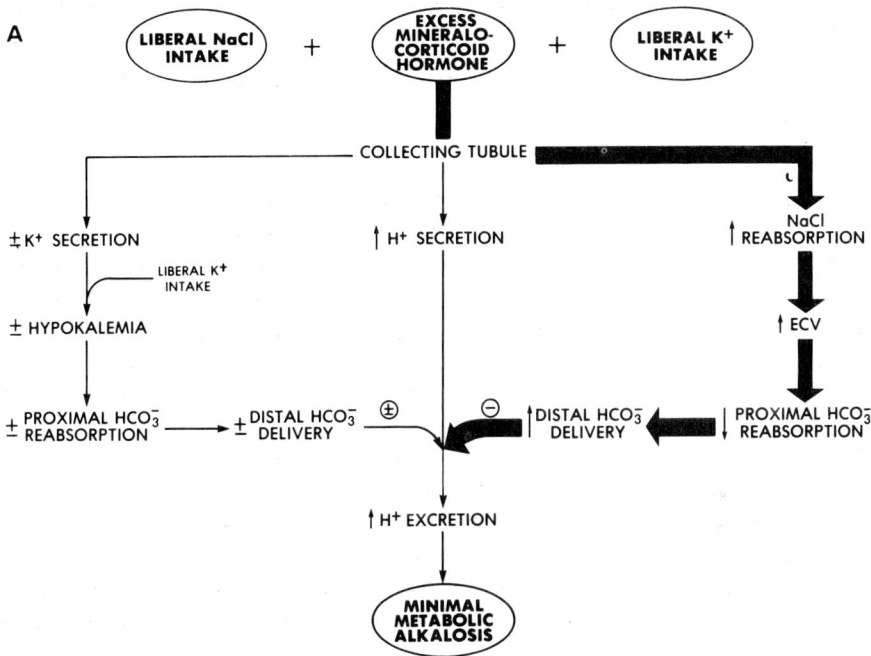

Figure 4.8a Interrelated effects of potassium, mineralocorticoid hormone, and extracellular fluid volume on net acid excretion: Mineralocorticoid excess and liberal K+ intake.

hibit renal chloride reabsorption. The mechanism whereby these agents stimulate renal hydrogen ion secretion has not been elucidated. Furosemide (Burg et al., 1973) and ethacrynic acid (Burg and Green, 1973) inhibit electrogenic chloride reabsorption in the rabbit loop of Henle and reduce the magnitude of the normally lumen positive transtubular potential difference. Furosemide is also known to reduce the transtubular potential difference in the human cortical collecting tubule (Jacobson et al., 1976), and the direction of change is consistent with an inhibitory effect on electrogenic chloride reabsorption. While it is not known whether the transepithelial potential in the normal human cortical collecting tubule is oriented with lumen positive or negative, elimination of chloride reabsorption by removal of chloride from the lumen has the same directional effect as addition of furosemide. Any reduction in luminal positivity, or equivalent increase in lumen negativity by furosemide, would provide a more favorable electrochemical gradient for hydrogen ion secretion. The rate of acidification is directly related to the magnitude of the favorable potential difference in toad bladder epithelia (Ziegler et al., 1976) and by inference in the human distal nephron (Clapp et al., 1962; Bank and Schwartz, 1960; Schwartz et al., 1955). Bicarbonate reabsorption appears not to occur in the

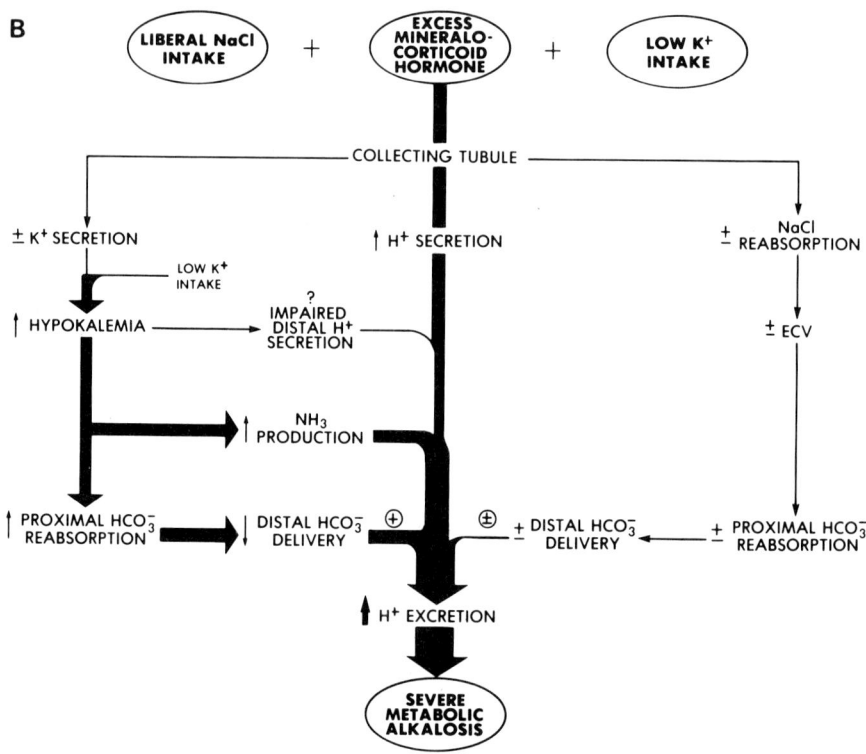

Figure 4.8b Interrelated effects of potassium, mineralocorticoid hormone, and extracellular fluid volume on net acid excretion: Mineralocorticoid excess and low K+ intake.

loop of Henle (Seldin, Rosin and Rector, 1975), but has been observed to occur in cortical collecting tubules (McKinney and Burg, 1977b). Accordingly, it seems reasonable to postulate that furosemide and other chloruretic diuretic agents induce metabolic alkalosis in part by increasing the electrical gradient favoring hydrogen ion secretion in the collecting tubules, thereby increasing net acid excretion and the ratio of HCO_3^-/Cl^- in the renal reabsorbate.

The increase in net acid excretion and subsequent occurrence of metabolic alkalosis during chronic administration of furosemide is not dependent on the development of negative chloride balance or on negative sodium or potassium balance (Bosch et al., 1977). The increase in plasma bicarbonate concentration is attenuated by prevention of electrolyte depletion with increased intake, but net acid excretion and plasma bicarbonate concentration, nevertheless, increase substantially. It has been similarly demonstrated that the stimulation of hydrogen ion secretion during ethacrynic acid administration is also not dependent on the occurrence of chloride depletion (Gyory and Lissner, 1977). These findings suggest that the stimulatory effect of these agents on renal hydrogen ion secretion occurs as a closely linked consequence of inhibition of the chloride reabsorptive mechanism. The normal operation of an electrogenic chloride reabsorptive mechanism in the collecting tubules may provide an inhibitory influence on hydrogen ion secretion.

The negative balance of chloride that ordinarily attends the administration of chloriuretic agents clearly contributes to the genesis and maintenance of metabolic alkalosis. Inhibition of net chloride reabsorption results in an inhibition of sodium reabsorption of nearly equal magnitude, and as a consequence, ECF volume decreases without proportionate reduction in extracellular bicarbonate stores (Cannon et al., 1965). So-called "contraction alkalosis" thereby results. This effect can be magnified in the treatment of patients with edema in whom massive losses of extracellular chloride, sodium, and water have been induced with chloriuretic agents. In such patients, the magnitude of the increase in plasma bicarbonate concentration is due largely to the ECF volume contraction and correlates with the degree of contraction (Fig. 4.2).

Acute administration of chloriuretic agents causes an increase in urinary potassium excretion, an increase in plasma renin levels, and an increase in aldosterone secretion. Although the occurrence of metabolic alkalosis is not dependent on negative potassium balance or on the presence of adrenal glands, the magnitude of the effect in clinical situations may be influenced by the degree of potassium depletion and hyperaldosteronism that ordinarily occur.

Hyperglucocorticoidism

Chronic administration of large doses of cortisone or hydrocortisone to man and dog results in metabolic alkalosis and hypokalemia (Grollman and Gamble, 1959; Gwinup, Gantt and Hamwi, 1964; Sprague et al., 1950) in association with an increased excretion of potassium and net acid (Seldin, 1976). In Cushing's syndrome the severity of metabolic alkalosis is directly related to the degree of hypokalemia and hypercortisolemia (Christy and Laragh, 1961). The alkalosis-producing effect of hydrocortisone is believed to reflect the "mineralocorticoid" properties of the hormone. Renal tubular reabsorption of sodium and chloride is also stimulated in response to hydrocortisone administration (Sprague et al., 1950; Seldin, 1976). Administration of a mineralocorticoid antagonist, spironolactone, at least partially reverses the mineralocorticoid effects of hydrocortisone (Gwinup et al., 1964). Metabolic alkalosis appears not to be a common consequence of the chronic administration of synthetic glucocorticoids that have lesser mineralocorticoid activity than hydrocortisone (prednisone, dexamethasone and triamcinolone).

The glucocorticoid effects of hydrocortisone might contribute to the genesis of metabolic alkalosis in patients with Cushing's syndrome. Glucocorticoids are known to increase glomerular filtration rate (Lindheimer, Lalone and Levinsky, 1967) and thereby possibly increase potassium and net acid excretion by virtue of an increase in distal delivery of sodium. In rats large doses of triamcinolone increase renal production and excretion of ammonia (Welbourne et al., 1976) in the absence of a reduction in plasma potassium concentration or urine pH. Evidence suggests that mitochondrial uptake and utilization of the major ammoniagenic precursor, glutamine, is enhanced by triamcinolone (Welbourne et al., 1976). It is conceivable that the glucocorticoid effect of hydrocortisone has a similar ammoniagenic effect, independent of its mineralocorticoid activity.

Hypercalcemia, hypoparathyroidism, hypervitaminosis D, and hyperphosphatemia

Metabolic alkalosis has been reported to occur in patients with hypercalcemia not caused by hyperparathyroidism (Heinemann, 1965) and in patients with hypoparathyroidism not caused by hypercalcemia (Barzel, 1969). Acute intravenous administration of calcium salts to man results in a reduction in urine pH and an increased excretion of titratable acid and ammonium (Richet et al., 1963). The acute bicarbonaturic response to bicarbonate infusion is lessened by calcium loading (Amiel et al., 1963). This enhancement of renal acidification may be due in part to inhibition of parathormone secretion, since parathyroidectomy results in an enhancement of renal bicarbonate reabsorptive capacity in the absence of hypercalcemia (Crumb et al., 1974; Karlinsky et al., 1974). Nevertheless, administration of calcium further increases renal bicarbonate reabsorption in hypoparathyroid animals (Crumb et al., 1974). Accordingly, hypercalcemia and hypoparathyroidism appear to be separate alkalosis-producing factors.

The occurrence of metabolic alkalosis in patients with vitamin D intoxication is presumably due at least in part to the attendant hypercalcemia and secondary hypoparathyroidism (Heinemann, 1965; Verner, Engel and McPherson, 1958). A preliminary report that intrarenal arterial injection of 25-hydroxy-vitamin D_3 in parathyroidectomized dogs acutely increases bicarbonate reabsorption raises the possibility that vitamin D may exert a "direct" effect on renal tubular secretion of hydrogen ion (Kumar et al., 1974). The results of micropuncture studies indicate that 25-hydroxyvitamin D_3 increases the reabsorption of sodium, calcium, and phosphorus in the proximal tubule (Puschett, Moranz and Kurnick, 1972). The antiphosphaturic effect of vitamin D appears to be mediated, in part, by antagonism of the tubular effect of parathyroid hormone (Popovtzer et al., 1974). The hydrogen ion stimulatory effect of vitamin D, therefore, may result from antagonism of the suppressive effect of parathyroid hormone on hydrogen ion secretion. The pathogenesis of metabolic alkalosis in vitamin D intoxication has not been systematically investigated.

ACKNOWLEDGMENT

This work was supported in part by grants from the National Institutes of Health (HL-06285, AM-16764) and the United States Public Health Service (SFO-63-74).

REFERENCES

Al-Awqati, Q., Norby, L. H., Mueller, A. & Steinmetz, P. R. (1976). Characteristics of stimulation of H+ transport by aldosterone in turtle urinary bladder. *Journal of Clinical Investigation*, 58, 351–358.

Amiel, C., Ardaillou, R., Lecestre, M. & Richet, G. (1963). Acidification de l'urine après injection intraveineuse de gluconate de calcium chez l'homme. *Revue Francaise d'Etudes Cliniques et Biologiques*, 8, 647–652.

Aquino, H. C. & Luke, R. G. (1973). Respiratory compensation to potassium-depletion and chloride-depletion alkalosis. *American Journal of Physiology*, 225, 1444–1448.

Atkins, E. L. & Schwartz, W. B. (1962). Factors governing correction of the alkalosis associated with potassium deficiency; the critical role of chloride in the recovery process. *Journal of Clinical Investigation*, 41, 218–229.

Bank, N. & Schwartz, W. B. (1960). The influence of anion penetrating ability on urinary acidification and the excretion of titratable acid. *Journal of Clinical Investigation*, 39, 1516–1525.

Bank, N. & Aynedjian, H. S. (1965). A micropuncture study of renal bicarbonate and chloride reabsorption in hypokalemic alkalosis. *Clinical Science*, 29, 159–170.

Barzel, U. S. (1969). Systemic alkalosis in hypoparathyroidism. *Journal of Clinical Endocrinology and Metabolism*, 29, 917–918.

Bear, R. A., Hammeke, M., Ho, M., Phillipson, E. A., Goldstein, M. B. & Halperin, M. L. (1977). Effect of metabolic alkalosis upon respiratory function in patients with chronic obstructive lung disease. *Kidney International*, 11, 379.

Berger, A. J., Mitchell, R. A. & Severinghaus, J. W. (1977). Regulation of respiration. *New England Journal of Medicine*, 297, 194–201.

Bieberdorf, F., Gorden, P. & Fordtran, J. S. (1972). Pathogenesis of congenital alkalosis with diarrhea. Implications for the physiology of normal ileal electrolyte absorption and secretion. *Journal of Clinical Investigation*, 51, 1958–1968.

Black, D. A. K. & Milne, M. D. (1952). Experimental potassium depletion in man. *Clinical Science*, 11, 397–415.

Bosch, J. P., Goldstein, M. H., Levitt, M. F. & Kahn, T. (1977). Effect of chronic furosemide administration on hydrogen and sodium excretion in the dog. *American Journal of Physiology*, 232, F397–F404.

Burg, M., and Green, N. (1973). Effect of ethacrynic acid on the thick ascending limb of Henle's Loop. *Kidney International*, 4, 301–308.

Burg, M., Stoner, L., Cardinal, J. & Green, N. (1973). Furosemide effect on isolated perfused tubules. *American Journal of Physiology*, 225, 119–124.

Burnell, J. M., and Dawborn, J. K. (1970). Acid-base parameters in potassium depletion in the dog. *American Journal of Physiology*, 218, 1583–1589.

Burnell, J. M., Teubner, E. J. & Simpson, D. P. (1974). Metabolic acidosis accompanying potassium deprivation. *American Journal of Physiology*, 227, 329–333.

Cannon, P. J., Heinemann, H. O., Albert, M. S., Laragh, J. H. & Winters, R. W. (1965). "Contraction" alkalosis after diuresis of edematous patients with ethacrynic acid. *Annals of Internal Medicine*, 62, 979–990.

Cannon, P. J., Ames, R. P. & Laragh, J. H. (1966). Relation between potassium balance and aldosterone secretion in normal subjects and in patients with hypertensive or renal tubular disease. *Journal of Clinical Investigation*, 45, 865–879.

Christy, N. P. & Laragh, J. H. (1961). Pathogenesis of hypokalemic alkalosis in Cushing's syndrome. *New England Journal of Medicine*, 265, 1083–1088.

Clapp, J. R., Rector, F. C., Jr. & Seldin, D. W. (1962). Effect of unreabsorbed anions on proximal and distal transtubular potentials in rats. *American Journal of Physiology*, 202, 781–786.

Cohen, J. J. (1968). Correction of metabolic alkalosis by the kidney after isometric expansion of extracellular fluid. *Journal of Clinical Investigation,* **47,** 1181–1192.

Cohen, J. J. (1970). Selective Cl retention in repair of metabolic alkalosis without increasing filtered load. *American Journal of Physiology,* **218,** 165–170.

Crumb, C. K., Martinez-Maldonado, M., Eknoyan, G. & Suki, W. N. (1974). Effects of volume expansion, purified parathyroid extract, and calcium on renal bicarbonate absorption in the dog. *Journal of Clinical Investigation,* **54,** 1287–1294.

Darrow, D. C., Schwartz, R., Ianucci, J. F. & Coville, F. (1948). The relation of serum bicarbonate concentration to muscle composition. *Journal of Clinical Investigation,* **27,** 198–208.

Elkinton, J. R., Squires, R. D. & Crosley, A. P., Jr. (1951). Intracellular cation exchanges in metabolic alkalosis. *Journal of Clinical Investigation,* **30,** 369–380.

Fencl, V., Miller, T. B. & Pappenheimer, J. R. (1966). Studies on the respiratory response to disturbances of acid-base balance, with deductions concerning the ionic composition of cerebral interstitial fluid. *American Journal of Physiology,* **210,** 459–472.

Fulop, M. (1976). Hypercapnia in metabolic alkalosis. *New York State Journal of Medicine,* **76,** 19–22.

Gabuzda, G. J. & Hall, P. W., III. (1966). Relation of potassium depletion to renal ammonium metabolism and hepatic coma. *Medicine,* **45,** 481–490.

Galla, J. H., Beaumont, J. E. & Luke, R. G. (1977). Effect of volume expansion with NaCl or $NaHCO_3$ on nephron fluid and Cl transport. *American Journal of Physiology,* **233,** F118–F125.

Garella, S., Chazan, J. A. & Cohen, J. J. (1970). Saline-resistant metabolic alkalosis or "chloride-wasting nephropathy." *Annals of Internal Medicine,* **73,** 31–38.

Goldring, R. M., Cannon, P. J., Heinemann, H. O. & Fishman, A. P. (1968). Respiratory adjustment to chronic metabolic alkalosis in man. *Journal of Clinical Investigation,* **47,** 188–202.

Grollman, A. P. & Gamble, J. L., Jr. (1959). Metabolic alkalosis, a specific effect of adrenocortical hormones. *American Journal of Physiology,* **196,** 135–140.

Gross, J. B., Imai, M., and Kokko, J. P. (1975). A functional comparison of the cortical collecting tubule and the distal convoluted tubule. *Journal of Clinical Investigation,* **55,** 1284–1294.

Gwinup, G., Gantt, C. L. & Hamwi, G. J. (1964). The production of hypokalemic alkalosis with hydrocortisone in subjects with adrenal insufficiency. *Metabolism, Clinical and Experimental,* **13,** 831–836.

Gyory, A. Z. & Lissner, D. (1977). Independence of ethacrynic acid-induced renal hydrogen ion excretion of sodium-volume depletion in man. *Clinical Science and Molecular Medicine,* **53,** 125–132.

Hanley, M. J. & Kokko, J. P. (1977). Characteristics of chloride transport across the rabbit cortical collecting tubule: response to desoxycorticosterone. *Clinical Research,* **25,** 506A.

Heinemann, H. O. (1965). Metabolic alkalosis in patients with hypercalcemia. *Metabolism, Clinical and Experimental,* **14,** 1137–1152.

Hulter, H. N., Ilnicki, L., Harbottle, J. & Sebastian, A. (1976a). On the mechanism of selective anion transport by the kidney during isometric volume expansion in metabolic acid base disorders. *Clinical Research,* **24,** 402A.

Hulter, H. N., Ilnicki, L., Harbottle, J. & Sebastian, A. (1976b). Pathogenetic role of aldosterone deficiency in the metabolic acidosis resulting from dietary potassium restriction. *Proceedings of the American Society of Nephrology,* **9,** 100A.

Hulter, H. N., Sigala, J. F. & Sebastian, A. (1977). Effects of preexisting dietary K^+ restriction on the renal action of mineralocorticoid hormone. *Clinical Research,* **25,** 436A. (also in press *American Journal of Physiology,* 1978)

Huth, E. J., Squires, R. D. & Elkinton, J. R. (1959). Experimental potassium depletion in normal human subjects. II. Renal and hormonal factors in the development of extracellular alkalosis during depletion. *Journal of Clinical Investigation,* **38,** 1149–1165.

Jacobson, H. R., Gross, J. B., Kawamura, S., Waters, J. D., and Kokko, J. P. (1976). Electrophysiological study of isolated perfused human collecting ducts. Ion dependency of the transepithelial potential difference. *Journal of Clinical Investigation,* **58,** 1233–1239.

Jarboe, T. M., Penman, R. W. & Luke, R. G. (1972). Ventilatory failure due to metabolic alkalosis. *Chest*, **61**, 61S–63S.

Karlinsky, M., Sager, D., Kurtzman, N. A. & Pillay, V. K. G. (1974). Effect of parathormone and cyclic adenosine monophosphate on renal bicarbonate reabsorption. *American Journal of Physiology*, **227**, 1226–1231.

Kassirer, J. P. & Schwartz, W. B. (1966a). The response of normal man to selective depletion of hydrochloric acid. The factors in the genesis of persistent gastric alkalosis. *American Journal of Medicine*, **40**, 10–18.

Kassirer, J. P. & Schwartz, W. B. (1966b). Correction of metabolic alkalosis in man without repair of potassium deficiency. A re-evaluation of the role of potassium. *American Journal of Medicine*, **40**, 19–26.

Kassirer, J. P., London, A. M., Goldman, D. M. & Schwartz, W. B. (1970). On the pathogenesis of metabolic alkalosis in hyperaldosteronism, *American Journal of Medicine*, **49**, 306–315.

Kassirer, J. P., Lowance, D. C. & Schwartz, W. B. (1971). Aldosterone-induced metabolic alkalosis in man. *Proceedings of the American Society of Nephrology*, **5**, 36A.

Kumar, R., Siegfried, J. D., Kurtzman, N. A. & Pillay, V. K. G. (1974). Effect of 25-hydroxycholecalciferol on bicarbonate reabsorption. *Clinical Research*, **22**, 618A.

Kunau, R. T. Jr., Frick, A., Rector, F. C. Jr. & Seldin, D. W. (1968). Micropuncture study of the proximal tubular factors responsible for the maintenance of alkalosis during potassium deficiency in the rat. *Clinical Science*, **34**, 223–231.

Kunau, R. T., Jr., Webb, H. L. & Borman, S. C. (1974). Characteristics of sodium reabsorption in the Loop of Henle and distal tubule. *American Journal of Physiology*, **227**, 1181–1191.

Kurtzman, N. A. (1970). Regulation of renal bicarbonate reabsorption by extracellular volume. *Journal of Clinical Investigation*, **49**, 586–595.

Kurtzman, N. A., White, M. G. & Rogers, P. W. (1973a). Pathophysiology of metabolic alkalosis. *Archives of Internal Medicine*, **131**, 702–713.

Kurtzman, N. A., White, M. G. & Rogers, P. W. (1973b). The effect of potassium and extracellular volume on renal bicarbonate reabsorption. *Metabolism, Clinical and Experimental*, **22**, 481–492.

Lemann, J., Piering, W. F. & Lennon, E. J. (1970). Studies of the acute effects of aldosterone and cortisol on the interrelationships between renal sodium, calcium and magnesium excretion in normal man. *Nephron*, **7**, 117–130.

Lennon, E. J., Lemann, J. & Litzow, J. R. (1966). The effects of diet and stool composition on the net external acid balance of normal subjects. *Journal of Clinical Investigation*, **45**, 1601–1607.

Leslie, B. R., Schwartz, J. H. & Steinmetz, P. R. (1973). Coupling between Cl- absorption and HCO_3^- secretion in turtle urinary bladder. *American Journal of Physiology*, **225**, 610–617.

Levine, D. Z., Walker, T. & Nash, L. A. (1973). Effects of KCl infusions on proximal tubular function in normal and potassium depleted rats. *Kidney International*, **4**, 318–325.

Lief, P. D. & Mutz, B. (1977). Thiazide stimulation of H^+ secretion in turtle urinary bladder: a possible explanation of diuretic-induced metabolic alkalosis. *Proceedings of the American Society of Nephrology*, **10**, 113A.

Lifschitz, M. D., Brasch, R., Cuomo, A. J. & Menn, J. (1972). Marked hypercapnia secondary to severe metabolic alkalosis. *Annals of Internal Medicine*, **77**, 405–409.

Lifschitz, M. D., Schrier, R. W. & Edelman, I. S. (1973). Effect of actinomycin D on aldosterone-mediated changes in electrolyte excretion. *American Journal of Physiology*, **224**, 376–380.

Lindheimer, M. D., Lalone, R. C. & Levinsky, N. G. (1967). Evidence that acute increase in glomerular filtration has little effect on sodium excretion in dog unless extracellular volume is expanded. *Journal of Clinical Investigation*, **46**, 256–265.

Luke, R. G. & Levitan, H. (1967). Impaired renal conservation of chloride and the acid-base changes associated with potassium depletion in the rat. *Clinical Science*, **32**, 511–526.

Luke, R. G., Wright, F. S., Fowler, N. B. & Giebisch, G. (1976). Effect of K-depletion on segmental chloride transport in the rat nephron. *Proceedings of the American Society of Nephrology*, **9**, 105A.

McKinney, T. D. & Burg, M. B. (1977a). Bicarbonate secretion by cortical collecting tubules. *Proceedings of the American Society of Nephrology*, 10, 115A.

McKinney, T. D. & Burg, M. B. (1977b). Bicarbonate transport by rabbit cortical collecting tubules. Effect of acid and alkali loads in vivo on transport in vitro. *Journal of Clinical Investigation*, 60, 766–768.

Mills, J. N., Thomas, S. & Williamson, K. S. (1960). The acute effect of hydrocortisone, deoxycorticosterone and aldosterone upon the excretion of sodium, potassium and acid by the human kidney. *Journal of Physiology*, 151, 312–331.

Mills, J. N., Thomas, S. & Williamson, K. S. (1961). The effects of intravenous aldosterone and hydrocortisone on the urinary electrolytes of the recumbent human subject. *Journal of Physiology*, 156, 415–423.

Moore, F. D., Boling, E. A., Ditmore, H. B. Jr., Sicular, A., Tetrick, J. E., Ellison, A. E., Hoye, S. J. & Ball, M. R. (1955). Body sodium and potassium, V. *Metabolism, Clinical and Experimental*, 4, 379–402.

Oliva, P. B. (1972). Severe alveolar hypoventilation in a patient with metabolic alkalosis. *American Journal of Medicine*, 52, 817–821.

Penman, R. W., Luke, R. G. & Jarboe, T. M. (1972). Respiratory effects of hypochloremic alkalosis and potassium depletion in the dog. *Journal of Applied Physiology*, 33, 170–174.

Petzel, R. A., Masler, D. S., Miller, T. C., Brown, D. C. & Mulhausen, R. O. (1976). Intravenous hydrochloric acid in the treatment of metabolic alkalosis. *Minnesota Medicine*, 59, 166–168.

Pitts, R. F. & Lotspeich, W. D. (1946). Bicarbonate in the renal regulation of acid-base balance. *American Journal of Physiology*, 147, 138–154.

Popovtzer, M. M., Robinette, J. B., DeLuca, H. F. & Holick, M. F. (1974). The acute effect of 25-hydroxycholecalciferol on renal handling of phosphorus. Evidence for a parathyroid hormone dependent mechanism. *Journal of Clinical Investigation*, 53, 913–921.

Purkerson, M. L., Lubowitz, H., White, R. W. & Bricker, N. S. (1969). On the influence of extracellular volume expansion on bicarbonate reabsorption in the rat. *Journal of Clinical Investigation*, 48, 1754–1760.

Puschett, J. B., Moranz, J. & Kurnick, W. (1972). Evidence for a direct action of cholecalciferol and 25-hydroxycholecalciferol on the renal transport of phosphate, sodium and calcium. *Journal of Clinical Investigation*, 51, 373–385.

Rector, F. C. Jr., Carter, N. W. & Seldin, D. W. (1965). The mechanism of bicarbonate reabsorption in the proximal and distal tubules of the kidney. *Journal of Clinical Investigation*, 44, 278–290.

Rector, F. C. Jr., (1973). Acidification of the urine. In *Handbook of Physiology, Section 8, Renal Physiology*, ed. Orloff J. & Berliner, R. W. Ch. 14. Washington D.C.: American Physiological Society.

Rector, F. C. Jr. (1976). Renal acidification and ammonia production; chemistry of weak acids and bases: buffer mechanisms. In *The Kidney*, ed. Brenner, B. M. & Rector, F. C. Jr. Ch. 9. Philadelphia: W. B. Saunders.

Relman, A. S. & Schwartz, W. B. (1952). The effect of DOCA on electrolyte balance in normal man and its relation to sodium chloride intake. *Yale Journal of Biology and Medicine*, 24, 540–558.

Relman, A. S., Lennon, E. J. & Lemann, J., Jr. (1961). Endogenous production of fixed acid and the measurement of the net balance of acid in normal subjects. *Journal of Clinical Investigation*, 40, 1621–1630.

Richet, G., Ardaillou, R., Amiel, C. & Lecestre, M. (1963). Acidification de l'urine par injection intraveineuse de sels de calcium. *Journal d'Urologie et de Nephrologie*, 69, 373–398.

Roberts, K. E., Poppell, J. W., Vanamee, P., Beals, R. & Randall, H. T. (1956). Evaluation of respiratory compensation in metabolic alkalosis. *Journal of Clinical Investigation*, 35, 261–266.

Roth, D. G. & Gamble, J. L., Jr. (1965). Deoxycorticosterone-induced alkalosis in dogs. *American Journal of Physiology*, 208, 90–93.

Saunders, N. A., Carter, J., Scamps, P. & Vandenberg, R. (1974). Severe hypercapnia associated with metabolic alkalosis due to pyloric stenosis. *Australian and New Zealand Journal of Medicine*, 4, 385–391.

Schwartz, W. B., Jenson, R. L. & Relman, A. S. (1955). Acidification of the urine and increased ammonium excretion without change in acid-base equilibrium: sodium reabsorption as a stimulus to the acidifying process. *Journal of Clinical Investigation,* **34,** 673–680.

Schwartz, W. B., van Ypersele de Strihou, C. & Kassirer, J. P. (1968). Role of anions in metabolic alkalosis and potassium deficiency. *New England Journal of Medicine,* **279,** 630–639.

Schwartz, W. B., and Cohen, J. J. (1978). The nature of the renal response to chronic disorders of acid-base equilibrium. *American Journal of Medicine,* **64,** 417–428.

Seldin, D. W., Rector, F. C. Jr., Carter, N. & Copenhaver, J. (1954). The relation of hypokalemic alkalosis induced by adrenal steroids to renal acid secretion. *Journal of Clinical Investigation,* **33,** 965–966.

Seldin, D. W., Welt, L. G. & Cort, J. H. (1956). The role of sodium salts and adrenal steroids in the production of hypokalemic alkalosis. *Yale Journal of Biology and Medicine,* **29,** 229–247.

Seldin, D. W. & Rector, F. C. Jr. (1972). The generation and maintenance of metabolic alkalosis. *Kidney International,* **1,** 305–321.

Seldin, D. W., Rosin, J. M. & Rector, F. C. Jr. (1975). Evidence against bicarbonate reabsorption in the ascending limb, particularly as disclosed by free-water clearance studies. *Yale Journal of Biology and Medicine,* **48,** 337–347.

Seldin, D. W. (1976). Metabolic alkalosis. In *The Kidney,* ed. Brenner, B. M. and Rector, F. C. Jr. Ch. 17. Philadelphia: W. B. Saunders.

Shear, L. & Brandman, I. S. (1973). Hypoxia and hypercapnia caused by respiratory compensation for metabolic alkalosis. *American Review of Respiratory Disease,* **107,** 836–841.

Slatopolsky, E., Hoffsten, P., Purkerson, M. & Bricker, N. S. (1970). On the influence of extracellular fluid volume expansion and of uremia on bicarbonate reabsorption in man. *Journal of Clinical Investigation,* **49,** 988–998.

Sprague, R. G., Power, M. H., Mason, H. L., Albert, A., Mathieson, D. R., Hench, P. S., Kendall, E. C., Slocumb, C. H. & Polley, H. F. (1950). Observations on the physiologic effects of cortisone and ACTH in man. *Archives of Internal Medicine,* **85,** 199–258.

Steinmetz, P. R. (1974). Cellular mechanisms of urinary acidification. *Physiological Reviews,* **54,** 890–956.

Struyvenberg, A., DeGraeff, J. & Lameijer, L. D. F. (1965). The role of chloride in hypokalemic alkalosis in the rat. *Journal of Clinical Investigation,* **44,** 326–338.

Tannen, R. L. (1970). The effect of uncomplicated potassium depletion on urine acidification. *Journal of Clinical Investigation,* **49,** 813–827.

Tannen, R. L. & Kunin, A. S. (1976). Effect of potassium on ammoniagenesis by renal mitochondria. *American Journal of Physiology,* **231,** 44–51.

Toussaint, C., Telerman, M. & Vereerstraeten, P. (1958). Effects of acute hypochloremia on glomerular filtration rate and electrolyte excretion in the dog. *Experientia,* **14,** 417–419.

Tuller, M. A. & Mehdi, F. (1971). Compensatory hypoventilation and hypercapnia in primary metabolic alkalosis. *American Journal of Medicine,* **50,** 281–290.

van Ypersele de Strihou, C., Brasseur, L. & DeConinck, J. (1966). The "carbon dioxide response curve" for chronic hypercapnia in man. *New England Journal of Medicine,* **275,** 117–122.

van Ypersele de Strihou, C. & Frans, A. (1973). The respiratory response to chronic metabolic alkalosis and acidosis in disease. *Clinical Science and Molecular Medicine,* **45,** 439–448.

Verner, J. V., Engel, F. L. & McPherson, H. T. (1958). Vitamin D intoxication: report of two cases treated with cortisone. *Annals of Internal Medicine,* **48,** 765–773.

Welbourne, T. C., Phenix, P., Thornley-Brown, C. & Welbourne, C. J. (1976). Triamcinolone activation of renal ammonia production. *Proceedings of the Society for Experimental Biology and Medicine,* **153,** 539–542.

Welbourne, T. C. & Francoeur, D. (1977). Influence of aldosterone on renal ammonia production. *American Journal of Physiology,* **233,** E56–E60.

Womersley, R. A. & Darragh, J. H. (1955). Potassium and sodium restriction in the normal human. *Journal of Clinical Investigation,* **34,** 456–461.

Wright, F. S., Streider, N., Fowler, N. B. & Giebisch, G. (1971). Potassium secretion by distal tubule after potassium adaptation. *American Journal of Physiology*, **221**, 437–448.

Yablon, S. & Relman, A. S. (1977). Excretion and total renal production of ammonia in K^+-depleted rats. *Clinical Research*, **25**, 452A.

Ziegler, T. W., Fanestil, D. D. & Ludens, J. H. (1976). Influence of transepithelial potential difference on acidification in the toad urinary bladder. *Kidney International*, **10**, 279–286.

5
Acid-base disorders of respiratory origin

JORDAN J. COHEN
NICOLAOS E. MADIAS

Introduction
Physiologic regulation of carbon dioxide (CO_2) tension
   CO_2 stores
   CO_2 input
   CO_2 transport
   CO_2 excretion
   Impact of CO_2 transport system on acid-base equilibrium
Respiratory acidosis
Pathophysiology
Secondary physiologic responses
   Acute hypercapnia
   Chronic hypercapnia
Cerebrospinal fluid acidity during respiratory acidosis
Intracellular acidity during respiratory acidosis
Other physiological and biochemical effects of respiratory acidosis
Diagnosis and clinical manifestations
Causes of respiratory acidosis
Treatment
Mixed acid-base disorders associated with respiratory acidosis
   Respiratory acidosis and metabolic acidosis
   Respiratory acidosis and metabolic alkalosis
   Acute respiratory and chronic respiratory acidosis
Respiratory alkalosis
Pathophysiology
Secondary physiologic responses
   Acute hypocapnia
   Chronic hypocapnia
Cerebrospinal fluid acidity during respiratory alkalosis
Intracellular acidity during respiratory alkalosis
Other physiological and biochemical effects of respiratory alkalosis
Diagnosis and clinical manifestations
Causes of respiratory alkalosis
Treatment
Mixed acid-base disorders associated with respiratory alkalosis
   Respiratory alkalosis and metabolic acidosis
   Respiratory alkalosis and metabolic alkalosis

INTRODUCTION

The level of plasma acidity, when viewed from a physiologic perspective, can be thought of as being totally dependent upon the prevailing levels of carbonic acid and plasma bicarbonate. Thus, changes in plasma acidity, whether they be physiologic adjustments or pathophysiologic disturbances, can occur only if carbonic acid concentration (i.e., $PaCO_2$) and/or bicarbonate concentration are altered. We will be concerned here with those acid-base disorders initiated by changes in $PaCO_2$, the so called "respiratory disorders."

Increases and decreases in $PaCO_2$, denoted by the terms hypercapnia and hypocapnia, result in disturbances referred to as respiratory acidosis and respiratory alkalosis, respectively. The degree to which plasma acidity is altered in response to primary changes in $PaCO_2$ is a reflection not only of the magnitude of the initiating change but also of the degree to which certain physiologic responses lead to secondary changes in plasma bicarbonate concentration. These adaptive adjustments in plasma bicarbonate concentration pursue a characteristic time course and tend to ameliorate the impact on acidity of the primary change in $PaCO_2$. As a general rule, however, they fall short of returning hydrogen ion concentration to completely normal levels. Since respiratory acid-base disturbances, which occur frequently in clinical medicine, often coexist with other disorders of acid-base equilibrium, it is necessary to know the response times and the magnitude of the anticipated secondary changes in bicarbonate concentration in order to make an intelligent appraisal of such mixed disorders.

It is our purpose here to summarize current concepts regarding the abnormalities in acid-base equilibrium initiated by changes in $PaCO_2$ and to define the limits of the unfettered physiologic response of the intact organism to both short-lived (acute) and persistent (chronic) deviations of $PaCO_2$ of graded degrees. As background, we will first review the physiologic regulatory system that controls the carbon dioxide tension of body fluids.

PHYSIOLOGIC REGULATION OF CARBON DIOXIDE (CO_2) TENSION

CO_2 stores

Man, as well as other mammals, contains a large reservoir of extractable CO_2, estimated to be approximately 1.8 liters per kg body weight at sea level (Farhi and Rahn, 1960; Rahn, 1962). The bulk of this extractable CO_2 is present in the body in chemical combination. Although bone carbonates account for most of the extractable CO_2, the bicarbonate ions of aqueous tissues and the carbamino groups on certain proteins (see below) play a larger physiologic role. A small fraction of the extractable carbon dioxide, approximately 1 percent, is dissolved as such in body water and a negligible fraction is present in the gaseous state within the lungs.

CO_2 input

Under basal conditions, cellular processes in the normal adult generate as much as 20,000 mmole of CO_2 (450 liters at sea level) daily. The rate of production is altered when energy requirements change but is remarkably constant from day to day in normal individuals if the level of exercise is reasonably stable.

CO_2 transport

The CO_2 produced in tissues is transported by the blood to the pulmonary capillaries for excretion (Roughton, 1964). Although the amount of CO_2 that can be transported is greatly enhanced by certain properties of hemoglobin, the process of diffusion is the fundamental driving force responsible for the steady flow of CO_2 from the tissues to the blood and from the blood to the alveolar spaces of the lungs. Accordingly, a progressive pressure gradient for CO_2 is required to transfer CO_2 across the several membranes between sites of production and alveoli. The normal pressure gradients for carbon dioxide are relatively small as compared with those for oxygen. Mean tissue pCO_2 is approximately 46 mmHg and alveolar pCO_2 approximately 40 mmHg. Transfer of CO_2 across the alveolar capillary membrane, which is some 20 times more permeable to CO_2 than to O_2, is almost instantaneous; equilibration between blood and alveolar pCO_2 is achieved well within 0.01 seconds.

As carbon dioxide is generated by cellular metabolism, it diffuses from cells into the bloodstream. The bulk of the CO_2 added from tissues diffuses into red blood cells where it is rapidly hydrated under the catalytic influence of carbonic anhydrase to form carbonic acid.*

$$CO_2 + H_2O \underset{\text{Anhydrase}}{\overset{\text{Carbonic}}{\rightleftarrows}} H_2CO_3 \rightleftarrows H^+ + HCO_3^-$$

Hydrogen ions relinquished by the dissociation of carbonic acid combine with available buffers, such as certain basic groups of hemoglobin. The buffering ability of hemoglobin is substantially enhanced by the reduced oxygen tension prevailing in venous blood because the pK' of reduced hemoglobin is higher than that of oxyhemoglobin. Consequently, a marked shift to the left in the CO_2 dissociation curve occurs during the desaturation of hemoglobin in the peripheral capillaries. That is to say, more CO_2 can be carried at the same pCO_2, thereby enhancing the removal of carbon dioxide from the tissues. This shift is referred to as the Haldane effect.

Bicarbonate ions generated in equivalent numbers during the process of dissociation of carbonic acid diffuse back into the plasma in exchange for chloride ions. A lesser but significant portion of the carbon dioxide entering the erythrocytes reacts directly with certain amino acid residues on hemoglobin to form carbamino compounds. The ability of hemoglobin to form carbamino compounds is also enhanced during the process of its oxygen

---

* It is imperative to take cognizance of the physiologic context within which the hydration of CO_2 and subsequent reversible dissociation of carbonic acid occurs. Because of the relatively high concentration of plasma bicarbonate maintained by the kidneys, the dissociation of carbonic acid formed when the carbon dioxide produced in tissues is hydrated must proceed in the face of a large quantity of its conjugate base. For this reason, the statistical likelihood of recombination is greatly enhanced, and equilibrium conditions are satisfied at a concentration of hydrogen ions that is correspondingly very low.

desaturation in the tissues. A small portion of CO_2 is carried as dissolved CO_2 within the erythrocytes.

The quantitative distribution of the CO_2 added to blood by the tissues is as follows: (1) Approximately 63 percent is carried in the form of bicarbonate ions; although red cell hemoglobin is responsible for generating this bicarbonate, the plasma compartment actually conveys most of it to the lungs. (2) Approximately 29 percent of the added CO_2 is conveyed as carbamino groups, virtually all of which are attached to hemoglobin. (3) The remaining 8 percent reaches the lungs in the form of dissolved CO_2, mostly in the plasma.

The CO_2 dissociation curve of blood approximates a straight line over the physiologic range of $PaCO_2$ (i.e., the quantity of CO_2 carried by the blood is proportional to the level of $PaCO_2$).

CO_2 excretion

When blood reaches the pulmonary capillaries, the entire sequence of events described above is reversed as CO_2 diffuses out of blood into the alveolar spaces and as the affinity of hemoglobin for hydrogen ions is lessened by the high oxygen tension prevailing in the pulmonary capillaries (i.e., the CO_2 dissociation curve is shifted to the right).

The rate at which carbon dioxide is excreted from the body is equal to the product of alveolar ventilation and alveolar carbon dioxide concentration. Since the concentration of carbon dioxide in the alveolar gas is proportional to its partial pressure, the following proportionality may be written:

$$CO_2 \text{ excretion} \sim P_ACO_2 \times \text{alveolar ventilation}$$

Since diffusion equilibrium for carbon dioxide is achieved with ease across the alveolar capillary wall, P_ACO_2 and $PaCO_2$ are virtually identical.* Furthermore, in the steady state, the rate of CO_2 excretion is equivalent to the rate of CO_2 production. Under steady-state circumstances, therefore, the following relationship governs:

$$PaCO_2 \sim \frac{CO_2 \text{ production}}{\text{alveolar ventilation}}$$

Thus, plasma carbon dioxide tension can be regarded as a dynamic physiologic variable, the value of which is determined at every instant by a continuous byplay between the rate of carbon dioxide production and the rate of alveolar ventilation.

The regulation of carbon dioxide excretion is extremely efficient under normal circumstances owing to the exquisite sensitivity of the medullary respiratory centers to changes in $PaCO_2$. Thus, moment-to-moment changes in the rate of carbon dioxide production in a healthy subject are responded to swiftly by appropriate changes in the rate of alveolar ventilation, the level of $PaCO_2$ being only minimally and transiently affected. As a result of

* P_ACO_2, alveolar CO_2 tension; $PaCO_2$ arterial CO_2 tension.

these prompt adjustments of respiratory control mechanisms, $PaCO_2$ is normally maintained within a narrow range of approximately ±2 mmHg around a mean value of 40 mmHg. Consequently, $PaCO_2$ is an excellent index of the adequacy of alveolar ventilation.

It follows from the above considerations that an increase in $PaCO_2$ above the normal range must signify alveolar hypoventilation and, conversely, a decrease in $PaCO_2$ must denote alveolar hyperventilation.

Impact of CO_2 transport system on acid-base equilibrium

The physiologic importance of the CO_2 transport system centers about the removal of carbon dioxide from solution and its temporary conversion to bicarbonate and carbamino compounds. This process permits the large quantity of carbon dioxide produced by the tissues to be transported to the lungs without requiring a large carbon dioxide gradient. The analogy between this system and the buffering of endogenous (fixed) acid is apparent. Just as body buffering permits large quantities of hydrogen ion to flow through body fluids without large changes in hydrogen ion concentration, so these mechanisms for transporting carbon dioxide permit large quantities of this substance to pass through the system with minimal changes in carbon dioxide tension. Moreover, just as body buffering makes no direct contribution to the external hydrogen ion *balance*, so the buffering of carbonic acid by hemoglobin makes no direct contribution to external carbon dioxide *balance*. External balance is determined by the rates of carbon dioxide production and elimination and not by the intermediate mode of transport.

RESPIRATORY ACIDOSIS

Respiratory acidosis refers to that acid-base disturbance initiated by an elevation in carbon dioxide tension.

Pathophysiology

An elevation in carbon dioxide tension (hypercapnia) occurs whenever CO_2 excretion lags behind CO_2 production (positive CO_2 balance). Overproduction of CO_2 by itself, however, is never responsible for significant hypercapnia; in the presence of a normal respiratory system, as noted above, increased CO_2 production results in a stimulus to alveolar ventilation sufficient to maintain $PaCO_2$ at nearly normal levels. Therefore, the presence of hypercapnia can be interpreted as signifying some impediment to alveolar ventilation. Alveolar hypoventilation results not only in failure of carbon dioxide elimination but, when breathing room air, in failure of oxygenation as well; alveolar and arterial pO_2 fall pari passu with the rise in alveolar and arterial pCO_2.

Increased venous admixture (shunting) as occurs in patients with cardiac abnormalities (congenital or acquired right-to-left shunts) or pulmonary disease (alveolar atelectasis, consolidation, or edema) invariably produces hypoxemia but does not usually lead to notable hypercapnia. First, since the difference in carbon dioxide tension between mixed venous blood and pulmonary capillary blood is only about 6 mmHg, a trivial increase in $PaCO_2$ is produced when venous admixture occurs. Second, whatever increment in $PaCO_2$ does occur is promptly sensed by the respiratory centers, and an appropriate increase in ventilation is elicited. Because of the essentially linear CO_2 dissociation curve over the physiologic range, the resulting hyperventilation of normally perfused areas easily offsets the CO_2 retention arising from areas of shunting. On the other hand, such hyperventilation does not increase the O_2 content of blood from normal areas of the lung because this blood is already maximally saturated; as a result, hypoxia persists.

For the very same reasons, hypoxemia but not hypercapnia is the usual consequence of marked ventilation-perfusion ($\dot{V}/\dot{Q}$) inequality. In fact, such $\dot{V}/\dot{Q}$ inequality is largely responsible for the hypoxemia in patients with chronic obstructive lung disease. A decrease in ventilation in relationship to perfusion (i.e., a low $\dot{V}/\dot{Q}$ ratio) modifies the composition of the alveolar gas towards that of mixed venous blood. The net effect is that the corresponding pulmonary capillary blood will feature a relatively low PaO_2 and a relatively high $PaCO_2$. The reverse is true for areas of high $\dot{V}/\dot{Q}$ ratio, in which the composition of alveolar gas is modified towards that of inspired air. A relative preponderance of areas of low $\dot{V}/\dot{Q}$ ratio tends to elevate $PaCO_2$ but, once again, increased ventilation restores $PaCO_2$ towards normal. Therefore, increased venous admixture and marked $\dot{V}/\dot{Q}$ inequality can lead to sustained hypercapnia *only* when alveolar ventilation cannot be augmented. From this analysis it can be concluded that the pathogenetic mechanism for all forms of elevated $PaCO_2$ is *inadequate alveolar ventilation.*

Secondary physiologic responses

It has been well established that increments in $PaCO_2$ elicit certain physiologic responses in the intact organism that result in secondary elevations in plasma bicarbonate concentration. These elevations in bicarbonate occur in two distinct steps (Fig. 5.1), the first related to the titration of nonbicarbonate body buffers and the second to renal adaptive mechanisms.

Acute hypercapnia

A small increase in bicarbonate concentration is observed within moments following the onset of hypercapnia as hydrogen ions derived from carbonic acid are removed from solution by nonbicarbonate buffers (Brackett, Cohen and Schwartz, 1965). Barring further changes in $PaCO_2$, this immediate adjustment is completed within five to ten minutes and is followed by a stable period of a few hours during which no further changes in acid-base equi-

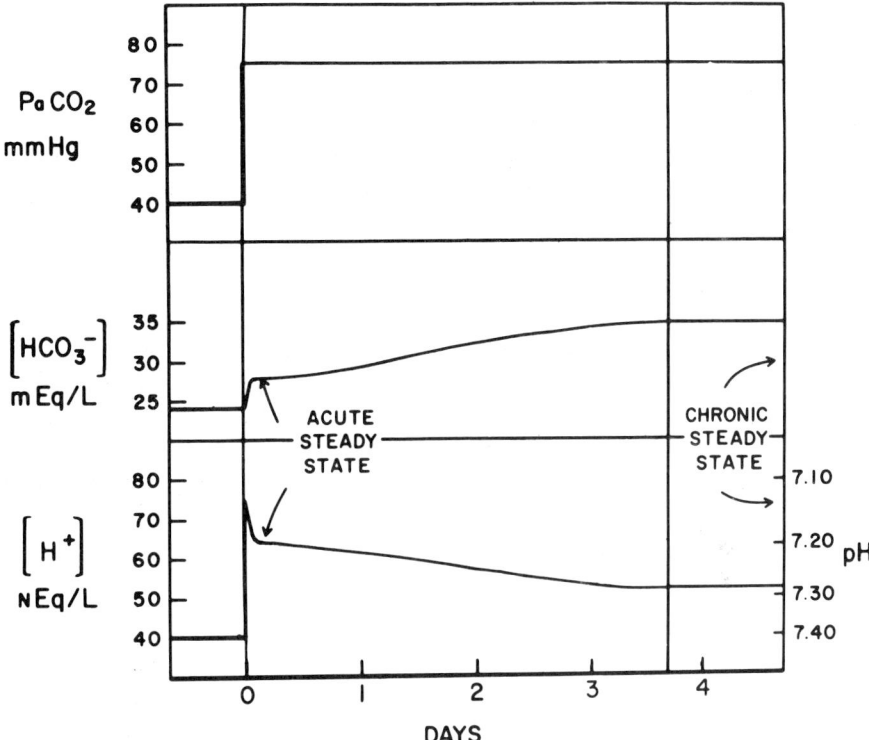

Figure 5.1 Schematic time course of the changes in plasma acid-base equilibrium during the development of respiratory acidosis. In this scheme, $PaCO_2$ is assumed to rise abruptly from 40 to 75 mmHg and to remain unchanged thereafter.

librium are detectable; this period can be defined as an "acute steady state." This immediate change is quite small, however, increasing the level of plasma bicarbonate concentration only 3 to 4 mEq/l above normal even when hypercapnia is extreme ($PaCO_2$ 80–90 mmHg).

This modest increase in plasma bicarbonate in vivo contrasts sharply with the large increase observed during in vitro titration of whole blood with carbon dioxide over the same range of carbon dioxide tensions (Fig. 5.2). The seemingly lower in vivo buffer capacity is attributable, at least in part, to the fact that some bicarbonate generated by blood buffers must diffuse into the poorly buffered interstitial space in the intact organism. It has been estimated that approximately 97 percent of the increment in extracellular bicarbonate stores during acute hypercapnia is accounted for by intracellular buffers (including hemoglobin), the remainder being attributed to plasma proteins.

Figure 5.3, lower panel, depicts the range of secondary changes in bicarbonate concentration seen in human subjects in response to graded degrees of acute respiratory acidosis. The upper panel depicts the corresponding range for hydrogen ion concentration. These data were obtained during

144 ACID BASE AND POTASSIUM HOMEOSTASIS

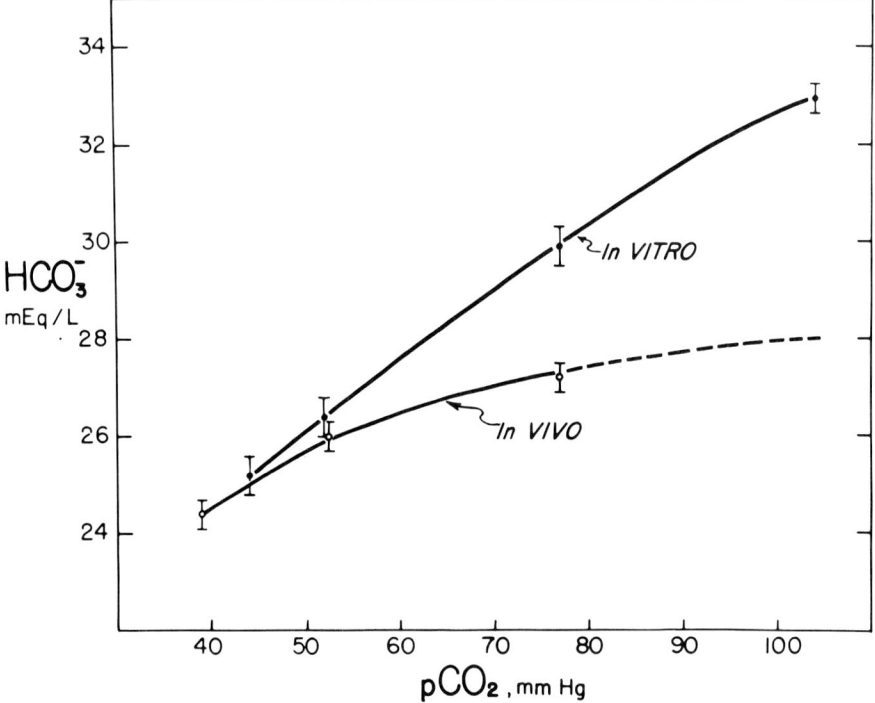

Figure 5.2 Comparison of the in vitro and in vivo carbon dioxide titration curves of human blood. The in vivo curve was drawn by inspection through the average plasma bicarbonate concentrations observed in normal human volunteers who were exposed to graded degrees of acute hypercapnia within an environmental chamber. (From Brackett et al., 1965, with permission of the publisher)

acute "whole-body" titration from conscious volunteer subjects exposed to high concentrations of inspired CO_2 within a large environmental chamber (Brackett et al., 1965). The net effect of the body buffer response to acute hypercapnia is that plasma hydrogen ion concentration is elevated by approximately 0.75 nEq/l for each mmHg increase in $PaCO_2$.

The 95 percent confidence interval constructed from these data (Fig. 5.3) represents the range of responses within which acid-base equilibrium would be expected to fall if acute hypercapnia were the only factor disturbing acidity.

Chronic hypercapnia

Several hours following the acute titration of body buffers, the renal response to persistent hypercapnia begins to be manifested by a further, gradual rise in plasma bicarbonate (Fig. 5.1). Under the influence of persistent hypercapnia, the rate of sodium-hydrogen exchange by the renal tubule is accelerated; as a consequence, the rate of net acid excretion (largely in the form of augmented ammonium excretion) transiently ex-

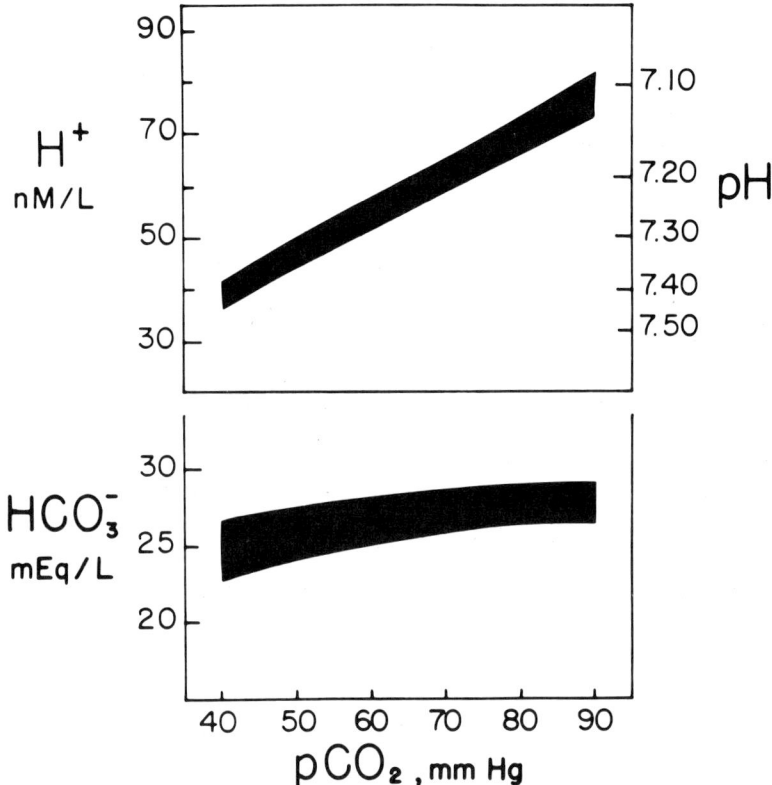

Figure 5.3 Ninety-five percent significance bands for plasma bicarbonate concentration (lower panel) and plasma hydrogen ion concentration (upper panel) in acute hypercapnia. The bands were calculated from data obtained in normal human volunteers. (From Brackett et al., 1965, with permission of the publisher)

ceeds the rate of endogenous acid production, negative hydrogen ion balance occurs, and new bicarbonate is generated (Schwartz et al., 1965). The persistent acceleration of sodium-hydrogen exchange during hypercapnia ensures the conservation of these new bicarbonate ions as they are recycled through the kidney by glomerular filtration. As bicarbonate stores are being augmented by this process, chloride stores are correspondingly depleted as a result of enhanced urinary chloride excretion in association with the increased excretion of ammonium. Based on the observations in the dog, it appears that these adjustments are completed in less than three to five days.

Studies in animals (Schwartz et al., 1965) indicate that a highly predictable relationship exists between the degree of chronic hypercapnia and the level at which plasma bicarbonate concentration stabilizes following full physiologic adaptation (Fig. 5.4). Observations in patients with

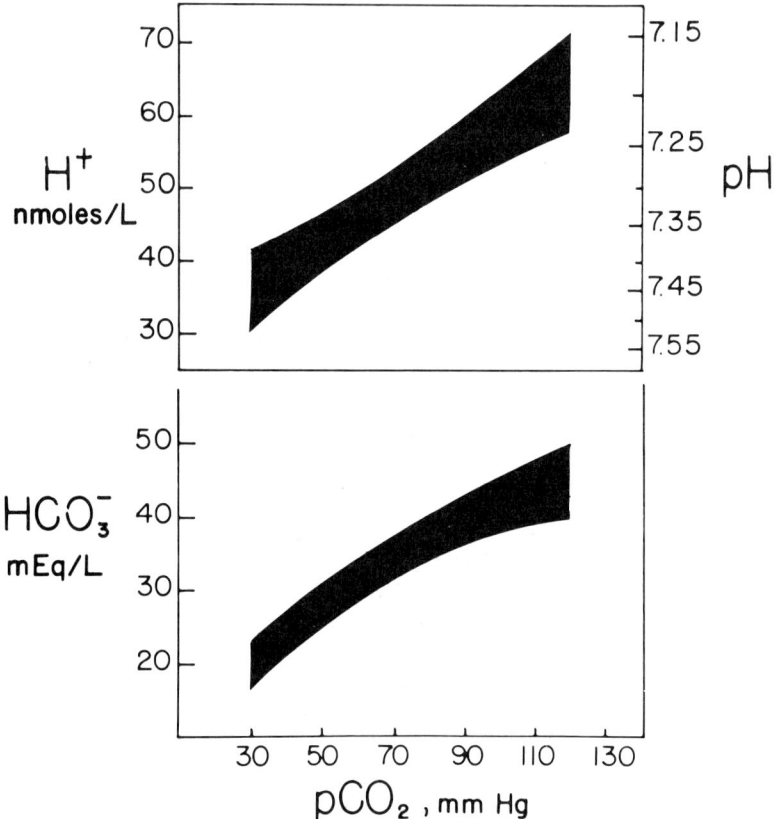

Figure 5.4 Ninety-five percent significance bands for plasma bicarbonate concentration (lower panel) and plasma hydrogen ion concentration (upper panel) in chronic hypercapnia. The bands were calculated from data obtained in dogs exposed to graded degrees of chronic hypercapnia within an environmental chamber. As noted in the text, observations of patients with relatively stable chronic hypercapnia have indicated that the human response is remarkably similar to that of the dog. (From Schwartz et al., 1965, with permission of the publisher)

chronic, stable respiratory acidosis (van Ypersele de Strihou, Brasseur and De Coninck, 1966; Brackett et al., 1969) appear to confirm the presence of a predictable pattern of response when no complicating acid-base disturbances are present. These findings in experimental animals and in man indicate that chronic hypercapnia is associated with an increment in hydrogen ion concentration of 0.25–0.30 nEq/l for each mmHg increment in $PaCO_2$ (Fig. 5.4).

It has been demonstrated that the renal response to chronic hypercapnia is not appreciably altered by the coexistence of such stresses as moderate hypoxia, sodium deprivation, potassium depletion, or alkali loading (Sapir, Levine and Schwartz, 1967; Polak et al., 1961; van Ypersele de Strihou, Gulyassy and Schwartz, 1962). However, it has been shown that failure to replenish the chloride ions lost during adaptation to respiratory acidosis severely hampers the ability of the kidneys to restore normal acid-base equilibrium when the hypercapnia abates (Schwartz et al., 1961); hence, if an elevated $PaCO_2$ is returned towards normal without adequate provision of exogenous chloride, plasma bicarbonate concentration cannot fall to normal resulting in a "posthypercapnic" alkalosis.

Cerebrospinal fluid acidity during respiratory acidosis

It has long been recognized that increases in $PaCO_2$ are rapidly reflected in the cerebrospinal fluid (CSF) producing a decrement in CSF pH. Experimental and clinical studies (Kety and Schmidt, 1948; Lambertsen, 1960; Merwarth, Sieker and Manfredi, 1961; Posner, Swanson and Plum, 1965; Pontén and Siesjö, 1966; Messeter and Siesjö, 1971) have demonstrated a narrowed CSF-arterial pCO_2 difference during respiratory acidosis which has been attributed to the associated increase in cerebral blood flow. With persistent hypercapnia, CSF bicarbonate concentration increases progressively so that the rise in CSF hydrogen ion concentration is ameliorated. Studies in dogs exposed to 12 percent CO_2 within an environmental chamber for a period of five days showed a gradual increment in CSF bicarbonate during the first 24 hours to a value approximately 13 mEq/l higher than control; CSF bicarbonate remained virtually unchanged thereafter (Bleich, Berkman and Schwartz, 1964). Plasma bicarbonate increased to almost an identical degree but, as noted above, rose somewhat more slowly. Similarly, the steady-state pH decrement was virtually identical in the two compartments.

In contrast, several clinical studies in patients with pulmonary insufficiency have indicated a relatively poor "protection" of CSF pH during hypercapnia, with pH shifts greater than those of plasma (Merwarth et al., 1961; Posner et al., 1965). The reason for these discrepant results is not apparent.

Intracellular acidity during respiratory acidosis

Several attempts have been made to define the changes in intracellular acidity (using indirect methods) in response to acute hypercapnia but the results are conflicting. Studies in human volunteers breathing a 7 percent CO_2 mixture in air for a period of three hours ($PaCO_2$ 55–60 mmHg) suggested a larger fall in "total-body" intracellular pH than in plasma pH (Manfredi, 1967). Similar results were obtained during in vivo studies of brain pH in the dog (Kibler, O'Neill and Robin, 1964). Conversely, in vitro experiments using the rat diaphragm have indicated a relative stability of intracellular acidity during moderate increases in $PaCO_2$ (Adler, Roy and Relman, 1965). In these studies, intracellular acidity was unaffected by increases in pCO_2 up to 70 mmHg despite the fact that the hydrogen ion concentration of the bathing media increased nearly twofold. When greater increments in pCO_2 were used, however, intracellular pH did fall in this model.

There is currently no information regarding the impact of chronic hypercapnia on intracellular acidity.

Other physiological and biochemical effects of respiratory acidosis

Several hemodynamic responses to acute hypercapnia have been described. Decreased myocardial contractility has been demonstrated consistently as a direct effect of acute elevations of $PaCO_2$ in the isolated cardiac muscle and in the perfused heart preparation in vitro (Vaughan Williams, 1955, McElroy, Gerdes and Brown, 1958; Ng, Levy and Zieske, 1967). Depressed ventricular performance during acute hypercapnia has been also demonstrated in intact animals (Noble, Trenchard and Guz, 1966; Horwitz, Bishop and Stone, 1968). Acute hypercapnia also produces a fall in peripheral vascular resistance due to a direct vasodilating effect on most systemic arteries (Price, 1960, Wendling et al., 1967; Kontos, Richardson and Patterson, 1968). Conversely, hypercapnia is thought to produce venoconstriction and a consequent redistribution of blood from the periphery to the pulmonary circulation (Mitchell, Wildenthal and Johnson, 1972).

Indirect effects on the circulation result from both central and peripheral reflex stimulation of the sympathetic nervous system and from a release of catecholamines (Downing, Mitchell and Wallace, 1963; DeGeest, Levy and Zieske, 1965a,b). Tending to offset this latter effect, however, hypercapnia is known to attenuate the response of the heart and vessels to these sympathetic mediators (Wood, Manley and Woodbury, 1963; Bendixen, Laver and Flacke, 1963). Finally, each of these effects is modified significantly by the degree of the accompanying acidosis and by the rapidity with which it develops.

Despite these many and varied effects, cardiac output and systemic blood pressure are usually well maintained during acute hypercapnia (Monroe, French and Whittenberger, 1960) and, in fact, may increase due to predominating sympathetic activity. Only rarely (e.g., during extreme acidemia, severe congestive heart failure, beta-adrenergic blockade) does acute hypercapnia result in a significant compromise in cardiovascular function.

The effect of acute hypercapnia on specific vascular beds has also received considerable study. Acute hypercapnia of mild to moderate degree results in renal vasodilation (Simmons and Olver, 1965) signifying the predominance of the direct effects of $PaCO_2$ on vascular smooth muscle. On the other hand, acute hypercapnia of greater severity has been shown in several studies to induce renal vasoconstriction, apparently the result of an overriding catecholamine effect (Brooker, Ansell and Brown, 1960; Bersentes and Simmons, 1967). Both responses have been shown to be independent of renal nerves.

Hypercapnia produces vasodilation and enhances blood flow in the brain (Kety and Schmidt, 1948; Reivich, 1964) but causes vasoconstriction in the mesenteric circulation (Epstein et al., 1961). Contradictory evidence has been obtained about the effects of hypercapnia on pulmonary vascular resistance and pulmonary artery pressure, some studies indicating an increase in these parameters (Nisell, 1950; Duke, 1951; Horwitz et al., 1968) and others failing to demonstrate any significant effect (Rahn and Bahnson, 1953; Stroud and Rahn, 1953; Fishman, Fritts and Cournand, 1961; Kato and Staub, 1966). Limited information in man appears to support the latter view (Twining et al., 1968; Housley et al., 1970).

Although cardiac arrhythmias have been described frequently during both acute and chronic hypercapnia (Hudson et al., 1973; Sideris et al., 1975), it is not certain that elevated $PaCO_2$, per se, is responsible. Other factors that might well contribute include hypoxia, abnormalities in plasma potassium concentration, the use of cardioactive medications, and, especially, the rapid appearance of alkalemia following sudden restoration of normal $PaCO_2$ (e.g., via mechanical ventilation).

Complex and unpredictable shifts in the oxyhemoglobin dissociation curve occur during hypercapnia since increased pCO_2 tends to shift the curve to the right (Bohr effect) and acidemia (by decreasing intracellular 2,3-DPG) tends to shift the curve to the left (Severinghaus, 1958; Naerraa et al., 1966; Lawson, 1966; Klocke, Bauer and Forster, 1970; Bellingham, Detter and Lenfant, 1971). Further complexity is introduced if chronic hypoxia is present since intracellular 2,3-DPG is augmented, which tends to shift the curve to the right (Oski et al., 1969).

Hyperphosphatemia is the only notable alteration in plasma electrolyte composition attributable to hypercapnia (Brackett, Jr. et al., 1965; Webb et al., 1977), other than those directly related to the acid-base changes.

Diagnosis and clinical manifestations

Clinical examination cannot be relied upon to assess the adequacy of alveolar ventilation. Hence, the diagnosis of respiratory acidosis inevitably depends upon accurate laboratory data. Arterial blood gases should be obtained whenever the clinical suspicion of hypoventilation arises.

The acid-base parameters themselves are diagnostic of some element of respiratory acidosis only when hypercapnia is associated with acidemia. When hypercapnia is associated with alkalemia, a separate element of respiratory acidosis may be diagnosed only if the degree of hypercapnia is clearly beyond the range anticipated in response to the coexisting metabolic alkalosis (Cohen and Schwartz, 1966). For moderate degrees of primary metabolic alkalosis (plasma bicarbonate no greater than 40 mEq/l), secondary hypoventilation would not be expected to raise $PaCO_2$ much above 50 mmHg.

Clinical manifestations of respiratory acidosis vary markedly according to the rapidity with which hypercapnia develops. Acute hypercapnia is often associated with marked anxiety, severe breathlessness, disorientation, confusion, combativeness and, in unusually severe instances, stupor or coma (Dulfano and Ishikawa, 1965; Kilburn, 1965). Chronic hypercapnia appears to be tolerated better but may also be responsible for confusion, loss of memory, and somnolence. Asterixis is a frequent accompaniment of both acute and chronic hypercapnia. Signs and symptoms of increased intracranial pressure (pseudotumor cerebri) are not uncommon and appear to be related to the vasodilating effects of carbon dioxide on cerebral blood vessels; frank papilledema may be found when hypercapnia is severe (Austen, Carmichael and Adams, 1957).

Concomitant hypoxemia may be an important contributing factor to the symptoms of respiratory acidosis. In the absence of significant hypoxemia, even severe chronic hypercapnia is well tolerated with minimal central nervous system dysfunction (Neff and Petty, 1972). In patients with longstanding hypercapnia, attempts to alleviate hypoxia by administering high-inspired oxygen concentrations should be made with considerable caution because such therapy may remove the hypoxic drive to respiration and result in even further reductions in alveolar ventilation.

Causes of respiratory acidosis

Alveolar hypoventilation may result from disease or malfunction within any element in the regulatory system controlling respiration, including the respiratory center; the central and peripheral nervous system; the respiratory muscles; the thoracic cage, pleural space, and lung parenchyma; and the airways. Table 5.1 separates the common causes of respiratory acidosis

Table 5.1 Causes of respiratory acidosis

Acute	Chronic
Airway obstruction	**Airway obstruction**
Aspiration of foreign body or vomitus	Chronic obstructive lung disease
Laryngospasm	(bronchitis, emphysema)
Generalized bronchospasm	**Respiratory center depression**
Respiratory center depression	Chronic sedative overdose
General anesthesia	Primary alveolar hypoventilation
Sedative overdosage	(Ondine's curse)
Cerebral trauma or infarction	Pickwickian syndrome
Circulatory catastrophies	Brain tumor
Cardiac arrest	**Neuromuscular defects**
Severe pulmonary edema	Poliomyelitis
Neuromuscular defects	Multiple sclerosis
High cervical cordotomy	Amyotrophic lateral sclerosis
Botulism, Tetanus	Diaphragmatic paralysis
Guillain-Barré syndrome	Myopathic diseases
Crisis in myasthenia gravis	**Restrictive defects**
Familial hypokalemic periodic paralysis	Kyphoscoliosis, spinal arthritis
Hypokalemic myopathy	Fibrothorax
Drugs or toxic agents (e.g., curare, succinylcholine, aminoglycosides)	Hydrothorax
	Interstitial fibrosis
Restrictive diseases	Decreased diaphragmatic movement
Pneumothorax	(e.g., ascites)
Hemothorax	Prolonged pneumonitis
Flail chest	Obesity
Severe pneumonitis	
Smoke inhalation	
Mechanical ventilators	

into two large groups in accordance with their usual mode of onset and duration. This classification also serves to underscore the biphasic time course that characterizes the secondary, biologic responses to hypercapnia.

A common cause of acute hypercapnia is sudden obstruction of the airways, secondary to *aspiration, laryngospasm,* or *generalized bronchospasm.* Acute hypercapnia is also seen in association with sudden suppression of the respiratory center, as may occur during *general anesthesia* or *sedative overdosage.* Acute respiratory failure frequently accompanies *cardiac arrest* and the resulting hypercapnia contributes significantly, along with the associated metabolic factors (e.g., lactic acid overproduction) to the severe acidemia commonly observed in this situation. An abrupt reduction in alveolar gas exchange, resulting in carbon dioxide retention, may occur during severe *pulmonary edema, pneumothorax,* and *chest injuries.* Several *neuromuscular defects* may lead to acute respiratory acidosis. Similarly, acute hypercapnia can result from improperly adjusted *mechanical ventilators.*

The great majority of patients with long-standing respiratory acidosis have underlying *chronic obstructive lung disease.* Some of the other less common causes of sustained hypercapnia are listed in Table 5.1.

Treatment

Treatment of *acute* respiratory acidosis, regardless of etiology, must be directed, if possible, at prompt removal of the underlying cause. It is imperative to restore adequate ventilation swiftly in order to prevent severe hypoxia and extreme elevations in plasma hydrogen ion concentration. Although circumstances may require the administration of bicarbonate to blunt the developing severe acidemia, such therapy is a temporizing maneuver at best and should not delay efforts to improve ventilation and thus hypercapnia and hypoxia.

Treatment of the acidemia of *chronic* respiratory acidosis is seldom necessary because only mild to moderate reductions of pH occur typically even with the most severe chronic elevations of $PaCO_2$. Treatment should be directed towards maximizing alveolar ventilation (e.g., bronchodilators, postural drainage, chest physiotherapy, diuretics for congestive failure, antibiotics for pulmonary infections). Acute exacerbation of chronic obstructive lung disease often requires periods of assisted ventilation. Furthermore, the superimposition of one or more additional acid-base disturbances may complicate the management of patients with respiratory acidosis. Such "mixed" disturbances are discussed in the following section.

Mixed acid-base disorders associated with respiratory acidosis

Complicating acid-base disorders occur with surprising frequency in patients with acute or chronic respiratory acidosis (Robin, 1963) and often present a considerable challenge to the clinician. In order to recognize the presence and reconstruct the pathogenesis of mixed acid-base disturbances, one must be aware of the degree to which secondary, physiologic responses alter acid-base equilibrium in the absence of complicating factors and of the time interval over which such adjustments run their full course (Cohen and Schwartz, 1966). What follows is a brief account of the most common "mixed" acid-base disorders associated with respiratory acidosis.

Respiratory acidosis and metabolic acidosis

The combination of respiratory acidosis and metabolic acidosis is commonly seen in patients with cardiorespiratory arrest and patients with severe pulmonary edema. In such settings, the interplay of poor tissue perfusion (leading to lactic acidosis) and inadequate ventilation (causing carbon dioxide retention) may produce profound falls in blood pH.

Since the contribution of body buffering during the acute phase of hypercapnia is so clearly delimited (see above), it is relatively easy under these circumstances to recognize the coincidental presence of a complicating metabolic acid-base disturbance. Thus, if hypercapnia is known to be of brief duration, and bicarbonate concentration is lower than 25–26 mEq/l, a cause for a coexistent metabolic acidosis should be sought. An element of metabolic acidosis superimposed on chronic hypercapnia may only suppress plasma bicarbonate partially towards the normal level. Under these circumstances, the acid-base parameters could be indistinguishable from those encountered during partial physiologic adaptation to hypercapnia alone and a valid judgment about the complex origin of the acid-base disturbance would depend upon clinical and other laboratory data.

Respiratory acidosis and metabolic alkalosis

This combination of disturbances is seen relatively frequently because patients with carbon dioxide retention secondary to severe pulmonary disease often develop metabolic alkalosis as a result of vomiting or administration of potent diuretics. In addition, patients with long-standing hypercapnia and secondary hyperbicarbonatemia due to renal adaptation are primed to develop "posthypercapnic" metabolic alkalosis if a sudden improvement in ventilation (e.g., tracheal suction, mechanical ventilation) returns $PaCO_2$ towards normal; such an alkalosis persists until the adaptive increase in bicarbonate stores is removed by renal excretion.

This mixed disturbance is readily identified when the plasma bicarbonate concentration is elevated beyond the range appropriate for full adaptation to the observed level of $PaCO_2$. In such cases, plasma pH will be inappropriately high; in fact, pH may be in the normal or even the alkalemic range.

Failure to recognize and correct superimposed metabolic alkalosis in patients with underlying pulmonary dysfunction can have important clinical consequences since relative or absolute alkalemia may abrogate an important stimulus to respiration in such patients. If plasma acid-base values are diagnostically inconclusive, a measurement of urine chloride concentration may help to exclude the presence of the most common forms of superimposed alkalosis. In the absence of recent diuretic administration, the finding of abundant urinary chloride provides reasonable assurance that a superimposed, chloride-responsive alkalosis is not present.

Acute respiratory acidosis and chronic respiratory acidosis

Patients fully adapted to an elevated level of $PaCO_2$ may experience a sudden deterioration of pulmonary function which causes $PaCO_2$ to rise even further. Under such circumstances, acid-base equilibrium will be intermediate between that characteristic of acute and that characteristic of chronic respiratory acidosis (Cohen and Schwartz, 1966). It should be noted, however, that a similar acid-base picture (i.e., an elevated $PaCO_2$ coupled with a level of plasma bicarbonate concentration intermediate between the ranges appropriate for acute and chronic hypercapnia) can also be seen (1) in patients with a stable degree of hypercapnia but in whom full adaptation has not yet occurred, (2) in patients with chronic respiratory acidosis in whom mild to moderate metabolic acidosis has supervened, and (3) in patients with acute respiratory acidosis complicated by metabolic alkalosis.

RESPIRATORY ALKALOSIS

Respiratory alkalosis refers to that acid-base disturbance initiated by a reduction in carbon dioxide tension.

Pathophysiology

Increased alveolar ventilation is the only process that can result in a period of negative carbon dioxide balance and, hence, in a reduction in carbon dioxide tension (hypocapnia). Primary hyperventilation and respiratory alkalosis are, therefore, synonymous terms.

Secondary physiologic responses

The changes in acid-base equilibrium associated with primary hyperventilation are shown in Figure 5.5 and are, for the most part, the obverse of those seen during respiratory acidosis. If $PaCO_2$ falls, the tendency for hydrogen ion concentration to be reduced is ameliorated by a secondary, adaptive reduction in plasma bicarbonate concentration. The decrements in bicarbonate induced by hypocapnia, like the analogous increments induced by hypercapnia, are the consequence of the immediate response of body buffering and of a more protracted response of renal adaptive mechanisms.

The biologic importance of these secondary responses is underscored by noting how extreme an effect even modest hyperventilation would have on hydrogen ion concentration if no change in bicarbonate were to occur. Thus, an increase in alveolar ventilation sufficient to reduce $PaCO_2$ from 40 to 20

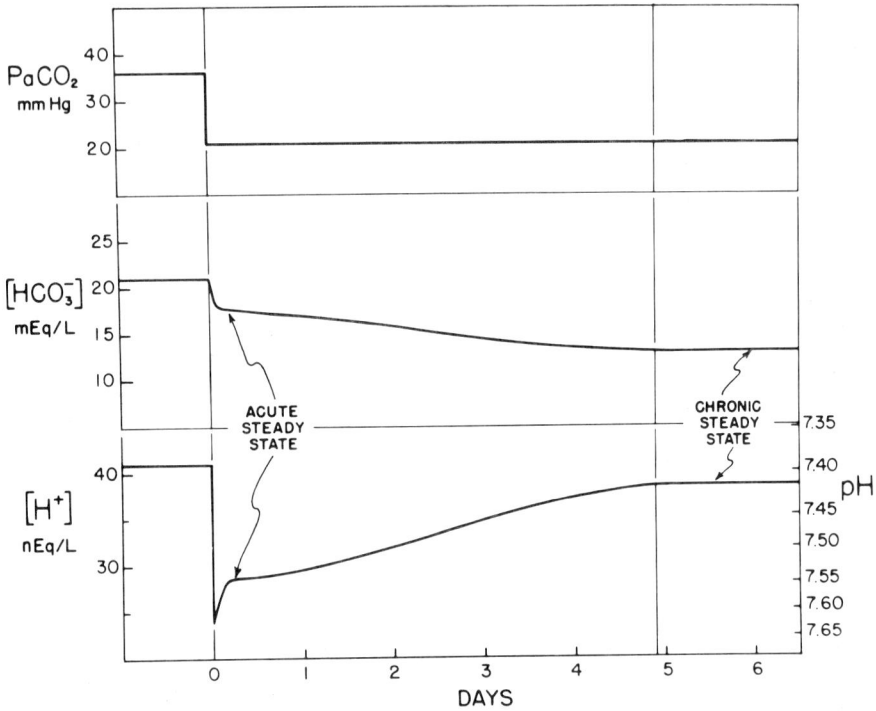

Figure 5.5 Schematic time course of the changes in plasma acid-base equilibrium during the generation of respiratory alkalosis. In this scheme, it is assumed that a $\Delta PaCO_2$ of 15 mmHg is produced abruptly and maintained unchanged thereafter.

mmHg would, in the face of a normal bicarbonate level, cause hydrogen ion concentration to fall to the dangerously low level of 20 nEq/l (pH 7.70). It is interesting to contrast this situation with that prevailing in metabolic alkalosis, in which such alarming reductions in hydrogen ion concentration would require extreme elevations in bicarbonate concentration even if secondary, adaptive hypercapnia failed to occur.

Acute hypocapnia

The adjustment in acid-base equilibrium following the induction of acute hypocapnia results totally from nonrenal mechanisms and appears to be accounted for virtually entirely by the titration of the nonbicarbonate buffers of the body. Assuming no further changes in $PaCO_2$, this immediate response is completed within 5 to 10 minutes; subsequently, no further, detectable change in acid-base equilibrium occurs for a period of several hours. Operationally, this period can be considered therefore as an "acute steady state." Figure 5.6 indicates the range of secondary changes in bicarbonate concentration and the corresponding range for hydrogen ion concentration seen in human subjects in response to graded degrees of acute hypocapnia

ACID-BASE DISORDERS OF RESPIRATORY ORIGIN 155

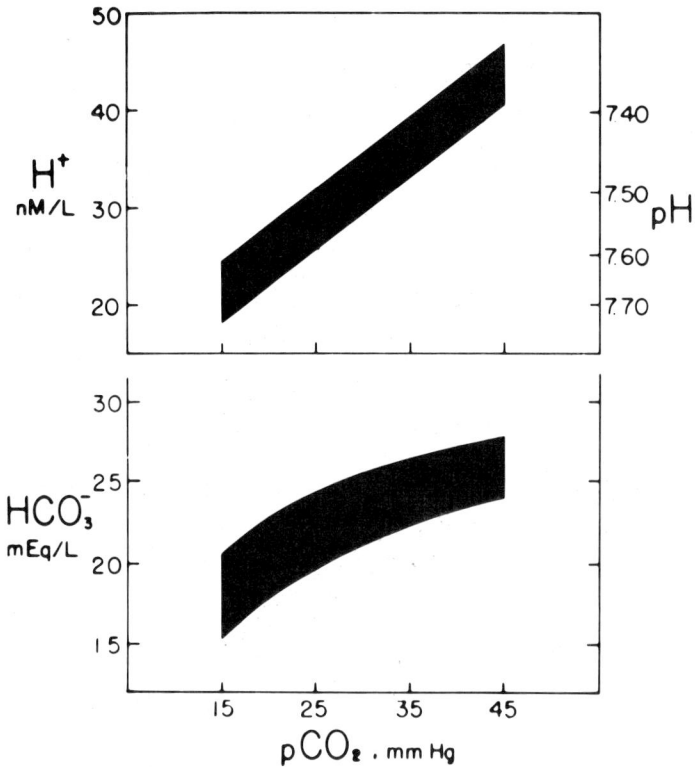

Figure 5.6 Ninety-five percent significance bands for plasma bicarbonate concentration (lower panel) and plasma hydrogen ion concentration (upper panel) in acute hypocapnia. The bands were calculated from data obtained in anesthetized patients who were passively hyperventilated during minor surgical procedures. (From Arbus et al., 1969, with permission of the publisher)

(Arbus et al., 1969). The data from which these ranges were calculated were obtained during the course of acute "whole-body" titration experiments utilizing passive hyperventilation of anesthetized subjects undergoing minor surgical procedures. The acute change in plasma bicarbonate concentration is substantially greater in magnitude than that observed during acute hypercapnia of comparable degree, a decrement in bicarbonate of some 3 to 4 mEq/l occurring within minutes after $PaCO_2$ is lowered to 20–25 mmHg. The resulting change in plasma hydrogen ion concentration, however, is strikingly similar to that observed during acute *elevations* in CO_2 tension. On the average, each mmHg reduction in $PaCO_2$ results in an immediate fall in hydrogen ion concentration of approximately 0.75 nEq/l. The 95 percent confidence intervals constructed from these data represent the range of responses within which acid-base equilibrium would be expected to fall if an acute reduction in $PaCO_2$ were the only factor disturbing plasma acidity. Several workers (Eichenholz et al., 1962; Zborowska-Sluis and Dossetor, 1967; Plum and Posner, 1967) have ascribed a major role to the accumulation of organic acids (notably lactic acid) in the reduction of bicarbonate concentration during acute hypocapnia. Other studies (Sykes and Cooke, 1965; Eldridge and Salzer, 1967; Arbus et al., 1969; Gledhill, Beirne and Dempsey, 1975), however, have failed to support this contention. It is possible that the large increase in lactate reported in some studies reflect a degree of circulatory insufficiency consequent to the experimental maneuvers required to produce the hyperventilation (e.g., mechanical ventilation, anesthesia, surgical stress).

Chronic hypocapnia

If hypocapnia persists, plasma bicarbonate concentration falls further as a consequence of renal adaptive responses. Studies in normal dogs (Gennari, Goldstein and Schwartz, 1972) have revealed that dampening of renal sodium-hydrogen exchange occurs, which causes in turn a transient suppression of net acid excretion and a persistent reduction in the rate of bicarbonate reabsorption. Approximately two to four days are required for the completion of these chronic physiologic responses (Fig. 5.5). These adaptive mechanisms serve to produce and maintain the reduction in plasma bicarbonate characteristic of the new, chronic steady state and result in a highly predictable relationship between the degree of chronic hypocapnia and the level at which plasma bicarbonate stabilizes; it has been shown that each mmHg chronic reduction in $PaCO_2$ is associated with a fall in plasma bicarbonate concentration averaging 0.4–0.5 mEq/l. As depicted in Figure 5.7, this decrement in bicarbonate is sufficient to limit the fall in hydrogen ion concentration dramatically. Thus, in the chronic steady state, plasma hydrogen ion concentration stabilizes at a level only slightly lower than control, despite a marked degree of hypocapnia. There is no appreciable change in plasma lactate during chronic hypocapnia (Gennari et al., 1972).

The regulation of acid-base equilibrium during chronic hypocapnia has not yet been well studied in man. Short-term studies (Gledhill et al., 1975) have revealed the early appearance of a renal response consisting of bicarbonaturia and suppression of ammonium and titratable acid excretion. Studies in high-altitude dwellers, as well as in volunteers at simulated altitude (Dill, Talbott and Consolazio, 1937; Houston and Riley, 1947; Chiodi, 1957; Lahiri and Milledge, 1967) have suggested changes in plasma bicarbonate of similar magnitude to those observed in the dog. If so, it would appear that plasma hydrogen ion concentration is maintained at more nearly normal levels during uncomplicated chronic hypocapnia than is the case during uncomplicated chronic hypercapnia.

Cerebrospinal fluid acidity during respiratory alkalosis

The relationship between arterial and CSF acid-base parameters during respiratory alkalosis has been studied in experimental hyperventilation, in human volunteers adapting to high altitudes, in high-altitude natives, and

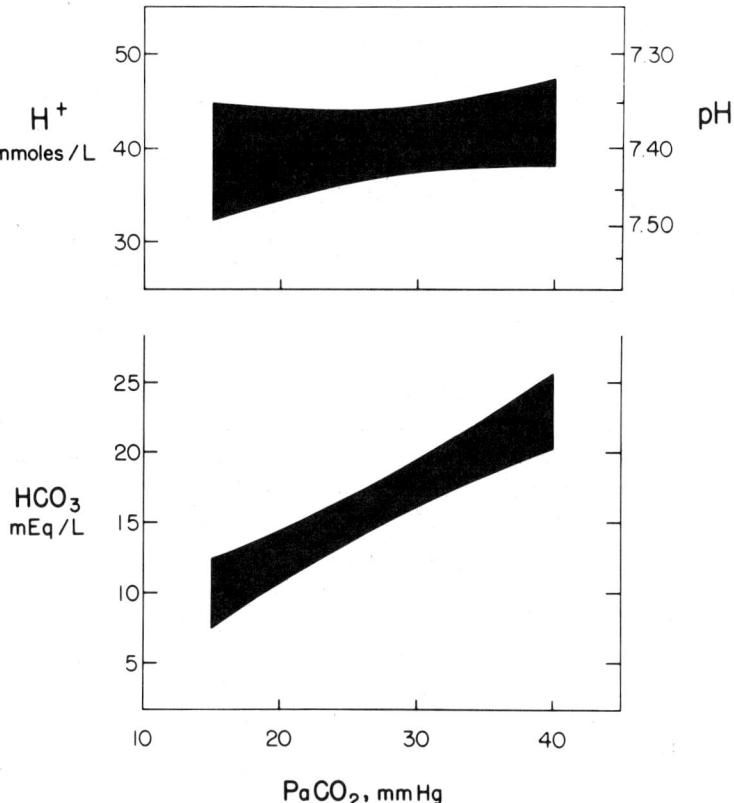

Figure 5.7 Ninety-five percent significance bands for plasma bicarbonate concentration (lower panel) and plasma hydrogen ion concentration (upper panel) in chronic hypocapnia. The bands were calculated from data obtained in dogs exposed chronically to an hypoxic atmosphere within an environmental chamber. (From Gennari et al., 1972, with permission of the publisher)

in certain clinical states associated with hyperventilation (liver disease, salicylate poisoning, head injuries, pregnancy).

The rise in plasma pH induced by acute experimental hyperventilation is followed by a significant elevation in CSF pH. Several studies (Kety and Schmidt, 1948; Lambertsen, 1960; Merwarth et al., 1961; Fisher and Christianson, 1963; Posner et al., 1965; Pontén and Siesjö, 1966) have emphasized, however, that the pCO_2 in CSF falls to a lesser extent than that in blood, resulting in a widening of the CSF-arterial pCO_2 difference; this is thought to be secondary to the decrease in cerebral blood flow associated with hypocapnia (Kety and Schmidt, 1948; Reivich, 1964). Whereas plasma bicarbonate, as noted above, falls abruptly in response to acute hypocapnia, CSF bicarbonate remains relatively stable for several minutes, falling gradually only over the subsequent several hours. As a consequence, CSF pH is gradually returned to its normal value. It has been suggested that a delayed increase in CSF lactate plays a role in this phenomenon (Van Vaerenbergh, Demeester and Leusen, 1965). Passive hyperventilation in dogs (Plum and Posner, 1967) to a $PaCO_2$ of approximately 20–22 mmHg resulted in a doubling of CSF lactate concentration (from 2.5 to 5.3 mmol/l by approximately five hours). It has been postulated that this gradual increase in CSF lactate reflects a change in CNS tissue metabolism. Two possibilities have been considered to account for this (Van Vaerenbergh et al., 1965). First, cerebral hypoxia resulting from the intense vasoconstricting effects of hypocapnia and second, increased synthesis of lactate resulting from alkalemia-induced acceleration of anaerobic glycolysis.

Several studies (Pauli, Vorburger and Reubi, 1962; Severinghaus et al., 1963; Mitchell et al., 1965) have demonstrated that CSF pH during *chronic* respiratory alkalosis deviates minimally from control. It is clear that the associated decrement in CSF bicarbonate concentration cannot be attributed to the barely detectable persistent increases in CSF lactate.

Intracellular acidity during respiratory alkalosis

It has been suggested (Manfredi, 1967) that intracellular hydrogen ion concentration, as assessed by indirect methods, falls in parallel with extracellular hydrogen ion concentration when healthy human subjects hyperventilate voluntarily to achieve a $PaCO_2$ of 15–20 mmHg. Similar results have been obtained from studies in isolated muscle preparations (Adler et al., 1965). The response of intracellular acidity to persistent hypocapnia has not been studied.

Other physiological and biochemical effects of respiratory alkalosis

The hemodynamic changes produced by hypocapnia are, in general, the obverse of those produced by hypercapnia. Acute hypocapnia produces a modest increase in myocardial contractility (Vaughan Williams, 1955; McElroy et al., 1958; Wead and Little, 1967) and direct vasoconstriction in most vascular beds. Cerebral vasoconstriction and reduced cerebral blood flow have been well documented during acute hypocapnia (Kety and Schmidt, 1948; Wasserman and Patterson, 1961). Studies on the vascular response to hypocapnia in the human forearm (Kontos et al., 1972) have shown an initial vasodilating effect (probably owing to transient histamine release), which is soon followed by persistent vasoconstriction. Acute hypocapnia causes an increase in renal vascular resistance and a decrease in renal blood flow; these changes are not mediated by the renal nerves (Simmons and Olver, 1965).

A decrease in $PaCO_2$ (and an increase in pH) cause a shift to the left in the oxyhemoglobin dissociation curve (Bohr effect) resulting in an impediment of oxygen unloading to the tissues but in an enhancement of oxygen loading in the lungs (Severinghaus, 1958; Lawson, 1966). During chronic alkalosis, these effects are offset by a shift of the curve to the right produced by an increased intracorpuscular concentration of 2,3-DPG (Klocke et al., 1970; Bellingham, 1971; Lenfant et al., 1968).

Hypophosphatemia is the only consistent change in plasma electrolyte composition (Okel and Hurst, 1961; Mostellar and Tuttle, 1964; Arbus et al., 1969) other than those related to the acid-base disorder itself.

Diagnosis and clinical manifestations

The acid-base parameters themselves are diagnostic of some element of respiratory alkalosis only when hypocapnia is associated with alkalemia. When hypocapnia is associated with acidemia, a separate element of respiratory alkalosis may be diagnosed only if the degree of hypocapnia is clearly beyond the range anticipated in response to the coexisting metabolic acidosis.

Acute hypocapnia is frequently accompanied by numbness and paresthesias of the lips and extremities and may be associated with light-headedness, mental confusion, muscle cramps, increased deep tendon reflexes, carpopedal spasm and, occasionally, seizures (Okel and Hurst, 1961; Saltzman, Heyman and Sieker, 1963; Kilburn, 1966). Electroencephalographic abnormalities have been reported (Brown, 1953). Pseudoangina and ischemic ST-T-wave changes have been observed in hyperventilating subjects (Jacobs, Battle and Ronan, 1974; Evans and Lum, 1977) with no evidence of coronary artery disease, as judged by clinical evaluation and coronary cineangiography. The pathogenesis of these findings remains unknown, although myocardial ischemia secondary to hypocapnia-induced vasoconstriction and alkalemia-induced shift to the left of the oxyhemoglobin dissociation curve have been proposed. Alkalemia may lead to serious cardiac arrhythmias (Ayres and Grace, 1969), especially in patients receiving cardioactive medications. Such arrhythmias are frequently resistant to standard forms of therapy.

Chronic hypocapnia is typically accompanied by few symptoms other than those associated with the underlying disease process. It is particularly noteworthy that marked hypocapnia can be sustained without a clinically evident increase in respiratory effort. For this reason, the physical examination should not be relied upon to diagnose respiratory alkalosis.

Causes of respiratory alkalosis

Respiratory alkalosis is perhaps the most frequent acid-base disorder encountered. Table 5.2 lists some of the common causes of this condition. Most are associated with the abrupt appearance of hypocapnia but in many instances the process may be sufficiently prolonged to permit full, chronic adaptation to occur. Consequently, no attempt has been made to separate these conditions into acute and chronic subgroups.

Table 5.2 Causes of respiratory alkalosis

Anxiety, Hysteria
Fever
Salicylate intoxication
Central Nervous System diseases
   Cerebrovascular accident
   Trauma
   Infection
   Tumor

Intrathoracic processes
   Congestive heart failure
   Pneumonitis
   Bronchial asthma attack
   Pulmonary fibrosis
   Pulmonary emboli
   Foreign bodies

Hypoxemia
   Lowered barometric pressure
   Increased venous admixture
   Marked ventilation-perfusion inequality
Hepatic insufficiency
Gram-Negative septicemia
Pregnancy
Mechanical ventilators

The mechanisms by which these processes lead to hyperventilation differ markedly. *Anxiety* and *hysteria* appear to act through cortical pathways, thus accounting for the improvement that often follows reassurance or sedation. Rebreathing into a closed system (e.g., a paper bag) may also be helpful by interrupting the vicious cycle that may result from the reinforcing effects of the symptoms of hypocapnia.

Fever and *salicylate intoxication* appear to cause a more direct stimulation of the respiratory center (Tenney and Miller, 1955; Cameron and Semple, 1968). Various *central nervous system lesions* of vascular, traumatic, infectious and neoplastic origin, if located in the vicinity of the medullary respiratory centers, may also enhance stimulatory inputs or interrupt normal inhibitory pathways.

Numerous intrathoracic processes such as *congestive heart failure, bronchial asthma, pneumonitis, pulmonary fibrosis* and *pulmonary emboli* may result in hyperventilation by stimulating the afferent limbs of certain neural reflex mechanisms. *Hypoxemia* stimulates ventilation through chemoreceptor mechanisms, and may result in hypocapnia if carbon dioxide excretion is unimpeded. Such situations of hypoxemia include high-altitude residence and conditions producing increased venous admixture and marked ventilation-perfusion inequality; diffusion limitation is considered as a subcategory of ventilation-perfusion inequality.

The mechanisms responsible for the primary hyperventilation seen commonly in severe *hepatic insufficiency* (Vanamee et al., 1956, Mulhausen,

Eichenholz and Blumentals, 1967; Karetzky and Mithoefer, 1967) and in *gram-negative septicemia* (Simmons, Nicoloff and Guze, 1960; MacLean et al., 1967; Mazzara, Ayres and Grace, 1974) are not yet certain. Since respiratory alkalosis may be an early manifestation of each of these processes, however, unexplained hyperventilation may provide an important clue to their presence.

Special attention should be made of the chronic hypocapnia known for many years to be associated with normal *pregnancy*. Measurable increases in alveolar ventilation appear early in pregnancy (Magnus-Levy, 1904; Goodland, Reynolds and Pommerenke, 1954; Marx and Orkin, 1958) and reach their maximal level by the twelfth week (approximately 75 percent above control). Several studies have shown that $PaCO_2$ falls to approximately 30–32 mmHg by the first trimester, a level that is maintained virtually unchanged throughout the remainder of pregnancy; pH remains at or near normal levels. An element of acute hypocapnia is often superimposed during parturition, $PaCO_2$ occasionally falling by additional 10–12 mmHg during uterine contractions. Ventilation and acid-base parameters return to control values several days post partum.

It was postulated many years ago (Hasselbalch and Gammeltoft, 1915) that the hyperventilation of pregnancy is the result of the physiologic effects of progesterone. Supportive evidence has been provided by the observation (Griffith et al., 1929; Goodland and Pommerenke, 1952) that hyperventilation occurs during the luteal phase of the normal menstrual cycle ($\Delta PaCO_2$ of approximately 5–6 mmHg). In addition, it has been shown that reversible increases in ventilation can be produced by the administration of exogenous progesterone in normal men, in postmenopausal women and in patients with chronic hypoventilation due to emphysema (Cullen, Brum and Reidt, 1959; Tyler, 1960) or Pickwickian syndrome (Lyons and Huang, 1968; Sutton et al., 1975).

Finally, *mechanical ventilators* may induce respiratory alkalosis if they are improperly adjusted. This complication can be avoided only by assessing acid-base equilibrium frequently in patients undergoing assisted or controlled ventilation.

Treatment

Treatment of respiratory alkalosis must be directed at removing the underlying cause. Attempts to return $PaCO_2$ towards normal by increasing the carbon dioxide content of inspired air have not been rewarding.

Mixed acid-base disorders associated with respiratory alkalosis

Respiratory alkalosis is frequently associated with other acid-base disorders. We will describe here some of the more common "mixed" disturbances seen in clinical practice.

Respiratory alkalosis and metabolic acidosis

This mixed acid-base disorder is commonly encountered during the rapid correction of severe metabolic acidosis. During treatment, plasma bicarbonate concentration often returns towards normal more swiftly than the secondary hypocapnia (originally induced by the acidosis) can abate. As a consequence, the degree of hypocapnia often remains inappropriately great with respect to the then-current decrement in plasma bicarbonate concentration for a period of several hours or more; hydrogen ion concentration during such intervals may be normal, or even frankly low.

Patients with salicylate intoxication commonly manifest elements of both metabolic acidosis and respiratory alkalosis, reflecting the independent effects of the salicylate molecule on both cellular metabolism and ventilation. Patients with gram-negative sepsis may also develop this mixed disturbance, reflecting the frequent occurrence of primary hyperventilation, on the one hand, and of lactic acidosis and/or renal failure, on the other. Similarly, this combination might occur in the setting of combined hepatic and renal failure.

It should be noted that patients who have made only a partial renal adaptation to hypocapnia (i.e., if hypocapnia has been present for only a day or so), may have acid-base values indistinguishable from those characteristic of patients with this mixed disturbance. Other laboratory data and relevant historical information should serve to differentiate between these alternatives.

Respiratory alkalosis and metabolic alkalosis

Although these two disturbances do not coexist often, when they do hydrogen ion concentration may, of course, be driven to extremely low levels. Patients with hepatic insufficiency, who frequently have persistent hyperventilation, may develop this combination of acid-base disturbances if they are treated with potent diuretics or if they lose gastric fluid. Similarly, patients with an underlying metabolic alkalosis may develop such a picture if ventilation is stimulated (e.g., pulmonary embolus, sepsis).

This mixed disturbance can be diagnosed readily if the level of $PaCO_2$ is less than normal and if plasma bicarbonate concentration is frankly elevated. This disorder should be treated as a medical emergency, because extreme alkalemia carries a grave prognosis.

REFERENCES

Adler, S., Roy, A. & Relman, A. (1965). Intracellular acid-base regulation. I. Response of muscle cells to changes in CO_2 tension or extracellular bicarbonate concentration. *Journal of Clinical Investigation,* **44,** 8–20.

Arbus, G. S., Herbert, L. A., Levesque, P. R., Etsten, B. E. & Schwartz, W. B. (1969). Characterization and clinical application of the "significance band" for acute respiratory alkalosis. *New England Journal of Medicine,* **280,** 117–123.

Austen, F. K., Carmichael, M. W. & Adams, R. D. (1957). Neurological manifestations of chronic pulmonary insufficiency. *New England Journal of Medicine,* **257,** 579-590.

Ayres, S. M. & Grace, W. J. (1969). Inappropriate ventilation and hypoxemia as causes of cardiac arrhythmias. *American Journal of Medicine*, 46, 495–505.

Bellingham, A. J., Detter, J. C. & Lenfant, C. (1971). Regulatory mechanisms of hemoglobin oxygen affinity in acidosis and alkalosis. *Journal of Clinical Investigation*, 50, 700–706.

Bendixen, H. H., Laver, M. B. & Flacke, W. E. (1963). Influence of respiratory acidosis on circulatory effect of epinephrine in dogs. *Circulation Research*, 13, 64–70.

Bersentes, T. J. & Simmons, D. H. (1967). Effects of acute acidosis on renal hemodynamics. *American Journal of Physiology*, 212, 633–640.

Bleich, H. L., Berkman, P. M. & Schwartz, W. B. (1964). The response of cerebrospinal fluid composition to sustained hypercapnia. *Journal of Clinical Investigation*, 43, 11–16.

Brackett, N. C., Jr., Cohen, J. J. & Schwartz, W. B. (1965). Carbon dioxide titration curve of normal man: effect of increasing degrees of acute hypercapnia on acid-base equilibrium. *New England Journal of Medicine*, 272, 6–12.

Brackett, N. C., Jr., Wingo, C. F., Muren, O. & Solano, J. T. (1969). Acid-base response to chronic hypercapnia in man. *New England Journal of Medicine*, 280, 124–130.

Brooker, W. J., Ansell, J. S. & Brown, E. B., Jr. (1960). Effect of respiratory acidosis on renal blood flow. *Surgical Forum*, 10, 869–872.

Brown, E. B., Jr. (1953). Physiological effects of hyperventilation. *Physiological Reviews*, 33, 445–471.

Cameron, I. R. & Semple, S. J. G. (1968). The central respiratory stimulant action of salicylates. *Clinical Science*, 35, 391–401.

Chiodi, H. (1957). Respiratory adaptations to chronic high altitude hypoxia. *Journal of Applied Physiology*, 10, 81–87.

Cohen, J. J. & Schwartz, W. B. (1966). Evaluation of acid-base equilibrium in pulmonary insufficiency: an approach to a diagnostic dilemma. *American Journal of Medicine*, 41, 163–167.

Cullen, J. H., Brum, V. C. & Reidt, W. U. (1959). The respiratory effects of progesterone in severe pulmonary emphysema. *American Journal of Medicine*, 27, 551–557.

DeGeest, H., Levy, M. N. & Zieske, H. (1965a). Reflex effects of cephalic hypoxia, hypercapnia and ischemia upon ventricular contractility. *Circulation Research*, 17, 349–358.

DeGeest, H., Levy, M. N. & Zieske, H. (1965b). Carotid chemoreceptor stimulation and ventricular performance. *American Journal of Physiology*, 209, 564–570.

Dill, D. B., Talbott, J. H. & Consolazio, W. V. (1937). Blood as a physicochemical system. XII. Man at high altitude. *Journal of Biological Chemistry*, 118, 649–666.

Downing, S. E., Mitchell, J. H. & Wallace, A. G. (1963). Cardiovascular responses to ischemia, hypoxia and hypercapnia of the central nervous system. *American Journal of Physiology*, 204, 881–887.

Duke, H. N. (1951). Pulmonary vasomotor responses of isolated perfused cat lungs to anoxia and hypercapnia. *Quarterly Journal of Experimental Physiology*, 36, 75–88.

Dulfano, M. J. & Ishikawa, S. (1965). Hypercapnia: mental changes and extrapulmonary complications. An expanded concept of the "CO_2 intoxication" syndrome. *Annals of Internal Medicine*, 63, 829–841.

Eichenholz, A., Mulhausen, R. O., Anderson, W. E. & MacDonald, F. M. (1962). Primary hypocapnia: a cause of metabolic acidosis. *Journal of Applied Physiology*, 17, 283–288.

Eldridge, F. & Salzer, J. (1967). Effect of respiratory alkalosis on blood lactate and pyruvate in humans. *Journal of Applied Physiology*, 22, 461–468.

Epstein, R. M., Wheeler, H. O., Frumin, M. J., Habib, D. V., Papper, E. M. & Bradley, S. F. (1961). The effect of hypercapnia on estimated hepatic blood flow, circulating splanchnic blood volume, and hepatic sulfobromophthalein clearance during general anesthesia in man. *Journal of Clinical Investigation*, 40, 592–598.

Evans, D. W. & Lum, L. C. (1977). Hyperventilation: an important cause of pseudoangina. *Lancet*, 1, 155–157.

Farhi, L. E. & Rahn, H. (1960). Dynamics of changes in carbon dioxide stores. *Anesthesiology*, 21, 604–614.

Fisher, V. J. & Christianson, L. C. (1963). Cerebrospinal fluid acid-base balance during a changing ventilatory state in man. *Journal of Applied Physiology*, 18, 712–716.

Fishman, A. P., Fritts, H. W., Jr. & Cournand, A. (1961). Effects of breathing carbon dioxide upon the pulmonary circulation. *Circulation.* **22**, 220–225.

Gennari, F. J., Goldstein, M. B. & Schwartz, W. B. (1972). The nature of the renal adaptation to chronic hypocapnia. *Journal of Clinical Investigation,* **51**, 1722–1730.

Gledhill, N., Beirne, G. J. & Dempsey, J. A. (1975). Renal response to short-term hypocapnia in man. *Kidney International,* **8**, 376–386.

Goodland, R. L. & Pommerenke, W. T. (1952). Cyclic fluctuations of the alveolar carbon dioxide tension during the normal menstrual cycle. *Fertility and Sterility,* **3**, 394–401.

Goodland, R. L., Reynolds, J. G. & Pommerenke, W. T. (1954). Alveolar carbon dioxide tension levels during pregnancy and early puerperium. *Journal of Clinical Endocrinology,* **14**, 522–530.

Griffith, F. R., Jr., Pucher, G. W., Brownell, K. A., Klein, J. D. & Carmer, M. E. (1929). Studies in human physiology. III. Alveolar air and blood gas capacity. *American Journal of Physiology,* **89**, 449–470.

Hasselbalch, K. A. & Gammeltoft, S. A. (1915). Die Neutralitätsregulation des graviden Organismus. *Biochemische Zeitschrift,* **68**, 206–264.

Horwitz, L. D., Bishop, Y. S. & Stone, H. L. (1968). Effects of hypercapnia on the cardiovascular system of conscious dogs. *Journal of Applied Physiology,* **25**, 346–348.

Housley, E., Clarke, S. W., Hedworth-Whitty, R. B. & Bishop, J. M. (1970). Effect of acute and chronic acidaemia and associated hypoxia on the pulmonary circulation of patients with chronic bronchitis. *Cardiovascular Research,* **4**, 482–489.

Houston, C. S. & Riley, R. L. (1947). Respiratory and circulatory changes during acclimatization to high altitude. *American Journal of Physiology,* **149**, 565–587.

Hudson, L. D., Kurt, T. L., Petty, T. L. & Genton, E. (1973). Arrhythmias associated with acute respiratory failure in patients with chronic airway obstruction. *Chest,* **63**, 661–665.

Jacobs, W. F., Battle, W. E. & Ronan, J. A. (1974). False-Positive ST-T-Wave changes secondary to hyperventilation and exercise. *Annals of Internal Medicine,* **81**, 479–482.

Karetzky, M. S. & Mithoefer, J. C. (1967). The cause of hyperventilation and arterial hypoxia in patients with cirrhosis of the liver. *American Journal of Medical Sciences,* **254**, 797–804.

Kato, M. and Staub, N. C. (1966). Response of small pulmonary arteries to unilobar hypoxia and hypercapnia. *Circulation Research,* **19**, 426–440.

Kety, S. S. & Schmidt, C. F. (1948). The effects of altered arterial tensions of carbon dioxide and oxygen on cerebral blood flow and cerebral oxygen consumption of normal young men. *Journal of Clinical Investigation,* **27**, 484–491.

Kibler, R. F., O'Neill, R. P. & Robin, E. D. (1964). Intracellular acid-base relations of dog brain with reference to the brain extracellular volume. *Journal of Clinical Investigation,* **43**, 431–443.

Kilburn, K. H. (1965). Neurological manifestations of respiratory failure. *Archives of Internal Medicine,* **116**, 409–415.

Kilburn, K. H. (1966). Shock, seizures and coma with alkalosis during mechanical ventilation. *Annals of Internal Medicine,* **65**, 977–984.

Klocke, R. A., Bauer, C. & Forster, R. E. (1970). The kinetics of the oxygen-hemoglobin reactions: influence of 2,3-diphosphoglycerate (2,3 DPG) and pH. *Physiologist,* **13**, 242.

Kontos, H. A., Richardson, D. W. & Patterson, J. L., Jr. (1968). Vasodilator effect of hypercapnic acidosis on human forearm vessels. *American Journal of Physiology,* **215**, 1403–1405.

Kontos, H. A., Richardson, D. W., Raper, A. J., Zubair-Ul-Hassan & Patterson, J. L., Jr. (1972). Mechanisms of action of hypocapnic alkalosis on limb blood vessels in man and dog. *American Journal of Physiology,* **223**, 1296–1307.

Lahiri, S. & Milledge, J. S. (1967). Acid-base in Sherpa altitude residents and lowlanders at 4880 M. *Respiration Physiology,* **2**, 323–334.

Lambertsen, C. J. (1960). Carbon dioxide and respiration in acid-base homeostasis. *Anesthesiology,* **21**, 642–651.

Lawson, W. H., Jr. (1966). Interrelation of pH, temperature and oxygen on deoxygenation rate of red cells. *Journal of Applied Physiology,* **21**, 905–914.

Lenfant, C., Torrance, J., English, E., Finch, C. A., Reynafarje, C., Ramos, J. & Faura, J. (1968). Effect of altitude on oxygen binding by hemoglobin and on organic phosphate levels. *Journal of Clinical Investigation,* **47**, 2652–2656.

Lyons, H. A. & Huang, C. T. (1968). Therapeutic use of progesterone in alveolar hypoventilation associated with obesity. *American Journal of Medicine*, 44, 881–888.

MacLean, L. D., Mulligan, G. W., McLean, A. P. H. & Duff, J. H. (1967). Alkalosis in septic shock. *Surgery*, 62, 655–662.

McElroy, W. T., Jr., Gerdes, A. J. & Brown, E. B., Jr. (1958). Effects of CO_2, bicarbonate and pH on the performance of isolated perfused guinea pig hearts. *American Journal of Physiology*, 195, 412–416.

Magnus-Levy, A. (1904). Stoffwechsel und Nahrungsbedarf in der Schwangerschaft. *Zeitschrift für Geburtshilfe und Gynäkologie*, 52, 116.

Manfredi, F. (1967). Effects of hypocapnia and hypercapnia on intracellular acid-base equilibrium in man. *Journal of Laboratory and Clinical Medicine*, 69, 304–312.

Marx, G. F. & Orkin, L. R. (1958). Physiological changes during pregnancy: a review. *Anesthesiology*, 19, 258–274.

Mazzara, J. T., Ayres, S. M. & Grace, W. J. (1974). Extreme hypocapnia in the critically ill patient. *American Journal of Medicine*, 56, 450–456.

Merwarth, C. R., Sieker, H. O. & Manfredi, F. (1961). Acid-base relations between blood and cerebrospinal fluid in normal subjects and patients with respiratory insufficiency. *New England Journal of Medicine*, 265, 310–313.

Messeter, K. & Siesjö, B. K. (1971). Regulation of the CSF pH in acute and sustained respiratory acidosis. *Acta Physiologica Scandinavica*, 83, 21–30.

Mitchell, R. A., Carman, C. T., Severinghaus, J. W., Richardson, B. W., Singer, M. M. & Shnider, S. (1965). Stability of cerebrospinal fluid pH in chronic acid-base disturbances in blood. *Journal of Applied Physiology*, 20, 443–452.

Mitchell, J. H., Wildenthal, K. & Johnson, R. L., Jr. (1972). The effects of acid-base disturbances on cardiovascular and pulmonary function. *Kidney International*, 1, 375–389.

Monroe, R. G., French, G. & Whittenberger, J. L. (1960). Effects of hypocapnia and hypercapnia on myocardial contractility. *American Journal of Physiology*, 199, 1121–1124.

Mostellar, M. E. & Tuttle, E. P., Jr. (1964). The effects of alkalosis on plasma concentration and urinary excretion of inorganic phosphate in man. *Journal of Clinical Investigation*, 43, 138–149.

Mulhausen, R., Eichenholz, A. & Blumentals, A. (1967). Acid-base disturbances in patients with cirrhosis of the liver. *Medicine*, 46, 185–189.

Naerraa, N., Strange Petersen, E., Boye, E. & Severinghaus, J. W. (1966). pH and molecular CO_2 components of the Bohr effect in human blood. *Scandinavian Journal of Clinical and Laboratory Investigation*, 18, 96–102.

Neff, T. A. & Petty, T. L. (1972). Tolerance and survival in severe chronic hypercapnia. *Archives of Internal Medicine*, 129, 591–596.

Ng, M. L., Levy, M. N. & Zieske, H. A. (1967). Effects of changes of pH and of carbon dioxide tension on left ventricular performance. *American Journal of Physiology*, 213, 115–120.

Nisell, O. I. (1950). The action of oxygen and carbon dioxide on the bronchioles and vessels of the isolated perfused lungs. *Acta Physiologica Scandinavica*, 21 (suppl. 73), 5–62.

Noble, M. I. M., Trenchard, D. & Guz, A. (1966). Effects of changes in $PaCO_2$ and PaO_2 on cardiac performance in conscious dogs. *Journal of Applied Physiology*, 22, 147–152.

Okel, B. B. & Hurst, J. W. (1961). Prolonged hyperventilation in man. Associated electrolyte changes and subjective symptoms. *Archives of Internal Medicine*, 108, 157–162.

Oski, F. A., Gottlieb, A. J., Delivoria-Papadopoulos, M. & Miller, W. W. (1969). Red-cell 2,3-diphosphoglycerate levels in subjects with chronic hypoxemia. *New England Journal of Medicine*, 280, 1165–1166.

Pauli, H. G., Vorburger, C. & Reubi, F. (1962). Chronic derangements of cerebrospinal fluid acid-base components in man. *Journal of Applied Physiology*, 17, 993–998.

Plum, F. & Posner, J. B. (1967). Blood and cerebrospinal fluid lactate during hyperventilation. *American Journal of Physiology*, 212, 864–870.

Polak, A., Haynie, G. D., Hays, R. M. & Schwartz, W. B. (1961). Effects of chronic hypercapnia on electrolyte and acid-base equilibrium. I. Adaptation. *Journal of Clinical Investigation,* **40,** 1223–1237.

Pontén, U. and Siesjö, B. K. (1966). Gradients of CO_2 tension in the brain. *Acta Physiologica Scandinavica,* **67,** 129–140.

Posner, J. B., Swanson, A. G. & Plum, F. (1965). Acid-base balance in cerebrospinal fluid. *Archives of Neurology,* **12,** 479–496.

Price, H. L. (1960). Effects of CO_2 on the cardiovascular system. *Anesthesiology,* **21,** 652–663.

Rahn, H. & Bahnson, H. T. (1953). Effect of unilateral hypoxia on gas exchange and calculated pulmonary blood flow in each lung. *Journal of Applied Physiology,* **6,** 105–112.

Rahn, H. (1962). The gas stores of the body with particular reference to carbon dioxide. In *Man's Dependence on the Earthly Atmosphere,* ed., Schaefer, K. E. pp. 297–304. New York: Macmillan.

Reivich, M. (1964). Arterial P_{CO_2} and cerebral hemodynamics. *American Journal of Physiology,* **206,** 25–35.

Robin, E. D. (1963). Abnormalities of acid-base regulation in chronic pulmonary disease, with special reference to hypercapnia and extracellular alkalosis. *New England Journal of Medicine,* **268,** 917–922.

Roughton, F. J. W. (1964). Transport of oxygen and carbon dioxide. In *Handbook of Physiology,* Section 3: Respiration, ed., Fenn, W. O. and Rahn, H. Vol. 1, Ch. 31, pp. 767–825. Washington, D.C.: American Physiological Society.

Saltzman, H. A., Heyman, A. & Sieker, H. O. (1963). Correlation of clinical and physiological manifestations of sustained hyperventilation. *New England Journal of Medicine,* **268,** 1431–1436.

Sapir, D. G., Levine, D. Z. & Schwartz, W. B. (1967). The effects of chronic hypoxemia on electrolyte and acid-base equilibrium: an examination of normocapneic hypoxemia and of the influence of hypoxemia on the adaptation to chronic hypercapnia. *Journal of Clinical Investigation,* **46,** 369–377.

Schwartz, W. B., Hays, R. M., Polak, A. & Haynie, G. D. (1961). Effects of chronic hypercapnia on electrolyte and acid-base equilibrium. II. Recovery, with special reference to the influence of chloride intake. *Journal of Clinical Investigation,* **40,** 1238–1249.

Schwartz, W. B., Brackett, N. C., Jr. & Cohen, J. J. (1965). The response of extracellular hydrogen ion concentration to graded degrees of chronic hypercapnia: The physiologic limits of defense of pH. *Journal of Clinical Investigation,* **44,** 291–301.

Severinghaus, J. W. (1958). Oxyhemoglobin dissociation curve: correction for temperature and pH variation in human blood. *Journal of Applied Physiology,* **12,** 485–486.

Severinghaus, J. W., Mitchell, R. A., Richardson, B. W. & Singer, M. M. (1963). Respiratory control at high altitude suggesting active transport regulation of CSF pH. *Journal of Applied Physiology,* **18,** 1155–1166.

Sideris, D. A., Katsadoros, D. P., Valianos, G. & Assioura, A. (1975). Type of cardiac dysrhythmias in respiratory failure. *American Heart Journal,* **89,** 32–35.

Simmons, D. H., Nicoloff, J. & Guze, L. B. (1960). Hyperventilation and respiratory alkalosis as signs of gram-negative bacteremia. *Journal of the American Medical Association,* **174,** 2196–2199.

Simmons, D. H. & Olver, R. P. (1965). Effects of acute acid-base changes on renal hemodynamics in anesthetized dogs. *American Journal of Physiology,* **209,** 1180–1186.

Stroud, R. C. & Rahn, H. (1953). Effect of O_2 and CO_2 tension upon the resistance of pulmonary blood vessels. *American Journal of Physiology,* **172,** 211–220.

Sutton, F. D., Jr., Zwillich, C. W., Creagh, C. E., Pierson, D. J. & Weil, J. V. (1975). Progesterone for outpatient treatment of Pickwickian syndrome. *Annals of Internal Medicine,* **83,** 476–479.

Sykes, M. K. & Cooke, P. M. (1965). The effect of hyperventilation on "excess lactate" production during anaesthesia. *British Journal of Anaesthesiology,* **37,** 372–379.

Tenney, S. M. & Miller, R. M. (1955). The respiratory and circulatory actions of salicylate. *American Journal of Medicine,* **19,** 498–508.

Twining, R. H., Lopez-Majano, V., Wagner, H. N., Jr., Chernick, V. & Dutton, R. E. (1968). Effect of regional hypercapnia on the distribution of pulmonary blood flow in man. *Johns Hopkins Medical Journal*, **123**, 95–103.

Tyler, J. J. (1960). The effect of progesterone on the respiration of patients with emphysema and hypercapnia. *Journal of Clinical Investigation*, **39**, 34–41.

Vanamee, P., Poppell, J. W., Glicksman, A. S., Randall, H. T. & Roberts, K. E. (1956). Respiratory alkalosis in hepatic coma. *Archives of Internal Medicine*, **97**, 762–767.

Van Vaerenbergh, P. J. J., Demeester, G. & Leusen, I. (1965). Lactate in cerebrospinal fluid during hyperventilation. *Archives Internationales de Physiologie et de Biochimie*, **73**, 738–747.

van Ypersele de Strihou, C., Gulyassy, P. F. & Schwartz, W. B. (1962). Effects of chronic hypercapnia on electrolyte and acid-base equilibrium. III. Characteristics of the adaptive and recovery process as evaluated by provision of alkali. *Journal of Clinical Investigation*, **41**, 2246–2253.

van Ypersele de Strihou, C., Brasseur, L. & De Coninck, J. (1966). The "carbon dioxide response curve" for chronic hypercapnia in man. *New England Journal of Medicine*, **275**, 117–122.

Vaughan Williams, E. M. (1955). The individual effects of CO_2, bicarbonate and pH on the electrical and mechanical activity of isolated rabbit auricles. *Journal of Physiology*, **129**, 90–110.

Wasserman, A. J. & Patterson, J. L., Jr. (1961). The cerebral vascular response to reduction in arterial carbon dioxide tension. *Journal of Clinical Investigation*, **40**, 1297–1303.

Wead, W. B. & Little, R. C. (1967). Effect of hypocapnia and respiratory alkalosis on cardiac contractility. *Proceedings of the Society of Experimental Biology and Medicine*, **126**, 606–609.

Webb, R. K., Woodhall, P. B., Tisher, C. C., Glaubiger, G., Neelon, F. A. & Robinson, R. R. (1977). Relationship between phosphaturia and acute hypercapnia in the rat. *Journal of Clinical Investigation*, **60**, 829–837.

Wendling, M. G., Eckstein, J. W., Abboud, F. M. & Hamilton, W. K. (1967). Cardiovascular responses to carbon dioxide before and after beta-adrenergic blockade. *Journal of Applied Physiology*, **22**, 223–226.

Wood, W. B., Manley, E. S., Jr. & Woodbury, R. A. (1963). The effects of CO_2-induced respiratory acidosis on the depressor and pressor components of the dog's blood pressure response to epinephrine. *Journal of Pharmacology and Experimental Therapeutics*, **139**, 238–247.

Zborowska-Sluis, D. T. & Dossetor, J. B. (1967). Hyperlactatemia of hyperventilation. *Journal of Applied Physiology*, **22**, 746–755.

6
Disorders of potassium balance

ALAN S. KLIGER
JOHN P. HAYSLETT

Potassium distribution and excretion
Factors that influence potassium distribution
   Extracellular potassium concentration
   Hydrogen ion balance
   Insulin
   Epinephrine
   Aldosterone
Regulation of potassium excretion
   Renal excretion of potassium
   Intestinal excretion of potassium
Hypokalemia
Factors and syndromes associated with redistribution of potassium
   Alkalosis
   Hypokalemic periodic paralysis
   Barium poisoning
Factors and syndromes associated with potassium deficiency
   Inadequate intake
   Excessive losses into urine
   Primary hyperaldosteronism
   Excessive deoxycorticosterone or corticosterone
   Ectopic ACTH syndrome
   Bartter's syndrome
   Excessive renin production
   Magnesium deficiency
   Renal tubular acidosis
   Liddle's syndrome
   Licorice ingestion
   Excessive losses from the gastrointestinal tract
      Loss of gastric fluid
      Diarrhea
      Villous adenoma
   Excessive losses from integument
      Sweat
      Postburn hypokalemia
Manifestations of hypokalemia
Therapy for hypokalemia and potassium deficiency
Hyperkalemia
Factors and syndromes associated with excessive tissue release or redistribution of potassium from intracellular to extracellular fluid
   Acidosis
   Hyperkalemic periodic paralysis
   Hemolysis
   Tissue necrosis
Factors and syndromes associated with excessive potassium retention
   Excessive intake
   Renal failure
   Hypoaldosteronism
   End-organ unresponsiveness
   Hyporeninemic hypoaldosteronism (SHH)
Manifestations of hyperkalemia
Therapy for hyperkalemia
Reduction of total body potassium
Artificial causes of hyperkalemia

POTASSIUM DISTRIBUTION AND EXCRETION

Potassium is known to play a key role in numerous biochemical processes and in the excitatory properties of membranes of certain cell types, such as muscle and nerve cells. It is not unexpected, therefore, that organ dysfunction should result from alterations in the total body stores of potassium or in its distribution across cell membranes. Since an imbalance in the metabolism of this ubiquitous cation is a common phenomenon, an understanding of the type of change in potassium homeostasis (whether due to a variation in distribution or overall balance) and of the cause of the abnormality underlie the management of many clinical disorders. In this chapter we have attempted to provide an approach to the diagnostic evaluation of patients with a change in potassium homeostasis. The two categories of hypokalemia and hyperkalemia have been used to identify the nature of the problem because of common clinical usage and because of the ease with which the concentration of potassium in extracellular fluid is measured.

On the basis of isotope dilution and total body counting of naturally occurring ^{40}K, (Aikawa et al., 1952; Anderson, 1963; Corsa et al., 1950; Forbes and Lewis, 1956), total body potassium content has been estimated at 31–57 mEq/kg body weight, or approximately 3,500 mEq. Figure 6.1 illustrates the distribution between extracellular and intracellular sites. Since the concentration of potassium in extracellular fluid is approximately 4 mEq/l, nearly 98 percent of total body potassium is located within cells.

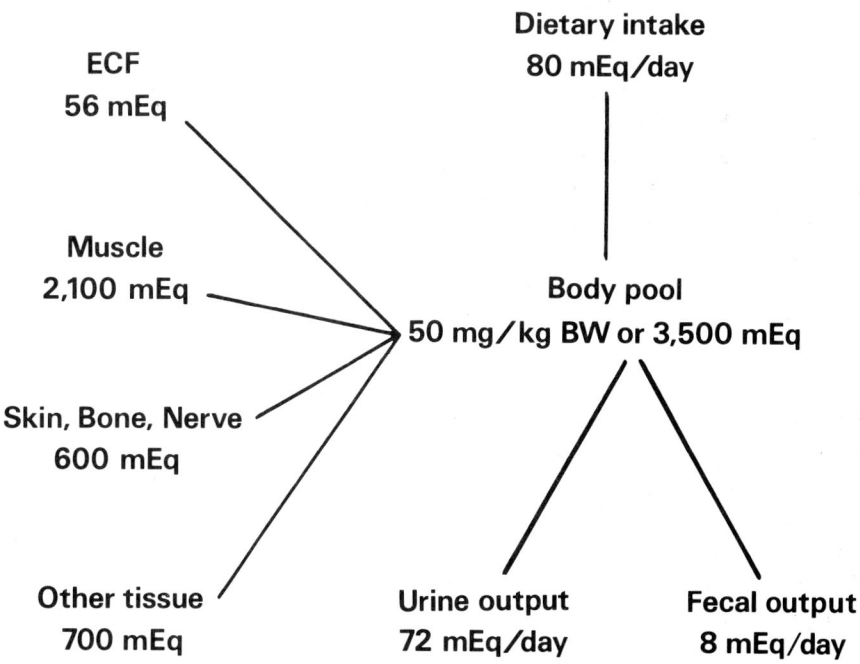

Figure 6.1 Distribution and balance of potassium in the body.

As a result of this distribution, potassium is the major intracellular cation, in contrast to extracellular fluid, where sodium predominates.

Cellular concentration of free potassium has been more difficult to ascertain, although it seems likely that the concentration of free potassium may not equal total potassium, determined by chemical analysis. In distal tubular cells, for example, potassium activity, estimated by ion sensitive electrodes, was reported to be 46.5 mM/l cell water, while 136 mM of total potassium content was found by chemical analysis of renal cortical slices (Khuri, 1972).

The characteristics of high potassium and low sodium, intracellularly, is common to all mammalian cells and has been shown to be regulated, in part, by a Na-K-ATPase pump mechanism in the cell membrane (Bonting and Caravaggio, 1963) and by electrochemical forces acting across the membrane. Since the cell membrane is more permeable to potassium than to sodium, the electrical asymmetry existing across the membranes plays an important role in modulating its distribution. These factors have important clinical implications since alterations in the Na-K exchange mechanism or passive forces responsible for alterations in potassium distribution may result in abnormally low or high levels in extracellular fluid.

Potassium homeostasis is characterized by the maintenance of total body stores within narrow limits, by achieving a zero net balance between input and loss, and by regulation of its distribution between cellular and extracellular sites. An understanding of the factors that regulate these processes is, therefore, essential to an analysis of clinical states in which homeostasis is changed in some way. A detailed evaluation of the handling of potassium by the kidney is provided in Chapter 1.

Factors that influence potassium distribution

Extracellular potassium concentration

Mechanisms that control the rate of cellular uptake of potassium are modulated, in part, by the concentration of potassium in extracellular fluid. During the acute infusion of potassium salts, a maneuver that raises the concentration of potassium in extracellular fluid and the load delivered to transporting epithelium, the rate of potassium secretion increases markedly in colon and in the distal portions of the nephron (Bastl, Hayslett and Binder, 1977; Schon, Silva and Hayslett, 1974). Under the same conditions, the rate of cellular uptake of potassium by nontransporting cells is probably also increased (Alexander and Levinsky, 1968).

Hydrogen ion balance

Changes in acid-base balance are known to influence the level of potassium in extracellular fluid, independently of changes in overall potassium balance. Plasma potassium levels fall in states characterized by alkalosis, with a fall in the level of hydrogen ion and rise during acidosis. Fenn and Cobb studied muscle in vitro and demonstrated that potassium diffusion into and out of muscle cells correlates inversely with the exchange of hydrogen ion. Mudge and Vislocky (1949) subsequently showed that metabolic acidosis was associated with a decrease in intracellular potassium in human striated muscle. On the basis of recent studies, evidence has been provided to suggest that bicarbonate levels, rather than hydrogen ion levels, are primarily responsible for regulating the movement of potassium across cell membranes. For example, in a report by Fraley and Adler (1976), potassium distribution was found to correlate with extracellular levels of bicarbonate, independently of pH, at high and low levels of plasma potassium. Plasma pH was maintained at constant levels by altering pCO_2.

Insulin

Harrop and Benedict (1923) first observed that insulin administration decreased the serum potassium concentration. Conversely, the acute infusion of KCl in man and experimental dogs results in increased levels of insulin in blood (Dluhy, Axelrod and William, 1972; Hiatt, Davidson and Bonorris, 1972). Further, insulin has been shown to increase potassium uptake in liver, muscle, and adipose tissue (Andres et al., 1962; Gourley and Bethea, 1964; Mortimore, 1961). The role of insulin in influencing the rate of cellular uptake of potassium, during acute loading with KCl, was studied by Hiatt (1973). He demonstrated a significant effect of insulin on net removal of potassium from extracellular fluid in dogs during infusion of KCl after renal excretion was eliminated by ligation of the ureters. In additional experiments, the rate of cell uptake of potassium was accelerated in pancreatectomized dogs after administration of insulin. Insulin was shown to augment both the rate of uptake and total amount of potassium in cellular stores.

While the mechanism of insulin-induced potassium cell uptake is not clear, there is evidence to suggest that such movement is independent, at least in part, of glucose entry into cells. In studies in vitro, insulin-induced potassium uptake into muscle cells occurred in the absence of glucose in bathing medium (Zierler, 1960). Moreover, net increase of muscle potassium may be induced by insulin levels that are too small to increase glucose uptake.

Epinephrine

The administration of epinephrine results in a biphasic response to plasma potassium, characterized by a brief rise and then a more prolonged fall (D'Silva, 1934; Vick, Todd and Leudke, 1972). The initial rise was shown to result from hepatic release of potassium (D'Silva, 1936) and to be dependent upon an interaction between both alpha- and beta-adrenergic receptors (Todd and Vick, 1971). Since epinephrine does not induce a kaliuresis, the sustained fall in plasma potassium produced by epinephrine infusion is presumably caused by increased intracellular potassium uptake. An enhanced uptake of potassium into both liver and skeletal muscle was demonstrated in dogs after epinephrine infusion (Vick et al., 1972). Since epinephrine-induced hypokalemia was reversed by propranalol, it has been suggested that the action of that sympathomimetic agent is mediated by adrenergic beta receptors.

Aldosterone
In addition to the well-known action of aldosterone to increase potassium excretion by distal portions of the nephron, salivary glands, and colon, aldosterone may influence the rate of cell uptake of potassium by nontransporting tissues. Studies by Alexander and Levinsky (1968) suggest that chronic potassium loading may increase the capacity of such tissue to extract potassium from extracellular fluids during acute loading with KCl. After nephrectomy, to obviate renal excretion, the rise in plasma potassium during acute loading experiments was significantly less in animals chronically adapted to a high potassium diet, compared with controls. The difference between experimental and control groups was abolished by adrenalectomy. In addition, the chronic administration of mineralocorticoids to intact animals that were on a normal potassium intake produced the same changes in capacity for cellular uptake of potassium as induced by a high potassium diet. The mechanism of this mineralocorticoid-mediated adaptation remains uncertain. One hypothesis proposes that chronic hyperaldosteronism induces a two or three percent depletion of muscle potassium content, creating a potential reservoir for potassium (Adler, 1970). Subsequent potassium loading may result in repletion of muscle potassium thus preventing extracellular hyperkalemia. Another study (Alexander and Levinsky, 1968), however, does not confirm an intracellular depletion of potassium during chronic aldosterone administration.

Regulation of potassium excretion

Renal excretion of potassium
Urinary potassium is derived primarily from potassium secreted by cells in the distal nephron. Berliner, Mudge and colleagues (Berliner, 1961; Davidson, Levinsky and Berliner, 1958; Mudge, Foulks and Gilman, 1948) demonstrated that urine potassium excretion is not dependent upon the filtered load of potassium and suggested that most potassium filtered at the glomerulus is reabsorbed in the proximal tubule. They further hypothesized that potassium secretion occurs at more distal sites, producing the bulk of urinary potassium, and that those secretory sites are responsible for maintaining absolute potassium excretion. Subsequent studies have confirmed their hypothesis. In stop flow experiments (Malvin, Wilde and Sullivan, 1958; Pitts et al., 1958), higher potassium concentrations were found in the more distal parts of the nephron, compared with the proximal portions. Further, following injection of radioactive ^{42}K, the specific activity in the urine approached the activity in tubular tissue rather than plasma, suggesting that urinary potassium excretion is more closely related to tubular

function than to glomerular filtration (Black et al., 1956). The relative importance of potassium secretion by distal nephron segments was finally established by an elegant series of micropuncture studies which demonstrated that changes in the fractional excretion of potassium delivered out of the proximal tubule have a minimal influence on the potassium content of final urine (Malnic, Klose and Giebisch, 1964, 1966). In the absence of diuretic agents or an osmotic agent, approximately 90 percent of filtered potassium is reabsorbed by the proximal tubule and loop of Henle. More distal sites exhibit a capacity for further reabsorption or secretion depending on potassium balance and other systemic conditions.

The relative importance of the secretory process in the distal tubule and collecting duct is unclear at the present time. Micropuncture studies have shown that under some conditions most potassium in the final urine can be accounted for by potassium secreted in the distal tubule segment. For example, secretion of potassium by distal tubule predominates during acute loading with KCl in both normal animals and in animals fed a high potassium diet (Malnic et al., 1964; Wright et al., 1971). Conversely, secretion by cells lining the collecting duct have been shown to play a primary role in determining urinary potassium in animals with hyperaldosteronism, induced by a diet chronically low in sodium (Wright et al., 1971), and in renal insufficiency, caused by surgical ablation of renal tissue (Bank and Aynedjian, 1973). The importance of medullary structures, presumably medullary collecting duct, in maintaining high levels of fractional excretion of potassium in renal insufficiency, has also been supported by studies in subjects with selective damage to the inner medulla. During acute infusion of KCl, the capacity to excrete potassium was blunted in rats with surgical papillectomy (Finkelstein and Hayslett, 1974). If the collecting duct and intact deep medullary structures are as important to potassium balance in azotemic humans as in the rat, individuals with renal disease involving primarily those deep medullary structures may be more likely to develop disorders in potassium homeostasis. Evidence to this effect in patients with sickle cell anemia has been described (DeFronzo et al., 1977).

Although a detailed analysis of the renal handling of potassium is provided in Chapter 1, it is pertinent to our discussion to delineate briefly the three major factors that increase renal potassium excretion.

Mineralocorticoids act to stimulate potassium excretion and reduce the urinary excretion of sodium (Williamson, 1963). Potassium retention and sodium wasting is a well-known consequence of adrenalectomy, a pattern that is reversed after supplemental doses of aldosterone are administered (Barger, Berlin and Tulenko, 1958). It is important to recall, however, that the action of mineralocorticoids on potassium and sodium transport may be dissociated. For example, the administration of small amounts of mineralocorticoid may stimulate sodium reabsorption, in the absence of a change in potassium secretion (Swingle et al., 1954). In addition, long-term administration of mineralocorticoids is characterized by persistent kaliuresis, while an 'escape' of the natriuretic effect occurs after a week or so (August, Nelson and Thorn, 1958).

The site of action of mineralocorticoids on specific segments of the nephron is less clear. Micropuncture studies have demonstrated an action to increase potassium secretion and sodium absorption in both distal tubule and collecting duct (Wiederholt et al., 1966; Uhlich, Baldamus and Ullrich, 1969). Recent studies, using different techniques, have been unable to show that aldosterone influences ion transport in early distal tubule, although prominent changes were demonstrable in cortical collecting duct (Gross and Kokko, 1977). It is possible, therefore, that the action of mineralocorticoids on renal tubular function is confined to the cortical and medullary collecting duct.

The second important influence on potassium excretion involves the acid-base status of the subject. Acute acidosis decreases and alkalosis increases the rate of potassium secretion by the renal distal tubule. These changes in secretory rate may correlate directly with similar changes in intracellular potassium activity (Giebisch, Berger and Pitts, 1955; Scribner, Fremont-Smith and Burnell, 1955; Swan and Pitts, 1955) although such a correlation cannot always be found (Burnell et al., 1956). More recent studies, however, have emphasized the complexity of the association between the pH of plasma and renal potassium excretion. For example, although potassium excretion immediately decreases after the induction of metabolic acidosis, the chronic administration of an acid load is associated with a kaliuresis (Gennari and Cohen, 1975).

Lastly, although potassium excretion does not rise during water diuresis (Ali, Cross and Pickford, 1958), maneuvers that increase the delivery of solute and water to distal nephron sites enhance both net distal tubular potassium secretion and final total potassium excretion (Morgan and Berliner, 1969). It seems likely that under the condition of volume expansion or urea loading (Khuri et al., 1975), both increased flow rate along the distal tubule and blunted potassium absorption by collecting duct are responsible for this kaliuresis. Conversely, measures that reduce distal delivery of solute and water result in significant reduction of potassium excretion and blunting or abolition of mineralocorticoid kaliuresis (Relman and Schwartz, 1952).

Intestinal excretion of potassium

The fecal excretion of potassium constitutes the second major pathway for the elimination of dietary potassium and for maintaining potassium balance. On a normal diet, in health, approximately 10 to 15 percent of dietary potassium is found in fecal excretion (Berger, 1960). The relative importance of the intestinal pathway, however, may increase when the capacity for urinary excretion is reduced. Hayes (1967) demonstrated that fecal potassium may represent as much as 35 percent of the oral intake in patients with severe renal failure.

Intestinal handling of potassium corresponds, in general, with the transport characteristics of the nephron. Potassium is absorbed in small bowel, presumably by passive forces, while the final regulation of intestinal excretion occurs in the distal segment and is determined by colonic secretion. The mechanism that modulates potassium secretion in colon, whether by an active or passive mechanism, has not been clearly established.

It is known, however, that mineralocorticoids act to increase potassium secretion and sodium absorption in colon and increase the luminal potential difference (Dolman and Edmonds, 1975). After adrenalectomy, these transport parameters are reduced below normal levels (Fisher, Binder and Hayslett, 1976). The intestinal route, therefore, becomes potentially important as a source for excessive losses of potassium in states characterized by increased aldosterone production. In addition, the high content of potassium in mucus and diarrheal stool are also important sources of excessive intestinal excretion of potassium.

HYPOKALEMIA

This condition is usually defined as serum potassium of less than 3.0 mEq/l. The low level of potassium in extracellular fluid can result from a redistribution of potassium, favoring intracellular sites, or may represent a reduction in total body potassium from excessive losses. The major factors or syndromes associated with hypokalemia from both causes are shown in Table 6.1.

Table 6.1 Causes of hypokalemia owing to alterations in potassium distribution and potassium deficiency

---

I. **Factors and syndromes associated with redistribution of potassium from extracellular fluid to intracellular fluid**
   A. Alkalosis—respiratory and metabolic
   B. Hypokalemic periodic paralysis
   C. Barium poisoning
II. **Factors and syndromes associated with potassium deficiency**
   A. Excessive losses into urine
      1. Diuretic drugs
      2. Excessive production of mineralocorticoids
      3. Bartter's syndrome
      4. Excessive renin production
      5. Magnesium deficiency
      6. Renal tubular acidosis
      7. Liddle's syndrome
      8. Licorice ingestion
   B. Excessive losses from gastrointestinal tract
      1. Loss of gastric fluid
      2. Diarrhea
      3. Villous adenoma
   C. Excessive losses from integument
      1. Sweat
      2. Postburn hypokalemia

Factors and syndromes associated with redistribution of potassium

Alkalosis

Both respiratory and metabolic alkalosis result in a fall in plasma potassium owing, in part, to an increased cellular uptake. Experimental evidence suggests that potassium enters cells in exchange for hydrogen ion. Recognition of this cause of hypokalemia is important since administration of potassium salts is not indicated to correct the laboratory finding and, indeed, may be contraindicated. It should also be noted that since hypokalemia is frequently found in critically ill patients with respiratory alkalosis, sensitivity of cardiac muscle to digitalis is increased and skeletal muscle and nervous tissue irritability is reduced in the presence of hypokalemia.

Hypokalemic periodic paralysis

This rare disorder, recently reviewed (Pearson and Kalyanaraman, 1972; Streeten, 1966), affects males more often than females. It is transmitted by autosomal dominance and is characterized by attacks of flaccid paralysis of the limbs and trunk lasting 6 to 24 hours. The plasma potassium level falls with onset of an attack, remains low during paralysis, and rises with recovery. During attacks the overall balance of potassium is positive, indicating that hypokalemia results from a shift of extracellular potassium into cells. Following the attack, potassium is released from cellular stores and a kaliuresis returns balance to the preattack level. Between attacks, total body potassium is low normal or low. Episodes of paralysis in predisposed patients can be precipitated by the administration of glucose and insulin, ACTH, some nonaldosterone mineralocorticoids, and a high carbohydrate–low potassium diet. Each of these regimens cause a fall in plasma potassium levels.

The underlying cause of this disorder and the factor responsible for precipitating attacks remain unknown. It is of interest that while ACTH and certain mineralocorticoids can induce an attack, the administration of glucocorticoids and aldosterone do not produce this same effect (Streeten and Speller, 1974). Furthermore, spironolactone can ameliorate an insulin-glucose–induced attack; and metyrapone and 11- and 18-hydroxylation blockers can reduce the severity of ACTH-induced attacks. On the basis of these observations, it has been suggested that a nonaldosterone mineralocorticoid, perhaps a precursor of aldosterone, may be responsible for inducing spontaneous attacks. Although the administration of potassium during an attack may accelerate improvement, the efficacy of prophylactic treatment to reduce the frequency of attacks has not been established.

Barium poisoning

Ingestion of soluble barium salts can result in a clinical picture that is similar to hypokalemic periodic paralysis. In addition to flaccid skeletal paralysis, however, barium poisoning may also result in respiratory paralysis.

Other symptoms include hypertension, vomiting, diarrhea, ventricular tachycardia, and fibrillation. Hypokalemia has been demonstrated in such patients (Diengott et al., 1964) and experimental work suggests that the low plasma level is due to a shift of potassium from the extra- to the intracellular sites (Roza and Berman, 1971).

Factors and syndromes associated with potassium deficiency

As indicated in Table 6.1, many conditions are associated with potassium deficiency because of excessive losses into the urine and feces or through the integument. As reported by Black (1967), a plasma level of potassium of 3.0 mEq/l, in the absence of acid-base imbalance, is usually the result of a deficiency of at least 100 to 200 mEq of potassium. When plasma potassium is lower than 3 mEq/l, each additional one mEq/l decrease of plasma potassium will reflect an additional deficit of 200–400 mEq (Scribner and Burnell, 1956). Hypokalemia, associated with electrocardiographic changes of that ionic disturbance, usually indicates a deficiency of at least 500 mEq.

Inadequate intake

While renal and gastrointestinal conservation of potassium is efficient, 5–10 mEq may be excreted daily even under hypokalemic conditions. Thus a chronic diet severely depleted of potassium could result in potassium deficiency and hypokalemia. However, in the absence of starvation or intractable vomiting, such dietary potassium restriction is unusual. Hypokalemia, therefore, usually represents redistribution or loss of total body potassium.

Excessive losses into urine

Diuretic drugs: Since 'diuretic drugs' are widely used to treat edematous conditions and are employed as antihypertensive agents, the kaliuresis associated with their use is a common clinical phenomenon. Thiazide derivates and the powerful 'loop' diuretics, furosemide and ethacrynic acid, may induce substantial urinary losses of potassium. The mechanism responsible for their kaliuretic effect is partially understood. Both categories of drugs inhibit NaCl reabsorption predominantly in the ascending limb of Henle's loop and increase the load of solute and water delivered to more distal segments of the nephron. Absolute potassium secretion is stimulated to increase in the distal convolution under these conditions (Morgan and Berliner, 1969). Further, these agents may directly influence the rate of potassium secretion by a mechanism independent of flow rate and solute delivery rate, although this action has not been established. Lastly, diuretic-induced metabolic alkalosis can be expected to enhance potassium secretion in the distal tubule.

While these agents may produce a substantial kaliuresis, their effect on plasma potassium and total body stores of potassium vary considerably amongst different groups of patients. In a study of 1,294 patients treated with potassium-depleting diuretics, less than five percent had reduced plasma potassium concentrations, and none developed serious complications of hypokalemia (Lawson, 1974). While edematous individuals have been shown to exhibit reduced stores of body potassium, potassium supplementation does not replete these potassium losses (Down et al., 1972). In these patients, treatment with loop diuretics produces an increase in urinary loss of potassium and thus, more significant potassium depletion may be induced. Despite substantial depletion of total potassium, plasma potassium levels are usually not decreased.

In contrast to the action of diuretic agents in edematous patients, nonedematous hypertensive patients treated with diuretics may more frequently develop a reduction of plasma potassium of from 0.3–1.2 mEq/l. However, total body potassium undergoes little or no reduction in this group treated with thiazides or furosemide (Healy et al., 1970; Wilkinson et al., 1975). The significance of diuretic-induced hypokalemia is further confused by reports that the induced hypokalemia may be transitory (Gifford et al., 1961).

Thus, dietary potassium supplements may not be necessary for most patients treated with diuretics. Nonedematous patients who maintain a plasma potassium above 3 mEq/l and have an adequate dietary intake of potassium should have plasma potassium levels measured at intervals, but they probably do not require potassium supplements (Dargie et al., 1974). In edematous patients, potassium supplementation may not increase total body stores. In patients receiving digitalis, in those with hyperaldosteronism, or in those prone to develop hepatic coma, potassium depletion may be present and should be treated even in the absence of reduced plasma levels. Oral potassium salt supplementation, or use of the aldosterone antagonist spironolactone, may be indicated to prevent further depletion of total body potassium.

Excessive production of mineralocorticoids: This category includes several conditions involving exogenous administration of mineralocorticoids or their excessive endogenous production by the adrenal gland or ectopic sites. Most of these conditions are associated with low plasma renin levels.

Primary hyperaldosteronism

Conn (1955) first described the syndrome in which hypertension was accompanied by significant hypokalemia and metabolic alkalosis, due to excessive production of aldosterone. Urinary aldosterone excretion was remarkably increased, while excretion of 17-ketosteroids and 17-hydroxycorticosteroids was normal. Surgical exploration revealed an adrenal adenoma, and following its removal, the elevated blood pressure and metabolic abnormalities returned to normal. Additional symptoms in patients with this disorder include muscular weakness, nocturia-polyuria, headache, polydypsia, and the occasional appearance of tetany, paresthesias, and paralysis. While this is a relatively uncommon cause of hypertension, the detection of primary hyperaldosteronism is important since improvement or remission in the hypertension has been reported in as many as 95 percent of patients following removal of the adenoma (Conn, Knopf and Nesbit, 1964).

The clinical diagnosis of the disorder should be suspected in any hypertensive patient with hypokalemia, not receiving diuretic agents. Indeed, a plasma potassium level should be determined in all hypertensive patients before diuretic therapy is begun. Once the diagnosis is suspected, evaluation includes eliminating possible causes of secondary hyperaldosteronism and eliciting evidence for high aldosterone production rates and low plasma renin levels. Urinary excretion rates of aldosterone and plasma renin levels are determined during chronic salt loading and are then compared with levels found in normal controls that are studied under similar conditions. The urinary excretion rates of 17-hydroxycorticosteroids and 17-ketosteroids are also measured since these values are usually elevated in the small percentage of individuals in whom the tumor is malignant.

Surgical treatment is preferred by most groups, although the aldosterone-blocking agent spironolactone may successfully control all manifestations of the disorder. Preoperative diagnosis has been aided by arteriography, adrenal venography, and selective blood sampling and the adrenal imaging with ^{131}I-19-iodocholesterol (Brillon, 1976).

In some patients with the clinical picture of Conn's syndrome, surgical exploration has revealed bilateral adrenal hyperplasia or normal appearing adrenal glands (Baer et al., 1970; Biglieri, Stockigt and Schambelan, 1972). Although uncommon, this group is important since bilateral adrenalectomy is usually unsuccessful in relieving hypertension. It is of interest that in some patients in this category, hyperaldosteronism can be suppressed to normal range by administration of deoxycorticosterone acetate (DOCA), although a similar suppression is not found in the adenomatous type of primary aldosteronism.

Excessive deoxycorticosterone or corticosterone

An excess of these mineralocorticoids may cause a syndrome characterized by hypertension, hypokalemia, and metabolic alkalosis (Biglieri et al., 1972). Plasma renin levels and aldosterone secretion rates are normal or low. This condition may result from the exogenous administration or excess secretion of desoxycorticosterone acetate (DOCA) and ACTH (Schambelan, Slaton and Biglieri, 1971). In addition, as will be noted below, an excess adrenal production of corticosterone may be found in adrenal hyperplasia.

Congenital adrenal hyperplasia: Two forms of congenital adrenal hyperplasia may cause hypokalemia.

1. *11-Hydroxylase deficiency*—In 1956, Eberlien and Bongiovanni (1956) described a female pseudohermaphrodite with systemic hypertension, elevated levels of 11 deoxycortisol (compound S) and decreased serum cortisol.

Replacement doses of cortisol suppressed the level of compound S and associated metabolites, hypertension improved, and virilization was suppressed. This patient was found to have 11β-hydroxylase deficiency. Other patients, with and without hypertension but with a similar defect, have subsequently been identified. Excess mineralocorticoid production in these patients is presumed to be responsible for the hypertension and hypokalemia.

2. *17-Hydroxylase deficiency*—Biglieri, Herron and Brust (1966) first described four females with hypogonadism, hypertension, hypokalemia, and metabolic alkalosis. Serum cortisol levels were negligible while levels of corticosterone and deoxycorticosterone were high. Since the excretion of urinary 17-ketosteroids and estrogens were absent, a defect in 17-hydroxylase was suggested. Subsequent reports of other patients with this defect have been published. The metabolic abnormality in potassium metabolism is probably due to the excess production of DOC and corticosterone.

Acquired adrenal hyperplasia—Cushing's syndrome: Cushing's syndrome is caused by increased production of 17-OH steroid and cortisol owing to primary adrenal hyperplasia or an increase in ACTH elaborated by the pituitary gland. Hypokalemia may be an associated feature. While the rise in urinary potassium excretion may result from excessive production of mineralocorticoids, recent reports indicate that hypertension and hypokalemic alkalosis may be caused by the high secretion rate of glucocorticoids (Christy and Laragh, 1961; Krakoff, Nicholis and Amsel, 1975).

It is of interest that in reported series of patients with Cushing's syndrome, hypokalemia, often in association with alkalosis, has been found in only 25 to 40 percent of patients (Christy and Laragh, 1961; Krakoff et al., 1975). Although some cases with alterations in potassium metabolism have ultimately been shown by surgery to have primary adrenal hyperplasia, the majority of patients had malignant adrenal neoplasms or nonadrenal tumors (Christy and Laragh, 1961).

Ectopic ACTH syndrome

ACTH produced by tumors of nonendocrine organs can result in gross elevation of plasma cortisol levels, with Cushingoid appearance, hypertension, and edema, and with a marked hypokalemic alkalosis (Rees, 1975). Although Liddle has coined the term "ectopic ACTH syndromes" to describe such cases (Liddle et al., 1965), subsequent studies have shown that Cushing's syndrome may result from the ectopic production of corticotrophin releasing factor (Upton & Amatruda, 1971). Evidence that the majority of tumors associated with Cushing's syndrome produce ACTH is based on the finding of (1) increased A-V gradient of ACTH levels across the tumor, (2) demonstration of ACTH in tumor by specific antisera, and (3) in vitro release of ACTH from clones of tumor cells.

In a review of the literature, Azzopardi and Williams (1968) found that ACTH-producing tumors fell into four general categories: (1) oat cell carcinoma of the bronchus; (2) endocrine tumors of foregut origin (bronchial carcinoids, islet cell pancreatic carcinomas, epithelial thymomas, medullary carcinomas of the thyroid, and possibly parathyroid carcinomas); (3) pheochromocytomas and related tumors; and (4) certain ovarian tumors (e.g., arrhenoblastomas). Among these, oat cell carcinoma of the lung most commonly causes this syndrome, with a reported incidence as high as 22 percent (Rees and Ratcliffe, 1974). It is pertinent that immunoreactive and biologically active ACTH has been found in tumor tissue of this type in patients without the overt features of Cushing's syndrome. In one study, for example, immunoreactive ACTH was present in 93 percent of extracts of primary lung tumors, and in 88 percent of extracts of metastatic tumor to the lung (Gerwirtz and Yalow, 1974). It is not clear whether the absence of the clinical features of Cushing's syndrome was due to lack of release of the polypeptide, or some other factor. The cause of hypokalemic alkalosis in the "ectopic" ACTH syndrome has not been established with certainty, but it is probably related to high levels of DOC and corticosterone (Schambelan, 1971). These tumors are usually autonomous and are rarely suppressed with dexamethasone.

Bartter's syndrome*

This clinical disorder is characterized by a normal blood pressure with markedly elevated production rates of aldosterone and renin. Patients with this syndrome also develop hypokalemia due to severe potassium deficiency and metabolic alkalosis; they are nonedematous (Bartter, 1962). Hyperplasia of the renal juxtaglomerular apparatus in the absence of an adrenal tumor is characteristic of this disorder. In recent studies, an abnormality in sodium metabolism has been described which results in excessively large urinary losses of sodium and a tendency toward volume contraction (Tomko, Yeh and Falls, 1976). Volume repletion induces an increased sodium excretion that tends to maintain the volume-depleted state (White, 1972). The hypokalemia in this group of patients is usually profound (plasma potassium often less than 2.5 mEq/l) and often accounts for presenting symptoms of weakness, muscle stiffness, and tetany. Excessive urinary losses of potassium persist despite marked hypokalemia (Cannon et al., 1968), adrenalectomy (Trygstad et al., 1969), and treatment with aminoglutethimide (Goodman, Vagnucci and Hartroft, 1969). A low sodium diet will reduce urinary potassium excretion, but only at the expense of volume contraction.

The pathogenesis of this disorder is unclear. Bartter and colleagues (1962) found vascular unresponsiveness to infusion of exogenous angiotensin II and suggested that the primary defect was related to that observation. In their schema, increased renin production was a compensatory response to maintain blood pressure and incidentally stimulated aldosterone production. Evidence to support this hypothesis includes the finding that volume expansion reduces renin and aldosterone levels but does not correct the vascular unresponsiveness (Solomon and Brown, 1975). In other studies, however,

---

* See also Chapter Nine.

aggressive salt and fluid loading resulted in volume expansion, fall in plasma renin activity, and return of normal angiotensin sensitivity (Fujita et al., 1977; White, 1972). In addition, impaired responsiveness to infused angiotensin has been observed in other high renin conditions.

A second proposed mechanism concerns the possibility of a primary renal tubular abnormality in sodium and potassium conservation (Cannon et al., 1968; Erkelens and Statius Van Eps, 1973; White, 1972). In most patients during balance studies, evidence for this hypothesis includes defective sodium conservation, associated with high rates of urinary potassium and acid excretion. Using clearance studies, a defect in sodium reabsorption in both proximal and distal nephron segments has been suggested (Bartter et al., 1974). Inappropriately high rates of sodium excretion may occur even after volume contraction has developed (White, 1972). It should be noted, however, that impaired sodium handling by the kidney is not found in all patients with Bartter's syndrome (Goodman et al., 1969; White, 1972). Moreover, some features of altered kidney function such as the defect in concentrating ability and maximal acidification may be caused by potassium deficiency itself, since correction of these transport defects may occur after potassium repletion. Furthermore severe potassium deficiency may decrease both renin and aldosterone activity, yet these may again become elevated and urinary potassium excretion increase with small doses of supplemental potassium that are insufficient to correct the hypokalemia (Cannon et al., 1968). In addition, other patients with Bartter's syndrome have been described with hypercalciuria (Fanconi et al., 1971), suggesting that patients with this disorder may not be a homogeneous group.

Recent studies have suggested a third possible mechanism involving prostaglandin production (Gill et al., 1976; Halushka et al., 1977; Verberckmoes et al., 1976). In a case report of a patient with Bartter's syndrome (Verberckmoes et al., 1976), treatment with indomethacin, an inhibitor of prostaglandin synthesis, returned the high renin and aldosterone levels to normal, restored responsiveness to angiotensin, and corrected the hypokalemia. Subsequently, Gill and colleagues (1976) demonstrated elevated levels of urinary prostaglandin E_2. After treatment with indomethacin, the reduced urinary excretion of prostaglandin was correlated with a fall in plasma renin levels and urinary excretion of aldosterone. It was of interest, however, that hypokalemia persisted in those patients despite a positive metabolic balance for sodium and potassium.

Excessive renin production

Since renin is a stimulus for adrenal production of aldosterone, it is not surprising that hypokalemia, because of excessive urinary losses of potassium, is found in several conditions characterized by high levels of plasma renin.

Renal vascular hypertension: Reduced renal blood flow, owing to partial obstruction of renal vessels, stimulates production of renin and may be associated with secondary hyperaldosteronism (Hayslett, Flinn & Mulrow, 1963). This sequence may result in hypokalemia and metabolic alkalosis. The clinical features in this group of patients, therefore, closely resemble primary hyperaldosteronism. In a cooperative study of 175 cases of surgically proven renovascular hypertension, hypokalemia (plasma potassium less than 3.4 mEq/l) was found in 16 percent, a value that was significantly greater than present in matched control subjects with essential hypertension (Simon et al., 1972). In addition, a 17 percent incidence of metabolic alkalosis was demonstrated in the group with renovascular hypertension.

A corrollary to the patient study has been established in an animal model of renovascular hypertension, produced by unilateral renal artery constriction (Möhring et al., 1975). Following the initial phase of sodium retention found in all animals, some exhibited persistent sodium retention, hypokalemia, and exaggerated rates of urinary potassium excretion. Plasma renin levels and aldosterone excretion rates were markedly elevated in the group with hypokalemia.

Malignant hypertension: Patients with malignant hypertension (diastolic pressure greater than 120 mmHg) have a high incidence (20 percent) of hypokalemia (plasma potassium lass than 3.5 mEq/l) (Wrong, 1961). The relative frequency of this complication apparently correlates with the severity of the hypertension, since the metabolic abnormality is found in 43 percent of subjects with papilledema. Although the cause of the disorder in potassium homeostasis has not been systematically analysed, it is probably related to the high rates of aldosterone production exhibited by patients with malignant hypertension (Holten and Petersen, 1956).

Renin-secreting tumor: An intrarenal renin-secreting tumor has been demonstrated (Conn et al., 1972) and is associated with hypertension, metabolic alkalosis, hypokalemia, and elevated production rates of renin and aldosterone. Postoperatively the release of large amounts of renin from tumor tissue has been demonstrated in vitro. Histologic examination revealed adenomatous transformation of juxtaglomerular tissue.

Preoperative evaluation involves selective renal arteriography to exclude the presence of renal artery stenosis. It may also reveal the intrarenal tumor. Moreover, a study of renin levels before and after volume expansion may help to distinguish renal vascular hypertension from hypertension owing to renin-secreting tumor, since renin production is partially suppressed in patients with vascular stenosis. In contrast, renin production is autonomous and is not influenced by salt loading in patients with a tumor. In a reported instance of an ectopic renin-secreting carcinoma (Genest et al., 1975), hypokalemia and hypertension were found, and renin secretion could not be suppressed by propranalol.

In patients with high, nonsuppressible renin hypertension, arteriography is often obtained to search for renal vascular stenosis. In the absence of such a lesion, bilateral renal vein renin determinations should follow since renin-producing tumors are not always observed roentgenographically (Conn et al., 1972).

Magnesium deficiency

The first report of hypokalemia associated with hypomagnesemia described three adult patients with metabolic alkalosis, impaired urinary conservation of potassium and magnesium, and a chronic dermatitis (Gitelman, Graham and Welt, 1966). Further study failed to demonstrate an increased excretion of mineralocorticoids, the presence of Fanconi's syndrome or an alteration in renal handling of sodium. Administration of neither potassium nor magnesium alone served to correct the excessive urinary losses of these cations. The authors suggested a specific tubular transport defect as an explanation for the clinical findings.

Subsequent reports have described a similar defect in children, which is associated with altered calcium excretion (Spencer and Voyce, 1976) and maximal concentrating ability (McCredie et al., 1971). It is of interest that in patients (Shils, 1969) and experimental animals (Whang, 1967) with primary magnesium deficiency, excessive urinary excretion of potassium is found and can be restored to normal only after correction of the magnesium defect.

Renal tubular acidosis*

Hypokalemia, due to excessive urinary losses of potassium, may occur in patients with a renal-acidifying defect involving both the proximal tubule (proximal RTA, type 2) and the distal tubule (distal RTA, type 1). Proximal RTA is often accompanied by a generalized impairment of proximal tubular function and results in hypokalemia, glucosuria, phosphaturia, and bicarbonaturia. Correction of systemic acidosis by the administration of alkali usually increases urinary potassium excretion (Sebastian, McSherry and Morris, 1971). Presumably, the mechanism responsible for high levels of urinary potassium involves a decrease in potassium reabsorption in the proximal tubule and an increase in secretion in distal portions of the nephron. Elevated rates of secretion are probably caused by the rise in sodium and bicarbonate loads delivered to the distal tubule. Administration of alkali increases distal delivery further.

An inappropriate urinary excretion of potassium may also be seen in distal RTA. In contrast to proximal RTA, correction of the acidosis with exogenous alkali usually reduces urinary potassium excretion, probably through correction of volume depletion and secondary hyperaldosteronism that is often found in untreated subjects with that disorder (Gill, Bell and Bartter, 1967). However, renal potassium wasting is not totally corrected by treatment of the acidosis with alkali (Sebastian et al., 1971).

---

* See also Chapter Two.

Liddle's syndrome

This uncommon condition was described (Liddle, Bledsoe and Cappage, 1963) in two siblings with the clinical findings of primary hyperaldosteronism. Both patients had hypertension and a hypokalemic metabolic alkalosis and exhibited a tendency for abnormal sodium retention and potassium excretion. Surprisingly, aldosterone excretion rate was markedly low and urinary excretion of sodium and potassium was not influenced by the administration of spironolactone or measures that inhibited aldosterone biosynthesis. Liddle and coworkers suggested that the disorder in these children was caused by an inherited defect in renal transport processes, since further studies showed the same disorder to be present in two additional siblings, the mother and two maternal uncles in the same kinship. Subsequent studies have suggested that the abnormality in ion transport represents a much more generalized disorder. The kinetics of sodium transport in erythrocytes of patients with Liddle's syndrome have been found to be markedly altered and to result in an increase in sodium influx and functional sodium efflux (Gardner et al., 1971).

Licorice ingestion

A high incidence of hypokalemia has been demonstrated in patients given licorice for the treatment of gastric ulcer, or in individuals drinking large volumes of a licorice-containing alcoholic beverage (Conn, Rovner and Cohen, 1968). Indeed, the ingestion of large amounts of licorice has been shown to cause severe symptoms of hypokalemia, including myopathy with myoglobinuria (Gross, Dexter and Roth, 1966), quadriplegia, tetany, and convulsions (Roussak, 1952). The causative factor in this syndrome has been shown to be glycyrrhizic acid, an ingredient in licorice, with mineralocorticoid properties. As expected, patients with hypokalemia owing to licorice ingestion have low levels of renin and aldosterone. The clinical manifestations of this disorder resolve after terminating licorice ingestion and can be simulated by the administration of ammonium glycyrrhizate.

Excessive losses from the gastrointestinal tract

Loss of gastric fluid

Hypokalemia may result from loss of gastric fluid, either by nasogastric suction or vomiting. The low plasma potassium level reflects both an alteration in potassium distribution between the cellular and extracellular space and an absolute potassium deficiency.

Although decreased intake and the loss of potassium in gastric fluid may play a role in the hypokalemia of gastric fluid loss, the primary pathogenic factors include volume contraction and induction of metabolic alkalosis. Removal of gastric acid induces a metabolic alkalosis that, as discussed above, favors potassium movement from the extracellular to intracellular space. Secondary hyperaldosteronism resulting from volume contraction stimulates renal potassium excretion.

Diarrhea

Fecal excretion of potassium normally amounts to approximately eight to 15 mEq per day. In diarrheal states, however, large potassium losses may be incurred since stool water has been shown to contain as much potassium as 100 mEq/l (Fordtran and Ingelfinger, 1968). Fortunately, as stool water increases in volume, potassium concentration falls to a level near that of plasma (Fordtran and Dietschy, 1966). A severe type of potassium deficiency has been reported in patients with non-insulin-secreting islet cell adenomas of the pancreas. In patients with this disorder, stool volume often exceeds six to eight liters per day and absolute fecal losses of potassium average 300 mEq/day. Total pancreatectomy has been suggested to avert the complications of fluid and potassium loss (Verner and Morrison, 1974).

Three factors appear to be important in governing the intestinal losses of potassium in diarrheal states. First, as noted, stool water may contain high levels of potassium. Secondly, colonic mucus is rich in potassium (100-140 mEq/l) and may be shed in large amounts (Crane, 1965). Lastly, volume depletion caused by diarrhea results in secondary hyperaldosteronism and increased urinary and colonic secretion of potassium (Shields and Miles, 1965).

In addition to diarrhea caused by acquired diseases of the gastrointestinal tract, laxative abuse may result in hypokalemia and metabolic alkalosis from high rates of fecal potassium excretion (Schwartz and Relman, 1953). In these cases with severe potassium depletion due to laxative abuse, stool water may contain as much as 50-55 mEq/l of potassium. This disorder is important because its clinical manifestations may be difficult to distinguish from Bartter's syndrome, and the use of laxatives is often concealed and may be difficult to identify. Elevated production rates of renin and aldosterone and juxtaglomerular hyperplasia, presumably due to chronic volume depletion, have been described in patients with laxative abuse (Fleischer et al., 1969).

Villous adenoma

Since the original description in 1954 (McKittrick and Wheelot), excessive potassium losses due to villous adenoma of colon and rectum have been widely confirmed. While less than one percent of patients with adenomas develop hypokalemia (Jahadi and Baldwin, 1975), those who do generally have long-standing diarrhea. The potassium content of stool in these patients is not higher than the level found in other types of diarrhea. Since the stool volume, however, may achieve levels of 1,500 to 3,500 ml/day, absolute excretion rate of potassium has been estimated to equal approximately 50 mEq/day.

The source of stool potassium has not been established but may result from increased secretion of colonic mucosal cells owing to secondary aldosteronism (Turnberg, 1970) or from a secretagogue elaborated by the tumor (Shields, 1966), or from excessive excretion of colonic mucus (Duthie and Atwell, 1963).

Excessive losses from integument

Sweat

The secretion of exocrine glands contains a low sodium concentration and a high potassium concentration (Gordon and Cage, 1966). Since volume is low under temperate conditions, electrolyte losses are negligible. During exercise and heat stress, however, flow rate may rise markedly to levels of several liters per day. Losses of potassium in sweat and concomitant urinary losses because of secondary hyperaldosteronism may result in potassium deficiency.

Postburn hypokalemia

Patients suffering full thickness burns over large areas may develop changes in potassium metabolism (Baxter, 1974). While plasma potassium in humans is usually in the normal range early in the postburn period, several factors contribute to an increased kaliuresis 36-48 hours later. Persistent respiratory alkalosis, hyperaldosteronism, and the obligatory excretion of large quantities of sodium contribute to a loss of 80 to more than 200 mEq of potassium daily in the urine. These factors, therefore, contribute to alterations in potassium distribution and large absolute losses, resulting in significant potassium deficiency if adequate replacement is not instituted.

MANIFESTATIONS OF HYPOKALEMIA

Sustained hypokalemia from any of these causes can produce widespread effects in the body (Welt, Hollander & Blythe, 1960). These may include impaired neuromuscular function, ranging in severity from weakness to paralysis, and myocardial abnormalities, including necrosis of cardiac muscle, conduction defects, arrhythmias (premature atrial and ventricular beats), atrial tachycardia and flutter, and altered sensitivity to digitalis. Intestinal dilatation and ileus are common manifestations and alterations in gastric secretions have been found. The kidney may be affected by hypokalemia, producing both structural and functional abnormalities. In 1956, Relman and Schwartz studied five patients with severe potassium deficiency with serial renal biopsies. A reversible disease was found, characterized, histologically by dilated proximal tubules with swollen epithelial cells cytoplasmic vacuoles or "foamy degeneration," and epithelial cell atrophy. Further studies (Schwartz and Relman, 1967) described osmophilic droplets in the cytoplasm of tubular epithelial cells, mitochondrial swelling, and thickening of tubular basement membrane. Significant abnormalities of renal function in potassium-deficient patients included a significant impairment of the ability to concentrate, abnormalities in diluting capacity, and impaired tubular secretion of the organic anion para-aminohippurate (PAH). Furthermore, a minor reduction in renal clearances of inulin, creatinine, and PAH was noted; and the ability to excrete a high sodium load and to maximally conserve sodium was impaired when dietary intake was restricted (Relman and Schwartz, 1956). These abnormalities were reversible and returned to normal when potassium was repleted. Long-standing hypokalemia may result in diffuse chronic abacteric interstitial nephritis and, eventually, in terminal renal failure (Cremer and Bock, 1977).

Hypokalemia may occasionally present as neuropsychiatric disturbances (Mitchell and Feldman, 1968). Patients may present with an acute brain syndrome, that is, with impaired memory, disorientation, and confusion. Psychoneuroses often characterized by a depressive reaction are also seen, and the neuromuscular abnormalities of hypokalemia may be misdiagnosed as a conversion reaction.

THERAPY FOR HYPOKALEMIA AND POTASSIUM DEFICIENCY

Since potassium is primarily an intracellular ion, plasma levels provide only a rough index to total body stores. Cellular potassium is stored in proportion to the protein and glycogen content, in the absence of systemic acid-base disturbances (Scribner and Burnell, 1956). In the absence of a disturbance of those factors, a total body deficit of 100–200 mEq of potassium usually results in a reduction in plasma levels of about 1.0 mEq/l below normal. After plasma potassium is decreased to a level of 3 mEq/l, further reduction in plasma levels reflect a deficit of 200–400 mEq per additional 1 mEq/l drop in plasma potassium.

The clinical approach to patients with alterations in potassium homeostasis is based upon whether the defect is due to a change in distribution or in total body stores. For example, in patients with hypokalemic alkalosis, therapy is directed towards correction of alkalosis rather than infusion of potassium. In patients with potassium deficiency, therapy requires both potassium repletion and identification and correction of the underlying disorder. Mild deficiencies, in the absence of complications, can usually be corrected by foods rich in potassium or supplemental KCl in liquid form or embedded in a waxy matrix ("slow K").

In severe potassium deficiency associated with complications, and in patients unable to take oral potassium, KCl may be administered intravenously. An estimate of the total amount required for repletion can be calculated by the rough guide outlined above. In the absence of acid-base disturbances, serial plasma levels can be used as a guide for adequate replacement. Since cellular uptake of potassium is not an immediate process, care must be exercised in arranging the rate of potassium administration to avoid severe hyperkalemia. As a general rule, the concentration of potassium should not exceed a level of 80 mEq/l of infusate and should not exceed an infusion rate of 30–40 mEq/hr.

HYPERKALEMIA

The term hyperkalemia is used to indicate a plasma potassium level greater than 5.5 mEq/l. As in the case of hypokalemia, alterations in potassium homeostasis, which result in elevated plasma levels, may be caused by changes in total body stores or in the relative distribution of potassium between cellular and extracellular sites. In addition, the clinical disorder of hyperkalemia may result from the release of tissue stores of potassium into extracellular fluid because of cellular damage. Major causes of hyperkalemia are listed in Table 6.2.

Table 6.2 Causes of hyperkalemia owing to alterations in potassium distribution and potassium retention

I. Factors and syndromes associated with excessive tissue release or redistribution of potassium from intracellular fluid to extracellular fluid
  A. Acidemia
  B. Hyperkalemic periodic paralysis
  C. Hemolysis
  D. Tissue necrosis
II. Factors and syndromes associated with excessive potassium retention
  A. Excessive intake
  B. Renal failure
  C. Hyporeninemic hypoaldosteronism
  D. End-organ unresponsiveness

Factors and syndromes associated with excessive tissue release or redistribution of potassium from intracellular to extracellular fluid

Acidosis

The plasma potassium level usually varies directly with the level of free hydrogen ion. Thus, hyperkalemia is an important complication of acidosis, and is therefore a common clinical disorder. Early studies demonstrated an elevation of plasma potassium concentration in both respiratory and metabolic acidosis (Scribner et al., 1955; Swan and Pitts, 1955), an elevation that was thought to result from the loss of cellular potassium in exchange for hydrogen ion. A recent report, however, has indicated that the plasma bicarbonate concentration, independent of blood pH, modulates the distribution of potassium across cell membranes (Fraley and Adler, 1976). Plasma potassium concentration was found to correlate inversely with variations in bicarbonate, at constant blood pH, an effect that could not be explained by differences in renal potassium excretion. It remains unclear whether the hyperkalemia of acidosis is always mediated by lower bicarbonate concentration, or if free hydrogen and bicarbonate concentrations independently vary the plasma potassium. A recent report of an acute, reversible lactic acidosis without hyperkalemia emphasizes the complex relationship between plasma potassium and acidosis (Orringer et al., 1977).

Hyperkalemic periodic paralysis

This uncommon disorder was first recognized two decades ago (Gamstorp, 1956) and is also presumably due to a shift of intracellular potassium into extracellular fluid. First described in two large kinships, hyperkalemia-induced attacks of paralysis occur episodically, although many affected individuals also experience muscle weakness between attacks.

Hyperkalemic periodic paralysis is inherited by autosomal dominant transmission and attacks usually first appear in childhood or early adulthood (Pearson and Kalyanaraman, 1972). Factors known to precipitate attacks include strenuous exercise, cold exposure, hunger, and the acute administration of potassium. In susceptible individuals, these factors may induce myotonia and paramyotonia, which peak in 30 to 40 minutes and then resolve in about 30 minutes. Since muscle stores of potassium have been shown to fall during attacks, it seems likely that the change in extracellular fluid potassium concentration is due to an alteration in distribution. The pathogenesis of potassium redistribution is unknown.

The frequency of attacks may be reduced by measures such as avoidance of vigorous exercise, a high carbohydrate diet and the administration of certain diuretics (thiazides and acetazolamide). It is of interest that Salbutamol, a drug that is thought to act directly on the reciprocal movement of sodium and potassium across the cell membrane, has been shown to have a beneficial effect in this disorder (Harris, 1976; Wang and Clausen, 1976).

Hemolysis

Rapid destruction of red blood cells may result in hyperkalemia owing to release of cellular stores of potassium at a rate that exceeds the capacity for urinary excretion and the rate of cell uptake by liver and muscle. This cause of hyperkalemia is usually transient although it may be life threatening in patients with renal insufficiency. The major causes of hemolysis associated with hyperkalemia include abnormal red cell morphology (as in hereditary spherocytosis and sickle cell disease) and antibody-mediated hemolytic anemia.

Tissue necrosis

In a manner similar to hemolysis, the sudden destruction of a large mass of non-hematogenous tissue may cause hyperkalemia. Under normal conditions, renal tubular cells respond to elevated plasma potassium levels by increasing the rate of secretion into tubular urine, preventing sustained hyperkalemia. Thus hyperkalemia caused by tissue necrosis is found predominantly in patients with impaired renal function. Disorders that may induce both tissue necrosis and acute renal failure often result in hyperkalemia; these include major trauma, rhabdomyolysis, and disseminated intravascular coagulation. In addition, destruction of anaplastic tumors with chemotherapy may cause plasma potassium levels to rise and has been reported in the treatment of Burkitt's lymphoma (Araseneau et al., 1973), acute lymphoblastic leukemia (Fennelly, Smyth and Muldowney, 1974), and lymphosarcoma (Muggia, 1973). Likewise, splenic irradiation in such patients may result in hyperkalemia (Kuslander, Stein and Roth, 1975). Probable predisposing factors to the development of hyperkalemia during chemotherapy include (1) a large amount of tumor tissue, (2) a high sensitivity of the tumor to therapy, and (3) a rapid destruction of cells after initiation of treatment.

Factors and syndromes associated with excessive potassium retention

Excessive intake

In the absence of impaired renal function, hyperkalemia owing to a high oral or parenteral intake of potassium is uncommon. The capacity for removal of excess potassium from extracellular fluid, however, may be exceeded during intravenous administration of potassium salts. Although the maximum rates of cellular uptake and rate of excretion by kidney and colon has not been systematically evaluated in man, clinical experience indicates that the rate of intravenous administration should not exceed 40 mEq/hr in patients with intact renal function.

Renal failure

Adaptive changes in the mechanism involved in regulating the rate of potassium excretion by colonic mucosal cells and renal tubular cells preserve potassium homeostasis until the glomerular filtration rate falls to a level of approximately 20 ml/min. In chronic renal insufficiency, the fraction of oral potassium excreted by feces increases from the normal level of 10 percent to a value of approximately 35 percent (Hayes et al., 1967) and may increase to even higher levels. Because of the rise in potassium excretion by individual surviving nephrons, fractional potassium excretion may exceed unity. It should be noted, however, that in types of renal disease in which medullary tissue is destroyed predominantly (Finkelstein and Hayslett, 1974), the adaptive capacity to increase secretion as the surviving nephron population falls, may be blunted.

In patients with moderate renal insufficiency, however, sudden elevations in potassium intake may result in hyperkalemia. As noted before, release of potassium from cellular stores due to hemolysis or tissue necrosis may exceed the rate at which potassium can be cleared from plasma. Other important causes of potassium overload in patients with moderate renal insufficiency include medications with a high potassium content, such as potassium penicillin (1.5 mEq per one million units for the parenteral and 3.1 mEq per one million units for the tablet preparation), 'salt' substitute (KCl), geophagia (chewing of clay) (Gelfand, Zarate, and Knepshield, 1975), inadequate mixing of intravenous fluids (Williams, 1973), and stored whole blood (which may contain levels of potassium as high as 10 mEq/l) (Bostic and Duvernoy, 1972).

Patients with severe renal failure (GFR $<$ 20 ml/min) eventually develop potassium retention and hyperkalemia, unless the rate of dietary intake is reduced. It is a common clinical observation that patients who become oligoanuric are particularly prone to develop hyperkalemia. The absolute potassium excretion is dependent on the concentration of potassium that can be generated in the distal tubule and collecting duct, and the volume of fluid excreted. Thus, marked alterations in volume can exert an important influence on potassium excretion, and patients with markedly reduced flow rate (oliguria) can be expected to have reduced potassium excretion. Conversely, high flow rates may significantly increase the capacity to excrete potassium, even under circumstances of impaired secretory ability of the distal nephron sites. Such is the case, for example, in the diuretic (recovery) phase following release of obstruction or acute tubular necrosis. Here, while urinary potassium concentration may not exceed 20–30 mEq/l, flow rates of more than 3 liters a day might permit excretion of more than 100 mEq potassium daily.

Hyperkalemia in patients with severe renal failure is caused by changes in distribution owing to metabolic acidosis as well as a positive potassium balance. Prophylactic measures designed to prevent alterations in potassium homeostasis, therefore, underlie conservative treatment in this group of patients.

Hypoaldosteronism

Since mineralocorticoids normally play an important role in modulating the rate of potassium excretion, potassium retention can result from a fall in production rate of aldosterone or in failure of target epithelium (distal nephron and colon) to respond to its action.

Primary adrenal insufficiency: In contrast to patients with pituitary disease and a fall in ACTH release, in whom mineralocorticoid production is relatively well maintained, impaired aldosterone production characterizes primary adrenal disease. Patients with Addison's disease, therefore, exhibit a fall in urinary potassium excretion and an inappropriate rise in the excretion of sodium (Thorn, Forsham and Emerson, 1949), despite normal or elevated production of renin.

As a general rule, hyperkalemia is not encountered in patients with Addison's disease unless overall renal function is reduced. Under the condition of volume contraction, owing to the large obligatory loss of solute, however, a reduction in renal blood flow and glomerular filtration rate may cause excessive retention of potassium and a rise in plasma levels of potassium. Hyperkalemic neuromyopathy with flaccid quadriplegia has been noted in rare patients (Pollen and Williams, 1960).

The capacity to maintain potassium balance is dependent, in large part, on the renal process for potassium secretion. This process is controlled by both aldosterone sensitive and nonsensitive mechanisms. Presumably, in the presence of normal renal function and a normal intake of potassium, the aldosterone-independent mechanism is capable of an adequate response to the excretory load. Harrison and Darrow (1939) studied adrenalectomized animals and showed an apparent maximum urine to plasma ratios (U/P) of potassium of approximately 10:1. These authors state that in normal individuals with intact adrenal function, the maximal U/P ratio of potassium is 50:1. If patients with adrenal insufficiency exhibit an impaired ability to achieve high concentrations of urinary potassium, their ability to maintain a rate of absolute excretion (U_kV) equal to intake would be critically dependent upon the level of renal blood flow and urinary flow rate. This is a fragile balance that may not always be achieved because of the important role of glucocorticoids in regulating renal blood flow and the permeability of renal tubular epithelium to water. In addition, recent studies have emphasized that elevated levels of antidiuretic hormone may reduce urinary flow rate in patients with glucocorticoid deficiency (Agus and Goldberg, 1971).

End-organ unresponsiveness

Studies in experimental animals have shown a reduced capacity to excrete potassium after selective damage to the medullary portion of the kidney (Finkelstein and Hayslett, 1974). Patients with interstitial nephritis may, therefore, have an impaired ability to secrete potassium and be especially prone to develop hyperkalemia during acute increases in potassium load. In the presence of normal or modestly reduced glomerular filtration rate and adequate amounts of aldosterone, the possibility of an impairment in the renal mechanism for potassium secretion has been amplified by recent studies in man. Children who appear to have a primary defect in tubular potassium secretion have been reported with hyperkalemia, despite normal glomerular filtration rates and adequate mineralocorticoid production (Weinstein, Allan and Mendoza, 1974). Low rates of urinary excretion of potassium have been reported in some patients (Perez et al., 1977), despite evidence that the renin-aldosterone mechanism was responsive to changes in extracellular volume. Four patients with a reversible defect in potassium excretion following successful renal transplantation have also been described (DeFronzo et al., 1977). Further, in a recent study of patients with sickle cell disease, characterized by normal glomerular filtration rate and reduced concentrating and renal acidifying ability, the rate of potassium excretion during acute administration of KCl was reduced, compared with normal control subjects (DeFronzo et al., 1977).

In addition to an intrinsic defect in the potassium secretory process, the rate of secretion may be inhibited by pharmacologic agents. The most commonly used therapeutic drugs with potassium retaining properties are spironolactone and triampterine. Since spironolactone is regarded as a specific inhibitor of the peripheral action of aldosterone, while triampterine is not, these agents presumably block potassium secretion by different mechanisms. Although treatment with either spironolactone or triampterine does not induce hyperkalemia in patients with intact renal function, their use may be associated with hyperkalemia in patients with renal insufficiency. For example, in one report 26 percent of patients with renal insufficiency were found to have an elevation of plasma potassium levels after administration of Diazide^R (hydrochlorthiazide and triampterine) (McDonald, 1976).

Hyporeninemic hypoaldosteronism (SHH)

A group of patients has been described in whom hyperkalemia occurs in association with hypoaldosteronism, in the absence of an impairment of glucocorticoid synthesis. Plasma renin activity, in contrast to the high levels in Addison's disease, is markedly depressed. Plasma potassium may exceed 7.0 mEq/l, and life-threatening arrhythmias and other complications of hyperkalemia may ensue. While most patients with this syndrome have evidence of primary renal disease with mild renal insufficiency and metabolic acidosis, all have hyperkalemia disproportionate to the degree of renal impairment and acidosis (Perez, Oster, and Vaamonde, 1974). Nearly half of the reported cases of SHH also have diabetes mellitus (Michelis and Murdaugh, 1975). In the initial description of this syndrome by Relman and colleagues (Hudson, Chronbanian and Relman, 1957), exogenous DOCA administration increased urinary excretion of potassium and corrected hyperkalemia, suggesting that despite renal insufficiency the kidney maintained its responsiveness to mineralocorticoid. This observation has been frequently confirmed, indicating that the pathogenesis of this disorder involves a disturbance in the metabolism of aldosterone. Further support for this concept comes from the observation that while correction of the metabolic acidosis with alkali may not improve the hyperkalemia (Szylman, 1976), mineralocorticoid administration may correct both the hyperkalemia and acidosis (Sebastian et al., 1977).

The etiology of SHH remains unclear. Two hypotheses have been proposed to explain the primary abnormality that leads to the clinical features of the syndrome: (1) a defect in renin; or (2) an impaired aldosterone metabolism. Since both renin and aldosterone secretion are depressed, some authors have suggested that a primary deficiency in renin secretion is responsible for SHH (Schambelan, Stockigt and Biglieri, 1972) since an isolated defect in aldosterone production should lead to increased renin levels. It is important to recall, however, that renin secretion might be depressed by hyperkalemia.

A primary defect in aldosterone production is suggested by the finding that the administration of neither angiotensin II nor ACTH stimulates aldosterone excretion, although excretion of 17-hydroxycorticosteroids rises in the usual way (Schambelan et al., 1972). This finding, however, may not exclude a primary defect in renin metabolism, since aldosterone responsiveness may be blunted in patients with an abnormal renin-angiotensin system (Biglieri et al., 1966).

Other mechanisms have been suggested to explain the hyperkalemia in SHH. In one report, renal tubular unresponsiveness to kaluretic stimuli was found (Arnold and Healy, 1969), suggesting a primary defect in renal tubular function in these patients. In addition, the syndrome has been reported with unusual frequency in patients with diabetes mellitus (Schambelan et al., 1972). Since insulin is an important mediator of potassium distribution, it is not surprising to find that defects in both insulin and aldosterone secretion predispose such patients to hyperkalemia. Some patients with diabetes mellitus demonstrate a paradoxical increase in plasma potassium concentration when hypertonic glucose is infused, a response abolished by pretreatment with insulin (Goldfarb et al., 1975). Since diverse mechanisms have been proposed on the basis of clinical observations, it is possible that patients grouped under the term SHH may represent heterogeneous disorders.

MANIFESTATIONS OF HYPERKALEMIA

Hyperkalemia can result in abnormalities of the neuromuscular, gastrointestinal, and cardiovascular systems. Neuromuscular manifestations include weakness and paresthesias, ascending paralysis and flaccid quadriplegia (Bull, Carter & Lowe, 1953). Gastrointestinal symptoms include nausea and vomiting, abdominal pain and ileus. The most immediate danger of hyperkalemia is its effect on cardiac conduction. While mild elevations of plasma potassium (5–7 mEq/l) may cause acceleration of cardiac conduction, marked hyperkalemia (> 8 mEq/l) induces a rapid depression of conduction and life-threatening arrhythmias (Ettinger, Regan and Oldewurtel, 1974). The EKG may serve as a sensitive indicator of potassium's effect on the heart. As the plasma potassium rises, the T waves become tall and peaked. As conduction is depressed, the PR interval prolongs, and then P waves disappear with atrial standstill. With worsening hyperkalemia, the QRS complexes widen and the R-R intervals become irregular. A conduction delay at the A-V junction causes the junctional pacemaker to accelerate, but conduction delay can also be measured in the His' bundle, the Purkinje network, and in the ventricular muscle. Finally, ventricular tachycardia that leads to ventricular fibrillation or to asystole is a terminal event.

THERAPY FOR HYPERKALEMIA

The two questions that should be answered in the rational approach to hyperkalemia are: (1) What is the cause? (2) How rapidly need it be corrected? In conditions characterized by a shift in potassium distribution from cellular to extracellular sites, such as acidosis, treatment directed towards correction of the underlying abnormality is indicated. Disorders that result in hyperkalemia owing to an increase in total potassium stores may require measures that rapidly induce a movement of potassium into cells, as well as procedures designed to increase the rate of absolute excretion. The clinical decision to introduce measures for rapid correction of elevated plasma levels is governed by the presence of cardiac conduction abnormalities and their severity.

The occurrence of cardiac conduction defects that are characteristic of hyperkalemia and of neuromuscular signs or symptoms are potentially fatal and require immediate reversal by treatment to reduce the plasma potassium concentration. Since changes in cardiac conduction are mirrored by the electrocardiogram, serial tracings serve as better guides to treatment than do determinations of plasma potassium levels.

Immediate treatment procedures include techniques to increase the rate of cellular uptake of potassium and measures to reduce the effect of high extracellular levels of potassium on cell membrane excitability. In the former category, administration of exogenous alkali, as $NaHCO_3$, and infusion of hypertonic glucose with crystalline insulin, usually cause a rapid but transient fall in plasma potassium levels. As noted previously, the mechanism by which bicarbonate administration increases the rate of cell uptake of potassium may involve a direct action of bicarbonate ion, rather than an exchange of extracellular potassium for intracellular hydrogen. The mechanism of the response to glucose and insulin treatment is not firmly established but probably involves insulin-dependent transport of glucose and potassium by cell membranes (Hiatt, 1973) and the role of intracellular glycogen in regulating cellular potassium stores (Scribner and Burnell, 1956). The action of high levels of extracellular potassium to depolarize cell membranes can be blocked by the administration of calcium and hypertonic NaCl. Both measures have been shown to cause an immediate improvement in hyperkalemia-induced conductive defects (Chamberlain, 1964). The action of hypertonic NaCl has been shown to result in an increase in the velocity of the action potential in cardiac fibrils (Ballantyne, David and Reynolds, 1975).

Reduction of total body potassium

Total body potassium can be reduced by hemo- or peritoneal dialysis, or with a cation exchange resin. These measures reduce total body stores of potassium relatively slowly, but are essential since the acute procedures, listed above, are only transiently effective in total body potassium excess. The exchange resin, sodium polystyrene sulfonate (Kayexalate), can be given orally or by enema, and exchanges sodium for potassium in the gastrointestinal tract (Scheer et al., 1961). Approximately 1 mEq of potassium is bound in each gram of resin when the resin is in the gut long enough for maximal exchange. Thus when given by enema, at least 30 minutes of retention is necessary for adequate exchange. When administered orally, 70 percent sorbitol is used to induce diarrhea, which may increase potassium exchange. Thus, 50 gm of Kayexalate can rapidly remove about 50 mEq of potassium. This can then be repeated at intervals.

Although dialysis can remove potassium, it is rarely more efficient than the use of cation exchange resins. Peritoneal dialysis in severely hyperkalemic patients can remove at most about 15 mEq of potassium per hour. Hemodialysis on the other hand can remove 50 mEq of potassium per hour when potassium-free dialysate is used and blood flow is more than 200 ml/min.

When severe hyperkalemia produces rhythm disturbances, transvenous pacemaker placement should be considered. Pacing, however, may be unsuccessful in severe hyperkalemia, as demonstrated by a case of iatrogenic hyperkalemia in which a transvenous cardiac pacemaker failed to effect a ventricular response. The pacemaker began to drive the ventricle normally when the plasma potassium began to fall towards normal (O'Reilly, Murnaghan and Williams, 1974).

ARTIFICIAL CAUSES OF HYPERKALEMIA

Since disturbances in potassium homeostasis are usually identified by changes in plasma potassium values, it is important to recall that several conditions may cause artificial elevations in measured potassium values. Blood samples taken from the same limb in which infusions of potassium salts are given is a well-known but easily forgotten cause of artificially elevated potassium levels in plasma.

Further, the term pseudohyperkalemia is used to describe the elevation of serum or plasma potassium values that result from the in vitro release of potassium from formed elements of blood. This phenomenon may occur when RBCs hemolyze in vitro, either due to excessive trauma or low threshold for hemolysis in some hematologic disorders. In addition, the release of potassium from platelets in cases with thrombocytosis, or from leukocytes, because of in vitro coagulation, may result in artificially high values. The determination of plasma potassium in anticoagulated blood may be necessary to exclude clot-induced cell lysis.

REFERENCES

Adler, S. (1970). An extrarenal action of aldosterone on mammalian skeletal muscle. *American Journal of Physiology,* 218, 616.

Agus, Z. S. & Goldberg, M. (1971). Role of antidiuretic hormone in the abnormal water diuresis of anterior hypopituitarism in man. *Journal of Clinical Investigation,* 50, 1478.

Aikawa, J. K., Harrell, G. T. & Eisenberg, B. (1952). The exchangeable potassium content of normal women. *Journal of Clinical Investigation,* 31, 367.

Alexander, E. A. & Levinsky, N. G. (1968). An extrarenal mechanism of potassium adaptation. *Journal of Clinical Investigation,* 47, 740.

Ali, M. H., Cross, R. B. & Pickford, M. (1958). Electrolyte excretion in diuretic and non-diuretic dogs. *Journal of Physiology* (London), **141**, 177.

Anderson, E. C. (1963). Three component body composition analysis based on K+ and water determinations. *Annals of the New York Academy of Sciences*, **110**, 189.

Andres, R., Baltzan, M. A., Cader, G. & Zierler, K. L. (1962). Effect of insulin on carbohydrate metabolism and on potassium in the forearm of man. *Journal of Clinical Investigation*, **41**, 108.

Araseneau, J. C., Bagley, C. L., Anderson, T. & Canellos, G. P. (1973). Hyperkalemia, a sequel to chemotherapy of Burkitt's lymphoma. *Lancet*, **1**, 10.

Arnold, J. E. & Healy, J. K. (1969). Hyperkalemia, hypertension and systemic acidosis without renal failure associated with a tubular defect in potassium excretion. *American Journal of Medicine*, **47**, 461.

August, J. T., Nelson, D. H. & Thorn, G. W. (1958). Response of normal subjects to large amounts of aldosterone. *Journal of Clinical Investigation*, **37**, 1549.

Azzopardi, J. G. & Williams, E. D. (1968). Pathology of non-endocrine tumors associated with Cushing's syndrome. *Cancer*, **22**, 274.

Baer, L., Sommers, S. C., Krakoff, L. R., Newton, M. A. & Laragh, J. H. (1970). Pseudo primary aldosteronism. *Circulation Research*, **26 & 27** (suppl. 1) I–203.

Ballantyne, F., III, David, L. D. & Reynolds, E. W., Jr. (1975). Cellular basis for reversal of hyperkalemic electrocardiographic changes by sodium. *American Journal of Physiology*, **229**, 935.

Bank, N. & Aynedjian, H. S. (1973). A micropuncture study of potassium excretion by the remnant kidney. *Journal of Clinical Investigation*, **52**, 1480.

Barger, A. C., Berlin, R. D. & Tulenko, J. F. (1958). Infusion of aldosterone, 9 α-fluorhydrocortisone, and antidiuretic hormone into the renal artery of normal and adrenalectomized dogs: Effect of electrolyte and water excretion. *Endocrinology*, **62**, 804.

Bartter, F. C., Pronove, P., Gill, J. R., Jr. & MacCardle, R. C. (1962). Hyperplasia of the juxtaglomerular complex with hyperaldosteronism and hypokalemic alkalosis. *American Journal of Medicine*, **33**, 811.

Bartter, F. C., Delea, C. S., Kawasaki, T. & Gill, J. R. (1974). The adrenal cortex and the kidney. *Kidney International*, **6**, 272.

Bastl, C., Hayslett, J. P. & Binder, H. J. (1977). Increased large intestinal secretion of potassium in renal insufficiency. *Kidney International*, **12**, 9.

Baxter, C. R. (1974). Fluid volume and electrolyte changes of the early postburn period. *Clinics in Plastic Surgery*, **1**, 693.

Berger, E. Y. (1960). Intestinal absorption and excretion. In *Mineral Metabolism—An Advanced Treatise*. ed. Comar, C. L. & Bronner, F. Vol. 1, pp. 249–286. New York: Academic Press.

Berliner, R. W. (1961). Renal mechanism for potassium excretion. *Harvey Lectures*, **55**, 141.

Biglieri, E. G., Herron, M. A. & Brust, N. (1966). 17-Hydroxylation deficiency in man. *Journal of Clinical Investigation*, **45**, 1946.

Biglieri, E. G., Slaton, P. E., Jr., Silen, W. S., Galante, M. & Forsham, P. H. (1966). Post-operative studies of adrenal function in primary aldosteronism. *Journal of Clinical Endocrinology and Metabolism*, **26**, 553.

Biglieri, E. G., Stockigt, J. R. & Schambelan, M. (1972). Adrenal mineralocorticoids causing hypertension. *American Journal of Medicine*, **52**, 623.

Black, D. A. K., Davies, H. E. F., Emery, E. W. & Wade, E. G. (1956). Renal handling of radioactive potassium in man. *Clinical Science*, **15**, 277.

Black, D. A. K. (1967). *Essentials of Fluid Balance*, 4/e. Philadelphia: F. A. Davis.

Bonting, S. L. & Caravaggio, L. L. (1963). Studies on sodium-potassium activated adenosine triphosphatase. V. Correlation of enzyme activity with cation flux in six tissues. *Archives of Biochemistry and Biophysics*, **101**, 37.

Bostic, O. & Duvernoy, W. F. C. (1972). Hyperkalemic cardiac arrest during transfusion of stored blood. *Journal of Electrocardiology*, **5**, 407.

Brillon, K. E. (1976). An approach to early adrenal visualization in Conn's and Cushing's syndromes. *Journal of Endocrinology*, **68**, 40P.

Bull, A. M., Carter, A. B. & Lowe, K. G. (1953). Hyperpotassaemic paralysis. *Lancet*, **2**, 60.

Burnell, J. M., Villamil, M. F., Uyeno, B. T. & Scribner, B. H. (1956). The effect in humans of extracellular pH changes on the relationship between serum potassium concentration and intracellular potassium. *Journal of Clinical Investigation*, **35**, 935.

Cannon, P. J., Leeming, J. M., Sommers, S. C., Winters, R. W. & Laragh, J. H. (1968). Juxtaglomerular cell hyperplasia and secondary hyperaldosteronism (Bartter's syndrome): a re-evaluation of the pathophysiology. *Medicine* (Baltimore), **47**, 107.

Chamberlain, M. J. (1964). Emergency treatment of hyperkalemia. *Lancet*, **1**, 464.

Christy, N. P. & Laragh, J. H. (1961). Pathogenesis of hypokalemic alkalosis in Cushing's syndrome. *New England Journal of Medicine*, **265**, 1083.

Conn, J. W. (1955). Primary aldosteronism: a new clinical syndrome. *Journal of Laboratory and Clinical Medicine*, **45**, 6.

Conn, J. W., Knopf, R. F. & Nesbit, R. M. (1964). Clinical characteristics of primary aldosteronism from an analysis of 145 cases. *American Journal of Surgery*, **107**, 159.

Conn, J. W., Rovner, D. R. & Cohen, E. L. (1968). Licorice-induced pseudoaldosteronism. *Journal of the American Medical Association*, **205**, 492.

Conn, J. W., Cohen, E. L., Lucas, C. P., MacDonald, W. J., Mayor, G. H., Blugh, W. M., Jr., Eveland, W. C., Bookstein, J. J. & Lapides, J. (1972). Primary reninism. *Archives of Internal Medicine*, **130**, 682.

Corsa, K., Jr., Olney, J. M., Jr., Sleenburg, R. W., Ball, M. R. & Moore, F. D. (1950). The measurement of exchangeable potassium in man by isotope dilution. *Journal of Clinical Investigation*, **29**, 1280.

Crane, C. W. (1965). Observations on the sodium and potassium content of mucus from the large intestine. *Gut*, **6**, 439.

Cremer, W. & Bock, K. D. (1977). Symptoms and course of chronic hypokalemic nephropathy in man. *Clinical Nephrology*, **7**, 112.

Dargie, H. J., Boddy, K., Kennedy, A. C., King, P. C., Read, P. R. & Ward, D. M. (1974). Total body potassium in long-term furosemide therapy: is potassium supplementation necessary? *British Medical Journal*, **4**, 316.

Davidson, D. G., Levinsky, N. G. & Berliner, R. W. (1958). Maintenance of potassium excretion despite reduction of glomerular filtration during sodium diuresis. *Journal of Clinical Investigation*, **37**, 548.

DeFronzo, R., Goldberg, M., Cooke, C. R., Barter, C., Grossman, R. & Agus, Z. (1977). Investigations into the mechanism of hyperkalemia following renal transplantation. *Kidney International*, **11**, 365.

DeFronzo, R. A., August, P., Black, H., McPhedran, P. & Cooke, C. R. (1977). Impaired renal tubular potassium secretion in sickle cell (SS) disease. *Proceedings of the American Society of Nephrology*, **10**, 13A.

Diengott, D., Roza, O., Levy, N. & Muammar, S. (1964). Hypokalemia in barium poisoning. *Lancet*, **2**, 343.

Dluhy, R. G., Axelrod, L. & William, G. H. (1972). Serum immunoreactive insulin and growth hormone response to potassium infusion in man. *Journal of Applied Physiology*, **33**, 22.

Dolman, D. & Edmonds, C. J. (1975). The effect of aldosterone and the renin-angiotensin system on sodium, potassium and chloride transport by proximal and distal rat colon in vivo. *Journal of Physiology* (London), **250**, 597–611.

Down, P. F., Polak, A., Rao, R. & Mead, J. A. (1972). Fate of potassium supplements in six outpatients receiving longterm diuretics for oedematous disease. *Lancet*, **2**, 721.

Duthie, H. L. & Atwell, T. D. (1963). The absorption of water sodium and potassium in the large intestine with particular reference to the effects of villous papillomas. *Gut*, **4**, 373.

D'Silva, J. L. (1934). The action of adrenaline on serum potassium. *Journal of Physiology* (London), **82**, 393.

D'Silva, J. L. (1936). The action of adrenaline on serum potassium. *Journal of Physiology* (London), **86**, 219.

Eberlien, W. R. & Bongiovanni, A. M. (1956). Plasma and urinary corticosteroids in the hypertensive form of congenital adrenal hyperplasia. *Journal of Biological Chemistry*, **223**, 85.

Erkelens, D. W. & Statius Van Eps, L. W. (1973). Bartter's syndrome and erythrocytosis. *American Journal of Medicine*, **55**, 711.

Ettinger, P. O., Regan, T. J. & Oldewurtel, H. A. (1974). Hyperkalemia, cardiac conduction and the EKG: a review. *American Heart Journal*, **88**, 360.

Fanconi, A., Schachenmann, G., Nussli, R. & Prader, A. (1971). Chronic hypokalemia with growth retardation, normotensive hyperrenin-hyperaldosteronism (Bartter's syndrome) and hypercalciuria. *Helvetica Paediatrica Acta*, **26**, 144.

Fenn, W. O. & Cobb, D. M. (1934). The potassium equilibrium in muscle. *Journal of General Physiology*, **17**, 629.

Fennelly, J. J., Smyth, H. & Muldowney, F. P. (1974). Extreme hyperkalemia due to rapid lysis of leukemic cells. *Lancet*, **1**, 27.

Finkelstein, F. O. & Hayslett, J. P. (1974). Role of medullary structures in the functional adaptation of renal insufficiency. *Kidney International*, **6**, 419.

Fisher, K. A., Binder, H. J. & Hayslett, J. P. (1976). Potassium secretion by colonic mucosal cells after potassium adaptation. *American Journal of Physiology*, **231**, 987.

Fleischer, N., Brown, H., Graham, D. Y. & Delenna, S. (1969). Chronic laxative-induced hyperaldosteronism and hypokalemia simulating Bartter's syndrome. *Annals of Internal Medicine*, **70**, 791.

Forbes, G. B. & Lewis, A. M. (1956). Total sodium, potassium and chloride in adult man. *Journal of Clinical Investigation*, **35**, 596.

Fordtran, J. S. & Dietschy, J. M. (1966). Water and electrolyte movement in the intestine. *Gastroenterology*, **50**, 263.

Fordtran, J. S. & Ingelfinger, F. J. (1968). Absorption of water, electrolytes and sugars from the human gut. In *Handbook of Physiology*, ed. Code, C. F. Vol. 3, Section 6: Alimentary Canal. Washington, D.C.: American Physiological Society.

Fraley, D. S. & Adler, S. (1976). Isohydric regulation of the plasma potassium by bicarbonate in the rat. *Kidney International*, **9**, 333.

Fujita, T., Sakaguchi, H., Shibagaki, M., Fukui, T., Nomura, M. & Sekiguchi, S. (1977). The pathogenesis of Bartter's syndrome. *American Journal of Medicine*, **63**, 467.

Gamstorp, I. (1956). Adynamia episodica hereditaria. *Acta Paediatrica* (Stockholm), **45**, (suppl. 108), 1.

Gardner, J. D., Lapey, A., Simapaoulos, A. P. & Bravo, E. L. (1971). Abnormal membrane sodium transport in Liddle's syndrome. *Journal of Clinical Investigation*, **50**, 2253.

Gelfand, M. C., Zarate, A. & Knepshield, J. H. (1975). Geophagia. A cause of life-threatening hyperkalemia in patients with chronic renal failure. *Journal of American Medical Association*, **234**, 738.

Genest, J., Rojo-Ortega, J. M., Kuchel, O., Boucher, R., Nawaczynski, W., Lefebvre, R., Chietien, M., Cantin, J. & Granger, P. (1975). Malignant hypertension with hypokalemia in a patient with renin-producing pulmonary carcinoma. *Transactions of the Association of American Physicians*, **88**, 192.

Gennari, F. J. & Cohen, J. J. (1975). Role of the kidney in potassium homeostasis: lessions from acid-base disturbances. *Kidney International*, **8**, 1.

Gerwirtz, G. & Yalow, R. S. (1974). Ectopic ACTH production in carcinoma of the lung. *Journal of Clinical Investigation*, **53**, 1022.

Giebisch, G., Berger, L. & Pitts, R. F. (1955). The extrarenal response to acute acid-base disturbances of respiratory origin. *Journal of Clinical Investigation*, **34**, 231.

Gifford, R. W., Mattox, V. R., Orvis, A. L., Sones, D. A. & Rosevear, J. W. (1961). Effect of thiazide diuretics on plasma volume, body electrolytes and excretion of aldosterone in hypertension. *Circulation*, **24**, 1197.

Gill, J. R., Bell, N. H. & Bartter, F. C. (1967). Impaired conservation of sodium and potassium in renal tubular acidosis and its correction by buffer anions. *Clinical Science*, **33**, 577.

Gill, J. R., Jr., Frolich, J. C., Bowden, R. E., Taylor, A. A., Keiser, H. R., Seyberth, H. W., Oates, J. A. & Bartter, F. C. (1976). Bartter's syndrome: a disorder characterized by high urinary prostaglandins and a dependence of hyperreninemia on prostaglandin synthesis. *American Journal of Medicine*, **61**, 43.

Gitelman, H. J., Graham, J. B. & Welt, L. G. (1966). A new familial disorder characterized by hypokalemia and hypomagnesemia. *Transactions of the Association of American Physicians*, **79**, 221.

Goldfarb, S., Strunk, B., Singer, I. & Goldberg, M. (1975). Paradoxical glucose-induced hyperkalemia. *American Journal of Medicine*, **59**, 744.

Goodman, A. D., Vagnucci, A. H. & Hartroft, P. M. (1969). Pathogenesis of Bartter's syndrome. *New England Journal of Medicine*, **281**, 1435.

Gordon, R. S., Jr. & Cage, C. W. (1966). Mechanism of water and electrolyte secretion by the eccrine sweat gland. *Lancet*, **1**, 1246.

Gourley, D. R. H. & Bethea, M. D. (1964). Insulin effect on adipose tissue sodium and potassium. *Proceedings of the Society for Experimental Biology and Medicine*, **115**, 821.

Gross, E. G., Dexter, J. D. & Roth, R. G. (1966). Hypokalemic myopathy with myoglobinuria associated with licorice ingestion. *New England Journal of Medicine*, **274**, 602.

Gross, J. B. & Kokko, J. P. (1977). Effects of aldosterone and potassium-sparing diuretics on electrical potential differences across the distal nephron. *Journal of Clinical Investigation*, **59**, 82.

Halushka, P. V., Wohitmann, H., Privitera, P. J., Hurwitz, G. & Margolios, H. S. (1977). Bartter's syndrome: urinary prostaglandin E-like material and kallikrein, indomethacin effects. *Annals of Internal Medicine*, **87**, 281.

Harris, P. M. (1976). Salbutamol in hyperkalemic familial periodic paralysis. *Lancet*, **1**, 427.

Harrison, H. E. & Darrow, D. C. (1939). Renal function in experimental adrenal insufficiency. *American Journal of Physiology*, **125**, 631.

Harrop, G. A. & Benedict, E. M. (1923). The role of phosphate and potassium in carbohydrate metabolism following insulin administration. *Proceedings of the Society for Experimental Biology and Medicine*, **20**, 430.

Hayes, C. P., McLeod, M. I. & Robinson, R. (1967). An extrarenal mechanism for the maintenance of potassium balance in severe chronic renal failure. *Transactions of the Association of American Physicians*, **80**, 207.

Hayslett, J. P., Flinn, R. B. & Mulrow, P. J. (1963). Hypertension and increased aldosterone excretion associated with unilateral renal artery stenosis. *Connecticut Medicine*, **27**, 8.

Healy, J. J., McKenna, J. J., Canning, B. V. J., Brien, T. G., Duffy, G. J. & Muldowney, F. P. (1970). Body composition changes in hypertensive subjects on long-term oral diuretic therapy. *British Medical Journal*, **1**, 716.

Hiatt, N., Davidson, M. B. & Bonorris, G. (1972). The effect of potassium chloride on insulin secretion in vivo. *Hormone and Metabolic Research*, **4**, 64.

Hiatt, N. (1973). Role of insulin in the transfer of infused potassium to tissue. *Hormone and Metabolic Research*, **5**, 84.

Holten, C. & Petersen, V. P. (1956). Malignant hypertension with increased secretion of aldosterone and depletion of potassium. *Lancet*, **2**, 918.

Hudson, J. B., Chronbanian, A. V. & Relman, A. S. (1957). Hypoaldosteronism. A clinical study of a patient with an isolated arenal mineralocorticoid deficiency, resulting in hyperkalemia and Stokes-Adams attacks. *New England Journal of Medicine*, **257**, 529.

Jahadi, M. R. & Baldwin, A. (1975). Villous adenomas of the colon and rectum. *American Journal of Surgery*, **130**, 729.

Khuri, R. H. (1972). Intracellular potassium in cells of the distal tubule. *Yale Journal of Biology and Medicine*, **45**, 384.

Khuri, R. N., Wiederholt, M., Strieder, N. & Giebisch, G. (1975). Effect of flow rate and potassium intake on distal tubular potassium transfer. *American Journal of Physiology*, **228**, 1249.

Krakoff, L., Nicholis, G. & Amsel, B. (1975). Pathogenesis of hypertension in Cushing's syndrome. *American Journal of Medicine*, **58**, 216.

Kuslander, R., Stein, R. S. & Roth, D. (1975). Hyperkalemia complicating splenic irradiation of chronic lymphocytic leukemia. *Cancer*, **36**, 926.

Lawson, D. H. (1974). Adverse reactions to potassium chloride. *Quarterly Journal of Medicine*, **43**, 433.

Liddle, G. W., Bledsoe, T. & Cappage, W. J., Jr. (1963). A familial renal disorder simulating primary aldosteronism but with negligible aldosterone secretion. *Transactions of the Association of American Physicians*, **79**, 199.

Liddle, G. W., Givens, J. R., Nicholson, W. E. & Island, D. P. (1965). The ectopic ACTH syndromes. *Cancer Research*, **25**, 1057.

Malnic, G., Klose, R. M. & Giebisch, G. (1964). Micropuncture study of renal potassium excretion in the rat. *American Journal of Physiology*, **206**, 647.

Malnic, G., Klose, R. M. & Giebisch, G. (1966). Micropuncture study of distal tubular potassium and sodium transport in rat nephron. *American Journal of Physiology*, **211**, 529.
Malvin, R. L., Wilde, W. S. & Sullivan, L. P. (1958). Localization of nephron transport by stop-flow analysis. *American Journal of Physiology*, **194**, 135.
McCredie, D. A., Blair-West, J. R., Scoggins, B. A. & Shipman, R. (1971). Potassium-losing nephropathy of childhood. *Medical Journal of Australia*, **1**, 129.
McDonald, C. J. (1976). Use of computer to detect and respond to clinical events: its effect on clinician behavior. *Annals of Internal Medicine*, **84**, 162.
McKittrick, L. S. & Wheelot, E. C., Jr. (1954). *Carcinoma of the Colon.* Springfield, Ill.: Thomas Publishers.
Michelis, M. E. & Murdaugh, H. V. (1975). Selective hypoaldosteronism. *American Journal of Medicine*, **59**, 1.
Mitchell, W. & Feldman, F. (1968). Neuropsychiatric aspects of hypokalemia. *Canadian Medical Association Journal*, **98**, 49.
Möhring, J., Möhring, B., Naumann, H. J., Philippi, A., Homsy, E., Orth, H., Dauda, G., Kazdd, S. & Gross, F. (1975). Salt and water balance and renin activity in renal hypertension of rats. *American Journal of Physiology*, **228**, 1847.
Morgan, T. & Berliner, R. W. (1969). A study by continuous microperfusion of water and electrolyte movements in the loop of Henle and distal tubule of the rat. *Nephron*, **6**, 388.
Mortimore, G. E. (1961). Effect of insulin on potassium transfer in isolated rat liver. *American Journal of Physiology*, **200**, 1315.
Mudge, G. H., Foulks, J. & Gilman, A. (1948). The renal excretion of potassium. *Proceedings of the Society for Experimental Biology and Medicine*, **67**, 545.
Mudge, G. H. & Vislocky, K. (1949). Electrolyte changes in human striated muscle in acidosis and alkalosis. *Journal of Clinical Investigation*, **28**, 482.
Muggia, F. M. (1973). Hypokalemia and chemotherapy. *Lancet*, **1**, 602.
O'Reilly, M. V., Murnaghan, D. P. & Williams, M. B. (1974). Transvenous pacemaker failure induced by hyperkalemia. *Journal of the American Medical Association*, **228**, 336.
Orringer, C. E., Eustace, J. C., Wunsch, C. D. & Gardner, L. B. (1977). Natural history of lactic acidosis after grand-mal seizures. *New England Journal of Medicine*, **297**, 796.
Pearson, C. M. & Kalyanaraman, K. (1972). Periodic paralysis. In *The Metabolic Basis of Inherited Disease*, 3/e. ed. Stanbury, J. B., Wyngaarden, J. B. & Frederickson, D. S. p. 1181. New York: McGraw Hill.
Perez, G. O., Oster, J. R. & Vaamonde, C. A. (1974). Renal acidosis and renal potassium handling in selective hypoaldosteronism. *American Journal of Medicine*, **57**, 809.
Perez, G. O., Lespier, L. E., Oster, J. R. & Vaamonde, C. A. (1977). Effect of alterations of sodium intake in patients with hyporeninemic hypoaldosteronism. *Nephron*, **18**, 259.
Pitts, R. F., Gurd, R. S., Kessler, R. H. & Hierholzer, K. (1958). Localization of acidification of urine, potassium and ammonia secretion and phosphate reabsorption in the nephron of the dog. *American Journal of Physiology*, **194**, 125.
Pollen, R. H. & Williams, R. H. (1960). Hyperkalemic neuromyopathy in Addison's disease. *New England Journal of Medicine*, **263**, 273.
Rees, L. H. & Ratcliffe, J. G. (1974). Ectopic hormone production by non-endocrine tumors. *Clinical Endocrinology* (New York), **3**, 263.
Rees, L. H. (1975). The biosynthesis of hormones by non-endocrine tumors—a review. *Journal of Endocrinology*, **67**, 143.
Relman, A. S. & Schwartz, W. B. (1952). The effects of DOCA on electrolyte balance in normal man and its relation to sodium chloride intake. *Yale Journal of Biology and Medicine*, **24**, 540.
Relman, A. S. & Schwartz, W. B. (1956). The nephropathy of potassium depletion: clinical and pathological entity. *New England Journal of Medicine*, **255**, 195.
Roussak, N. J. (1952). Fatal hypokalemic alkalosis with tetany during liquorice and P.A.S. therapy. *British Medical Journal*, **1**, 360.
Roza, O. & Berman, L. B. (1971). The pathophysiology of barium: hypokalemic and cardiovascular effects. *Journal of Pharmacology and Experimental Therapeutics*, **177**, 433.

Schambelan, M., Slaton, P. E., Jr. & Biglieri, E. G. (1971). Mineralocorteroid production in hyperadenocorticism. *American Journal of Medicine*, **51**, 299.

Schambelan, M., Stockigt, J. R. & Biglieri, E. G. (1972). Isolated hypoaldosteronism in adults. *New England Journal of Medicine*, **287**, 573.

Scheer, L., Ogden, D. A., Medd, A. W., Spritz, N. & Rubin, A. (1961). Management of hyperkalemia with a cation exchange resin. *New England Journal of Medicine*, **264**, 115.

Schon, D. A., Silva, P. & Hayslett, J. P (1974). Mechanism of potassium excretion in renal insufficiency. *American Journal of Physiology*, **227**, 1323.

Schwartz, W. B. & Relman, A. S. (1953). Metabolic and renal studies in chronic potassium depletion resulting from over use of laxatives. *Journal of Clinical Investigation*, **32**, 258.

Schwartz, W. B. & Relman, A. S. (1967). Effects of electrolyte disorders on renal structure and function. *New England Journal of Medicine*, **276**, 383.

Scribner, B. H., Fremont-Smith, K. & Burnell, J. M. (1955). The effect of acute respiratory acidosis on the internal equilibrium of potassium. *Journal of Clinical Investigation*, **34**, 1276.

Scribner, B. H. & Burnell, J. M. (1956). Interpretation of the serum potassium concentration. *Metabolism, Clinical and Experimental*, **5**, 468.

Sebastian, A., McSherry, E. & Morris, R. C., Jr. (1971). Renal potassium wasting in renal tubular acidosis. *Journal of Clinical Investigation*, **50**, 667.

Sebastian, A., Schambelan, M., Lindenfeld, S. & Morris, R. C., Jr. (1977). Amelioration of metabolic acidosis with fluorocortisone therapy in hyporeninemic hypoaldosteronism. *New England Journal of Medicine*, **297**, 576.

Shields, R. & Miles, J. B. (1965): Absorption and secretion in the large intestine. *Postgraduate Medical Journal*, **41**, 435.

Shields, R. (1966). Absorption and secretion of electrolytes and water by the human colon with particular reference to benign adenoma and papilloma. *British Journal of Surgery*, **53**, 893.

Shils, M. E. (1969). Experimental human magnesium depletion. *Medicine* (Baltimore), **48**, 61.

Simon, N., Franklin, S. S., Bleifer, K. H. & Maxwell, M. H. (1972). Clinical characteristics of renal vascular hypertension. *Journal of the American Medical Association*, **220**, 1209.

Skou, J. C. (1965). Enzymatic basis for active transport of sodium and potassium across cell membranes. *Physiological Reviews*, **45**, 596.

Solomon, R. J. & Brown, R. S. (1975). Bartter's syndrome. New insights into pathogenesis. *American Journal of Medicine*, **59**, 575.

Spencer, R. W. & Voyce, M. A. (1976). Familial hypokalemia and hypomagnesemia. *Acta Paediatrica Scandinavica*, **65**, 505.

Streeten, D. H. P. (1966). Periodic paralysis. In *The Metabolic Basis of Inherited Disease*, 2/e., ed. Stanbury, J. B., Wyngaarden, J. B. & Frederickson, D. S. p. 905. New York: McGraw-Hill.

Streeten, D. H. P. & Speller, P. J. (1974). The role of mineralocorticoids in the pathogenesis of hypokalemic periodic paralysis. *Journal of Clinical Endocrinology and Metabolism*, **39**, 326.

Swan, R. C. & Pitts, R. F. (1955). Neutralization of infused acid by nephrectomized dogs. *Journal of Clinical Investigation*, **34**, 205.

Swingle, W. W., Maxwell, R., Ben, M., Baker, C., LeBrie, S. J. & Eisler, M. (1954). A comparative study of aldosterone and other adrenal steroids, in adrenalectomized dogs. *Endocrinology*, **55**, 813.

Szylman, P. (1976). Role of hyperkalemia in the metabolic acidosis of isolated hypoaldosteronism. *New England Journal of Medicine*, **294**, 361.

Thorn, G. W., Forsham, D. H. & Emerson, K., Jr. (1949). *The diagnosis and treatment of adrenal insufficiency*. P. 182. Springfield, Ill.: Thomas Publishers.

Todd, E. P. & Vick, R. L. (1971). Kalemotropic effect of epinephrine: analysis with adrenergic agonists and antagonists. *American Journal of Physiology*, **220**, 1963.

Tomko, D. H., Yeh, B. P. Y. & Falls, W. F., Jr. (1976). Bartter's syndrome. Study of a 52 year old man with evidence for a defect in proximal tubular sodium reabsorption and comments on therapy. *American Journal of Medicine*, **61**, 111.

Trygstad, C. W., Mangos, J. A. & Bloodworth, J. M. B., Jr. & Lobeck, C. C. (1969). A sibship with Bartter's syndrome: failure of total adrenalectomy to correct the potassium wasting. *Pediatrics*, **44**, 234.

Turnberg, L. A. (1970). Electrolyte absorption from the colon. *Gut*, **11**, 1049.

Uhlich, E., Baldamus, C. A. & Ullrich, K. J. (1969). The effect of aldosterone on sodium transport in the collecting ducts of the mammalian kidney. *Pfluegers Archiv*, **308**, 111.

Upton, G. U. & Amatruda, T. T. (1971). Evidence for presence of tumor peptides with corticotropin releasing factor like activity in the ectopic ACTH syndrome. *New England Journal of Medicine*, **285**, 419.

Verberckmoes, R., VanDamme, B., Clement, J., Amery, A. & Michielsen, P. (1976). Bartter's syndrome with hyperplasia of renomedullary cells: successful treatment with indomethacin. *Kidney International*, **9**, 302.

Verner, J. V. & Morrison, A. B. (1974). Endocrine pancreatic islet disease with diarrhea. *Archives of Internal Medicine*, **133**, 492.

Vick, R. L., Todd, E. P. & Leudke, D. W. (1972). Epinephrine induced hypokalemia—relation to liver and skeletal muscle. *Journal of Pharmacology and Experimental Therapeutics*, **181**, 139.

Walker, W. G., Jost, L. J., Kowarski, A. & Dunn, M. J. (1968). Aldosterone secretion and sodium balance in salt-losing nephropathy. *Johns Hopkins Medical Journal*, **122**, 45.

Wang, P. & Clausen, T. (1976). Treatment of attacks in hyperkalemic familial periodic paralysis by inhalation of salbutamol. *Lancet*, **1**, 221.

Weinstein, S. F., Allan, D. M. & Mendoza, S. A. (1974). Hyperkalemic acidosis and short stature associated with a defect in potassium excretion. *Journal of Pediatrics*, **85**, 355.

Welt, L. G., Hollander, W., Jr. & Blythe, W. B. (1960). The consequences of potassium depletion. *Journal of Chronic Diseases*, **11**, 213.

Whang, R., Morosi, H. J., Rodgers, D. & Reyes, R. (1967). The influence of sustained magnesium deficiency on muscle potassium repletion. *Journal of Laboratory and Clinical Medicine*, **70**, 895.

White, M. G. (1972). Bartter's syndrome: a manifestation of renal tubular defects. *Archives of Internal Medicine*, **129**, 41.

Wiederholt, M., Stolte, H., Brecht, J. P. & Hierholzer, K. (1966). Mikropunklionsuntersuchungen über den Einfluβ von Aldosterone, Cortison und Dexamethason auf die renale Natriumresorption adrenalektomiester Ratten. *Pfluegers Archiv*, **292**, 316.

Wilkinson, P. R., Hesp, R., Issler, H. & Rattey, E. B. (1975). Total body and serum potassium during prolonged thiazide therapy for essential hypertension. *Lancet*, **1**, 759.

Williams, R. H. P. (1973). Potassium overdosage—a potential hazard of non-rigid parenteral fluid containers. *British Medical Journal*, **1**, 714.

Williamson, H. E. (1963). Mechanism of the antinatriuretic action of aldosterone. *Biochemical Pharmacology*, **12**, 1449.

Wright, F. S., Strieder, N., Fowler, H. B. & Giebisch, G. (1971). Potassium secretion by the distal tubule after potassium adaptation. *American Journal of Physiology*, **221**, 437.

Wrong, O. (1961). Incidence of hypokalemia in severe hypertension. *British Medical Journal*, **2**, 419.

Zierler, K. L. (1960). Effect of insulin on potassium efflux from rat muscle in the presence and absence of glucose. *American Journal of Physiology*, **198**, 1066.

7

The consequences of potassium deficiency

ROBERT E. CRONIN
JAMES P. KNOCHEL

Introduction
Etiology of potassium depletion
Effects of potassium depletion on the kidney
Acid-base disturbances
Impaired urinary concentration
Other tubular defects
Hyponatremia
Histologic alterations

Other physiologic effects of potassium depletion
On the skeletal muscles
On the myocardium
On the vascular system
Metabolic factors in potassium deficiency
Carbohydrate metabolism
Nitrogen metabolism
Conclusions

INTRODUCTION

Potassium (K) deficiency is a frequent clinical problem. Although it occurs in a wide variety of clinical settings, its diagnosis may be difficult. Of the methods available for assessing body potassium stores, the serum potassium concentration is the most readily available but can be the least reliable. While serum potassium concentration correlates with total body potassium (Boddy et al., 1976), under certain conditions it may be within the normal range despite potassium depletion. For example, during diabetic ketoacidosis, the rise in hydrogen ion concentration (H^+) of extracellular fluid (ECF) is buffered by the movement of hydrogen ion into cells and balanced by the movement of potassium into the ECF. In this situation, serum K may be normal or even elevated even though total body K is reduced. Therefore, K-deficiency might be overlooked or underestimated if serum potassium concentration were the only measure of potassium stores evaluated.

The methods used to estimate total body potassium include: (1) total body counting of the naturally occurring radioactive potassium isotope ^{40}K; (2) measurement of exchangeable potassium using the radioactive isotopes ^{42}K or ^{43}K; (3) chemical analysis of potassium in tissues; and (4) neutron activation analysis.

Estimation of total body K by counting the natural isotope ^{40}K is unquestionably the most advantageous, but it requires a whole body counter. Unfortunately, there are few of these expensive instruments in existence. An acceptable substitute for whole body counting is the isotope dilution technique using either ^{42}K or ^{43}K. These isotopes are readily available, can be employed without undue hazard of radiation injury, and where properly used, yield acceptably reproducible data. The major disadvantage of ^{42}K is its short life (viz., 12.5 hours). The half life of ^{43}K is 22 hours. In healthy subjects whose total body potassium averaged 3,250 mEq, values obtained in our laboratory were reproducible within a range of 200 mEq employing ^{42}K and allowing 24 hours for equilibration (Knochel, Dotin and Hamburger, 1972). In disease states it may be advantageous to allow 36 hours for equilibration of the isotope with total body K. By these methods, exchangeable ^{42}K will equilibrate with at least 95 percent of total body potassium.

In a healthy 70 kg man, the total quantity of potassium in the body is approximately 3,500 mEq or about 50 mEq/kg body weight. In terms of body weight, there is slightly less in women because of their increased partition of body fat. Most of the potassium in the body is contained in skeletal muscle and viscera. In these tissues, its concentration in intracellular water and its content in dried, defatted tissue is remarkably constant at about 150 mEq/l and 44.0 mEq/100 g fat-free dry solids, respectively. Body weight is often an unreliable index of potassium stores because of the wide individual variation in the content of adipose tissue, which contains little potassium. In health, more reproducible indices of total body K include lean body mass, total body water, or red cell mass. However, to quantitate one of these entities is not only impractical as a clinical tool, but in the presence of serious disease, each may become deranged independently of total body potassium. Thus, many serious diseases are associated with an increase of both intracellular and extracellular water. Since water is the major component of the lean, fat-free body, lean mass might be elevated although no actual deficit of potassium exists. A good example of this is the patient with expansion of total body water due to inappropriate secretion of antidiuretic hormone (SIADH). An illustration of this is as follows:

Table 7.1

	Body weight Kg	Body water L	LBM[a] Kg	^{42}Ke[b] mEq	^{42}Ke mEq/kg Body weight	Ke/LBM mEq/kg
Normal	70	42	57	3,500	50	61
SIADH	75	47	64	3,500	46	54

[a] LBM is estimated by dividing total body water by 0.732 (RP).
[b] Ke denotes exchangeable ^{42}K.

Equally spurious estimates of total body K may result from similar techniques in patients with dehydration.

In investigative work, one proven and reliable method to assess tissue potassium deficiency is by analysis of skeletal muscle. This technique is moderately tedious but nevertheless rewarding. It takes advantage of the fact that (1) defatted, dried solids in skeletal muscle are about 80 percent protein, and (2) the potassium/nitrogen ratio in muscle tissue is very close to 3.3 mEq/gm.* Thus the potassium content of normal muscle is about 40–45 mEq/100 gm fat-free dry solids. In potassium deficiency, muscle potassium content is invariably below normal.

Estimation of K content in certain other tissues (e.g., erythrocytes) tends to yield erratic values, perhaps resulting from the diverse age of red cells in circulating blood. Recent evidence suggests that K content of leukocytes may more accurately reflect total body stores (Edmondson et al., 1976). Even in advanced potassium deficiency, K content of myocardium and liver remains nearly normal, falling only in the preterminal state.

The high concentration gradient of potassium ions between the intracellular and extracellular fluids is thought to result from several phenomena. First, inside cells, there are several species of anionic macromolecules, including proteins and organic phosphates, that electrically attract positively charged potassium ions. Second, active transport of sodium (Na) ions from the inside to the outside of cells increases intracellular electronegativity by removal of positive charges. This outward movement of sodium is roughly coupled to the inward transport of potassium ions from plasma into cells. Third, most cellular membranes are less permeable to K than Na ions. Thus, electrogenic forces, active transport, and differential permeability contribute to the maintenance of an approximately 37:1 concentration gradient of K between cell water and ECF. This high intracellular concentration of potassium ions participates in several important functions including impulse conduction, regulation of certain enzymes, glycogen synthesis, muscle contraction, vascular tone, and rhythmicity of the heart.

Mild to moderate potassium depletion may go unnoticed since overt clinical abnormalities may be absent. Severe potassium depletion, however, usually produces serious derangements that may involve any organ system. Following a brief description of factors that predispose to potassium depletion, this chapter will review the effects produced by potassium depletion with particular emphasis on changes occurring in the kidney and muscle.

ETIOLOGY OF POTASSIUM DEPLETION

To maintain a high concentration gradient of K between intracellular and extracellular water requires work and expenditure of energy. Quite obviously, a variety of factors might interfere with energy utilization, impair cellular function, and thus permit K ions to escape from the cell. A classical example of such a defect is severe congestive heart failure where ischemia prevents delivery of both energy substrates as well as oxygen.

---

* One gm of N may be converted to protein by multiplying by 6.25.

Similar potassium deficits may occur in other disease states in which energy utilization is impaired (e.g., uncontrolled diabetes mellitus, severe infections, thermal burns, starvation, malnutrition, various chemical intoxications, and others).

In this type of potassium deficiency (i.e., where K ions leave the cell because of defective ion transports), the deficiency state can be repaired only if the primary disorder is corrected first. In these circumstances, the concentration of potassium inside the cell is only moderately depressed, and its concentration in extracellular water usually is normal. When the cause of the abnormality is removed, transient, modest hypokalemia may occur as a result of intracellular movement of potassium. Examples of this include recovery from respiratory acidosis, recovery from severe infection such as lobar pneumonia, and recovery from protein-calorie malnutrition.

Reduced potassium intake is an unusual cause for potassium depletion. However, it may occur in settings where nutritional needs are grossly neglected (e.g., chronic alcoholism). During protracted vomiting, the potassium lost each day in gastric juice approximates only about 10 mEq/l, but this does not represent the major part of the total potassium loss. The bulk of potassium loss consequent to vomiting is into the urine. Since the renal mechanism for potassium conservation is slow during states of gastrointestinal potassium loss, renal potassium excretion may result in a large net deficit during the first two or three days. In time, renal conservation of potassium will occur, but renal losses may still be as high as 20 mEq/day after two weeks. In addition, hydrogen ion loss during vomiting will generate an extracellular alkalosis that further contributes to the renal loss of potassium. In contrast to states of protracted vomiting, during diarrheal states or with laxative abuse, intestinal losses are rich in potassium bicarbonate, and potassium depletion may develop in the face of metabolic acidosis. Geophagia or clay ingestion causes potassium binding in the gut and in effect represents potassium malabsorption. With the rare villous adenoma of the colon, the large surface area of the tumor may allow the secretion of one or more liters of fluid containing as much as 100 mEq of potassium per day.

Excessive production of mineralocorticoid, either primary or secondary, may also be responsible for potassium deficiency. Unsuspected adrenal tumors that produce aldosterone may be discovered because of their propensity to cause hypokalemia after diuretic therapy for hypertension. In such cases, diuretics lead to enhanced potassium excretion by (1) augmenting sodium delivery to the aldosterone-dependent distal sodium-potassium exchange site and by (2) creating a metabolic alkalosis. Diuretic agents used to control uncomplicated hypertension uncommonly lead to significant potassium depletion. In contrast, when used in conjunction with mineralocorticoid agents, serious potassium depletion may occur. (This will be discussed again in a later section.) In renal tubular acidosis, the presence of mild volume contraction with secondary hyperaldosteronism combined with excess quantities of the sodium salt of a nonreabsorable anion (bicarbonate) in distal tubular fluid leads to enhanced potassium secretion.

High urine flows associated with osmotic diuresis (mannitol, glucose) enhance potassium secretion by lowering potassium concentration in distal tubular urine. This produces a more favorable chemical gradient for potassium secretion. In addition, with the osmotic diuresis of diabetic ketoacidosis, the presence of nonreabsorbable anions (keto acids) in the distal tubule augments the electrical gradient for potassium excretion by increasing luminal negativity. The profound depletion often experienced by these patients may only be appreciated following insulin therapy when correction of the acidosis leads to a precipitous fall in serum potassium.

EFFECTS OF POTASSIUM DEPLETION ON THE KIDNEY

Since the kidney is the primary organ responsible for excreting normal as well as excessive daily potassium loads, it understandably plays a central role in potassium conservation during states of deprivation. Several studies during potassium depletion indicate that the renal handling of potassium and sodium are linked and altered in such a way that potassium losses are reduced. During potassium depletion, proximal fractional sodium reabsorption is increased (Bank and Aynedjian, 1964) and distal sodium reabsorption is reduced. This latter effect is in large part due to reduced aldosterone levels resulting from potassium depletion (Cannon, Ames and Laragh, 1966). As a direct consequence of the reduced distal sodium reabsorption, the potential difference (PD) across the distal tubule falls (Finkelstein and Hayslett, 1975), making the electrical gradient for potassium diffusion into tubular urine less favorable. Also, during states of potassium depletion, the renal response to stimuli that normally increase potassium excretion appears to be blunted. For example, during potassium depletion, increases in urinary flow are not accompanied by the usual increase in potassium secretion (Khuri et al., 1975). Using data such as these, Sealey and Laragh (1974) proposed a cybernetic system wherein alterations in potassium balance were compensated by a homeostatic mechanism combining the interaction of the renin-angiotensin-aldosterone system with renal handling of potassium and sodium. Thus, during potassium depletion, potassium is conserved through diminished distal tubular sodium delivery and reabsorption, owing to increased proximal fractional sodium reabsorption and lower aldosterone levels, respectively. However, when these adaptive mechanisms fail and serious potassium depletion occurs, derangements in several important renal functions take place. The changes described below occur almost exclusively at the tubular level confirming earlier reports that potassium depletion has little effect on glomerular filtration rate in man (Schwartz and Relman, 1967) or dog (Beskind and Mudge, 1959) although a fall was reported in the rat (Muntwyler and Griffin, 1953).

Acid-base disturbances

The close inverse relationship between cellular potassium concentration and the level of intracellular acidity has been appreciated for some time (Fenn and Cobb, 1934). Although the exact relationship existing between these ions appears to be extremely complex and incompletely understood many of the metabolic and physiologic consequences of potassium depletion described in this chapter may occur because of potassium-induced alterations of intracellular pH (Adler and Fraley, 1977).

The close reciprocal relationship between the renal secretion of hydrogen and potassium ions is well-known and was responsible for the theory that potassium and hydrogen ions compete for a common carrier-mediated secretory mechanism in the distal tubules (Berliner, Kennedy and Orloff, 1951). Supporting this view was the observation that hydrogen ion secretion was reduced during potassium loading in man and experimental animals (Loeb, Atchley and Richards, 1932; Fuller, MacLeod and Pitts, 1955) and increased by potassium depletion (Roberts, Magida and Pitts, 1953). In addition, inhibition of hydrogen ion secretion with a carbonic anhydrase inhibitor caused a marked augmentation of potassium excretion (Berliner et al., 1951). Conversely, human subjects depleted of potassium develop an acid urine ("paradoxical aciduria") despite extracellular alkalosis if certain conditions exist: (1) alkalosis usually secondary to vomiting (2) vomiting has stopped (3) volume contraction is severe (Roberts et al., 1953). Therefore, the effect of potassium depletion on hydrogen ion secretion must, according to such a theory, occur at the distal tubule, the site of potassium secretion. Studies by Rector, Bloomer and Seldin (1964), however, demonstrated that during potassium depletion, secretion of hydrogen ion in the proximal tubule was also increased, a segment of the nephron where potassium is being reabsorbed rather than secreted. Clearly then, potassium depletion can augment hydrogen ion secretion independently of a carrier-mediated mechanism. The view generally offered to explain these observations is that intracellular losses of potassium are partially replaced by hydrogen ions with resultant intracellular acidosis (Adler, Zett and Anderson, 1972; Wilson and Simmons, 1970). Thus, the secretion of hydrogen ion would be favored at distal as well as proximal secretory sites. However, this explanation of renal hydrogen ion secretion remains a theoretical consideration since the pH of renal tubular cells has not been directly measured in vivo during potassium depletion. Despite the reservation, the bulk of the data indicates that rather than a tightly linked potassium-hydrogen exchange system, distal potassium secretion is best explained as a passive phenomenon generated by a transepithelial potential difference. The latter results from active sodium reabsorption. In addition, K secretion would be favored by the concentration gradient for potassium between tubular cells and the tubular fluid (Gennari and Cohen, 1975).

While the association between potassium deficiency and enhanced hydrogen ion secretion seems clear, its relationship to development of metabolic alkalosis is more complex and may not simply represent cause and effect. Since volume depletion per se can increase proximal tubular hydrogen ion secretion (Rector, Carter and Seldin, 1967), alternatively, then, metabolic alkalosis may result from chloride depletion out of proportion to sodium loss and not depletion of potassium per se (Aitkins and Schwartz, 1962; Schwartz, van Ypersele de Strihou and Kassirer, 1968). Atkins and Schwartz (1962) showed that the alkalosis associated with potassium depletion can be corrected by administration of saline alone. Moreover, the degree of the potassium depletion does appear to have an effect on the development of the alkalosis. The alkalosis in rats with a severe potassium deficiency maintained by DOCA is only partially corrected by the administration of saline (Seldin and Rector, 1972). Likewise, in patients with severe potassium depletion, saline alone may not correct the metabolic alkalosis until the large potassium deficit is repaired (Garella, Chazan and Cohen, 1970). Moreover, when extracellular volume is held constant, potassium deficiency clearly increases hydrogen ion secretion as indicated by increased renal bicarbonate reabsorption (Kurtzman, White and Rogers, 1973). In rats with potassium deficiency induced by a diet low in potassium but normal in chloride, alkalosis develops and is accompanied by a decrease in muscle pH (Adler, Zett and Anderson, 1974). Thus, the mechanism of enhanced bicarbonate generation leading to metabolic alkalosis in potassium depletion appears to depend upon at least two factors: (1) ECF contraction, which increases the apparent renal Tm for bicarbonate excretion and (2) intracellular acidosis as a result of potassium depletion, which increases hydrogen ion secretion. Species variation has lead to some confusion regarding the acid-base response to potassium depletion. The observations in rats that isolated potassium deficiency leads to metabolic alkalosis (Adler et al., 1974) does not appear to hold for the dog, where chronic dietary potassium restriction without ECF contraction results in diminished net acid excretion and a hyperchloremic metabolic acidosis (Burnell, Teubner Simpson, 1974; Hulter, Ilnicki and Sebastian, 1976; van Ypersele de Strihou and Dieu, 1977). The mechanism of this metabolic acidosis may be hypokalemia-induced aldosterone deficiency, since dietary potassium restriction in the adrenalectomized dog maintained on fixed dose of aldosterone prevented development of metabolic acidosis (Hulter et al., 1976). However, in examining the potassium deficiency acidosis of the dog in response to a chronic acid load, van Ypersele de Strihou and Dieu (1977) were unable to show that changes in aldosterone secretion mediated the effect. They proposed that potassium deficiency in addition to enhancing ammoniagenesis also impaired the pH gradient achieved across the tubular cells and by this mechanism reduced hydrogen ion excretion. Such a defect would in effect represent a form of acquired renal tubular acidosis. The development of acidosis rather than alkalosis following potassium restriction in these dogs may also be explained in part by the mild degree of potassium deficiency achieved and the maintenance of a normal extracellular fluid volume. Clearly, the dog is capable of responding to combined potassium and chloride deficiency with metabolic alkalosis (Wilson and Simmons, 1970). But in contrast to the marked potassium depletion developed in this and earlier studies (Kassirer et al., 1956; Aitkins and Schwartz, 1962), potassium depletion that caused metabolic acidosis was modest as reflected by balance data and the small reduction in plasma potassium during the potassium depletion period. In addition, ECF contraction was prevented during potassium depletion by substituting potassium losses with sodium chloride.

Several lines of evidence indicate that ammonia production is closely related to potassium homeostasis (Tannen, 1977). In 1970 Tannen clearly showed that with potassium depletion, ammonium excretion was elevated in the absence of changes in systemic pH, pCO_2, bicarbonate, or fixed acid excretion. Also, a primary increase in renal ammonia production is accompanied by a reciprocal decrease in potassium excretion (Tannen and Terrien, 1975). Moreover, potassium-deficient human subjects respond to an acute acid load with an increase in net acid excretion accounted for mainly by enhanced ammonium excretion. In contrast, van Ypersele de Strihou and Dieu (1977) studied potassium deficient dogs challenged with a chronic acid load and concluded that potassium deficiency acidosis resulted from an impaired ability to lower urinary pH with only a secondary effect on renal ammoniagenesis. Also, the potassium-depleted human kidney failed to reduce urinary pH in response to an acid load (Clarke et al., 1955; Wrong and Davies, 1959). These studies suggested that potassium depletion induced a state of acquired renal tubular acidosis. In view of the data cited earlier showing that potassium depletion enhances proximal tubular bicarbonate reabsorption (Rector et al., 1964), the site of this proposed renal tubular acidosis must necessarily be the distal tubule. However, a more plausible explanation for these findings appears to be that potassium depletion leads to a primary increase in ammonia production with a subsequent increase in urine pH (Tannen, 1977). Nevertheless, a hydrogen ion gradient defect in the distal tubule has not been conclusively excluded. How this linkage between potassium and ammonia production benefits the organism is unclear. Tannen (1977) suggested that potassium responsive changes in renal ammonia production serve as a mechanism for maintaining relatively normal hydrogen ion homeostasis during manipulation of potassium intake. The biochemical explanations for this reciprocal relationship between potassium levels and ammonia production are not clear but a phosphate-dependent glutaminase is increased in the rat kidney by potassium depletion (Ching, Rogoff and Gabuzda, 1973; Pagliara and Goodman, 1970), an observation consistent with an increased rate of ammonia production. The evidence for glutaminase induction in the dog is conflicting (Pollak et al., 1965; Gabuzda and Hall, 1966), but using in vivo techniques, potassium depletion increased renal ammonia production in normal dogs

(Gabuzda and Hall, 1966) and humans with cirrhosis (Baertle, Sancetta and Gabuzda, 1963).

Enhancement of ammonia production following potassium depletion may lead to hepatic encephalopathy when it occurs in advanced liver disease (Gabuzda and Hall, 1966). Under normal conditions, ammonia generated within renal tubular cells freely diffuses into the tubular urine. In direct proportion to the degree of urinary acidity, ammonia is "trapped" and excreted as the poorly diffusable ammonium ion. With severe liver disease, maximum urinary acidifying capability may be defective, which raises urinary pH and thus impairs the ammonia "trapping" mechanism. Thus potassium depletion enhances renal tubular ammonia production, which through failure of the trapping mechanism, backdiffuses into the peritubular circulation and thus to the systemic circulation. In the presence of severe liver disease, ammonia conversion to urea may be impaired and thus potassium deficiency in the cirrhotic may be responsible indirectly for hepatic coma. The clinical observation that hepatic coma often follows diuretic administration may have its origin in a further reduction of already marginal potassium stores. However, when carefully given, diuretics reportedly assist in mobilization of ascites in these patients without an added risk of encephalopathy (Gregory et al., 1977).

In summary, potassium depletion decreases intracellular pH in muscle and most likely in renal tubular cells. The metabolic alkalosis seen with moderate potassium depletion in part results from this intracellular acidosis, which favors renal hydrogen ion secretion. However, more important in its generation and correction is the accompanying chloride depletion (volume contraction). In some cases of severe potassium depletion, especially in association with mineralocorticoid excess, metabolic alkalosis cannot be corrected simply with sodium chloride (volume expansion) until the potassium deficit has been corrected. Mild degrees of simple potassium depletion seem to have little effect on overall acid-base balance in man although ammonia generation is enhanced. In liver disease, however, augmented ammonia production resulting from potassium depletion may precipitate hepatic coma.

Impaired urinary concentration

Impaired urinary concentrating ability and polyuria are consistent features of the nephropathy of potassium depletion (Relman and Schwartz, 1956; Relman and Schwartz, 1958). Kaliopenic nephropathy usually implies serious potassium deficiency. Kassirer and Harrington (1977) could find no evidence that potassium depletion nephropathy develops secondary to diuretic therapy, presumably reflecting the generally mild potassium deficit seen in this setting. In 17 patients with a negative potassium balance of 380 mEq after two to four months of diuretic treatment, none was unable to concentrate urine normally (Healy et al., 1970). The pathophysiologic mechanism responsible for this concentrating defect is unknown but several studies appear to shed some light on the answer. Potassium depletion impairs the water permeability of the toad bladder (Finn, Handler and Orloff, 1966) and the collecting tubules of the rat (Bank and Aynedjian, 1959). Moreover, Beck and Webster (1976) showed that potassium depletion in the rat reduces responsiveness of the renal cyclic AMP system to vasopressin stimulation. With potassium repletion this impairment of cyclic AMP generation was reversed. In addition, impairment of maximal urinary osmolality may also be related to reduced production of ATPase in the kidney. This enzyme is activated only when medullary tonicity exceeds 600 m0sm (Kannegiesser and Lee, 1971). Potassium depletion may also cause polyuria by stimulating the thirst mechanism (Berl et al., 1976). In this study, potassium-deficient rats had an increased water intake, which in large measure accounted for the increased urine flow. In the dog, the polyuria of hypokalemia is associated with increased urinary excretion of prostaglandins, and the defect is reversed by an inhibitor of prostaglandin synthesis (Galvez, et al., 1976). This suggests that potassium depletion increases renal prostaglandin synthesis, which antagonizes the effect of ADH at the level of the collecting tubules.

Other tubular defects

In addition to the concentrating defect associated with potassium depletion, a number of other tubular defects also may occur. Potassium depletion can impair parathyroid hormone-dependent cyclic-AMP generation in the renal cortex (Beck and Davis, 1975) and the phosphaturic response to PTH. The clearance of para-aminohippurate (PAH) may be reduced due to a decreased extraction and secretion (Schwartz and Relman, 1953). Aminoaciduria has been associated with potassium depletion but the relationship is not entirely clear (Stanbury and MaCaulay, 1957). Tubular handling of glucose in the dog appears to be normal (Beskind and Mudge, 1959).

Hyponatremia

Potassium depletion may lead to hyponatremia in patients treated with diuretics in the absence of an edema-forming state or extracellular fluid volume depletion. In a report by Fichman et al. (1971) hyponatremia was associated with only slight reductions in mean exchangeable sodium but a marked decrease in measured exchangeable potassium, suggesting that sodium had moved from an extracellular location to compensate for the intracellular potassium depletion. Furthermore, the hyponatremia was reversed with administration of potassium chloride despite continued use of the diuretic. Sodium retention is a known consequence of potassium depletion (Bay et al., 1976). Moreover, edema may occur in states of potassium depletion as a result of this sodium retention and may increase somewhat with potassium replacement, suggesting that intracellular sodium returns to the extracellular compartment during potassium repletion (Bland and Bassett, 1953; Fourman and Hervey, 1955).

Histologic alterations

The characteristic renal lesion of potassium depletion in man consists of vacuolization of tubular epithelial cells, most prominently those of the proximal tubule (Relman and Schwartz, 1956, Muehrcke and Rosen, 1964). The nature of the material in the vacuoles is unknown, and while its appearance is characteristic of potassium depletion, it is not specific since similar vacuolization of proximal tubular cells may occur following certain osmotic diuretics (Di Scala et al., 1965). Glomerular, vascular, and interstitial structures usually appear normal when the duration of the potassium depletion is relatively short. Although interstitial fibrosis with glomerular and vascular changes is reported in cases of chronic potassium depletion (Cremer and Bock, 1977), a clear etiologic relationship is disputed (Schwartz and Relman, 1967). Potassium depletion in experimental animals, however, does suggest an association between potassium depletion and the development of pyelonephritis (Muehrcke and Rosen, 1963; Woods et al., 1959), but these histological changes in the rat may have little relevance to man. The anatomical defect that occurs in the rat occurs in the cells of the papillary tip in which multivesicular bodies are seen (Toback et al., 1976). Later, in the outer medulla and at the corticomedullary junction, hyperplasia and swelling of the epithelial cells of the collecting tubules occurs. Unlike human potassium depletion, vacuolization of proximal tubular cells does not develop in the rat. Thus, a causal relationship between potassium depletion and interstitial nephritis in man had not been proven, and the interstitial nephritis that does occur in potassium-depleted experimental animals may have little relevance to the human disease.

OTHER PHYSIOLOGIC EFFECTS OF POTASSIUM DEPLETION

On the skeletal muscles

Skeletal muscle contains the largest reservoir of potassium in the body. It accounts for 94 percent of total body potassium and as such, serves a dual role. Normal muscle potassium content is required for proper muscle function but in addition, this large potassium reservoir under the control of several hormonal and nonhormonal regulators serves as a buffer pool preventing large potentially fatal swings in ECF potassium concentration. This section describes the consequences of potassium depletion on skeletal muscle.

In the clinical setting, muscle weakness is probably the most common symptom of potassium depletion, but frank paralysis and rhabdomyolysis also may occur (Campion, Arias and Carter, 1972). Profound hypokalemia

(1.7 mEq/l), leading to total paralysis was recently described in a patient with renal tubular acidosis following toluene "sniffing" (Taher et al., 1974). When potassium deficiency becomes sufficiently severe to induce frank rhabdomyolysis, many patients become paralyzed. This clinical situation has been mistaken for the Guillain-Barré syndrome (Roystan and Prout, 1976). Development of muscular weakness or paralysis is usually associated with a serum potassium concentration less than 2 mEq/l. One of the most devastating clinical consequences of potassium depletion is the occurrence of rhabdomyolysis. Clinically, most cases occur as the result of excess mineralocorticoids such as following ingestion of licorice or carbenoxolone, agents with mineralocorticoid activity. In contrast, chronic diuretic therapy will generally produce a mild stable reduction in total body potassium of approximately 250 mEq (McFarland and Carr, 1977), a level very unlikely to produce overt rhabdomyolysis. In those cases where frank rhabdomyolysis has followed diuretic therapy, the agent implicated has been a long-acting diuretic with a propensity toward potassium wasting (e.g., chlorthalidone) (Oh, Douglas and Brown, 1971; Cohen and Hills, 1959). In other cases, rhabdomyolysis occurred following ingestion of long-acting diuretics in conjunction with a kaliuretic agent (e.g., chlorthalidone and carbenoxolone or licorice) (Descamps, Vandenbroucke and van Ypersele de Strihou, 1977; Gross, Dexter and Roth, 1966).

Experimental studies have cast some light on the mechanism of rhabdomyolysis in potassium deficiency (Knochel and Schlein, 1972; Knochel and Carter, 1976; Bilbrey et al., 1973). In early deficiency, the fall of extracellular potassium concentration $[K]o$ is proportionally greater than that of intracellular potassium concentration $[K]i$. As predicted from the Nernst equation, the resting transmembrane electrical potential differences (Em) of the muscle rises. With more advanced potassium deficiency (i.e., approximately 25 percent or more), the Em declines. This suggests either impairment of sodium transport mechanisms or increased leakiness of the sarcolemmal membrane to sodium. The reduction in Em is associated with an increased intracellular content of sodium, chloride, and water. When the Em declines, muscle creatine phosphokinase activity in serum rises, suggesting decomposition of the muscle cell. Histologic studies of muscle under such conditions show myofibrillar vacuolization, patchy hyalinization, and isolated fiber necrosis.

Inferential observations suggest that cells such as those described above are in jeopardy of dissolution (rhabdomyolysis) when a stress is superimposed that critically depresses energy stores (e.g., ATP) to very low values. Examples of such stresses are muscular exercise or acute starvation, which can produce anoxia or sufficiently limit fuel to prevent adequate supplies of ATP to sustain cellular integrity (Sweetin and Thomson, 1973).

In support of the foregoing, experimental studies in the dog suggest that potassium depletion leads to rhabdomyolysis through failure of the vasodilatory response to exercise (Knochel and Schlein, 1972). In the normal state, exercise hyperemia of skeletal muscle results from a localized vasodilatation in response to hyperkalemia. The quantity and/or quality of this localized hyperemia in muscle appears to be defective in potassium-depleted states. The clinical counterpart of these experimental studies is the untrained military recruit in whom acute renal failure from rhabdomyolysis is most likely to develop after approximately two weeks of training, a time when potassium depletion is maximum (Schrier et al., 1970; Knochel et al., 1972). Why some species develop rhabdomyolysis after potassium depletion and others do not is unclear, but it may be due to the effect of potassium on membrane integrity. In man and the dog, rhabdomyolysis occurs following severe potassium depletion, and membrane potential prior to developing rhabdomyolysis is low. This does not seem to be the case in the rat, where rhabdomyolysis of skeletal muscle in potassium deficiency does not occur (Bilbrey et al., 1973).

Although the myopathy of alcoholism is commonly associated with hypokalemia and a decreased muscle potassium content is reported, (Martin et al., 1971), Knochel and coworkers (1975) were unable to confirm a decrease in intracellular muscle potassium. In this myopathy a deficiency of phosphorus caused by the chronic alcoholism is a more likely explanation for the death of muscle cells.

Hypokalemic periodic paralysis is a familial disorder characterized by weakness and flaccid paralysis of skeletal muscle. For a few days prior to attacks the excretion of potassium and phosphorus may fall (Graeff and Lameijer, 1965) and during attacks potassium appears to move into muscle cells (Grob, Johns and Liljestrand, 1957; Zierler and Andres, 1957). In the period between paralytic episodes, total body potassium in these patients is reported to be low or low normal (Coppen and Reynolds, 1966; Talso et al., 1963). Theoretically, a sudden increase in intracellular potassium and a fall in extracellular potassium would be expected to hyperpolarize muscle cells making them less excitable and would provide a logical explanation for the muscular weakness and paralysis seen with this disorder. Unfortunately, direct measurements of resting membrane potentials during paralysis have shown rather significant depolarization of the muscle membrane (Hofmann and Smith, 1970; Creutzfeldt et al., 1963). The precise relationship of high carbohydrate meals to precipitation of paralytic episodes is not clear. Carbohydrate-induced release of insulin, glycogen formation, and the associated movement of potassium to an intracellular location may provide part of the explanation. Muscle glycogen content is abnormally elevated in hypokalemic periodic paralysis. Our own observations have shown that muscle potassium content increases to abnormally high levels during paralytic episodes (unpublished observations, Fuller and Knochel). Although intermittent adrenocorticol hypersecretion has been proposed as a pathogenic mechanism in hypokalemic periodic paralysis, little substantial evidence can be marshalled to support this viewpoint (Streeten, 1966).

At least part of skeletal muscle uptake of potassium is a beta-adrenergic receptor function and the successful use of drugs that act at this site may have important etiologic implications in these diseases. Propranalol, an agent that blocks potassium uptake by skeletal muscle cells, has been used effectively in three patients with hypokalemic periodic paralysis associated with thyrotoxicosis (Conway, Seibel and Eaton, 1974). Salbutamol, a beta-adrenergic stimulator, causes hypokalemia apparently by enhancing muscular uptake of potassium and has been used successfully in the treatment of hyperkalemic familial periodic paralysis (Wang and Clausen, 1976).

For the present, the etiology of hypokalemic periodic paralysis remains unclear. Potassium deficiency per se does not appear to play a role in this fascinating disease.

On the myocardium

Cardiac abnormalities resulting from potassium depletion are primarily electrocardiographic. Older, equivocal observations suggest that congestive heart failure may occur (Perkins, Petersen and Riley, 1950; Harrison, Pilcher and Ewing, 1930). The human myocardium may undergo severe histological changes characterized by loss of muscle striation, myofibrovascularization, interstitial cellular infiltration and varying degrees of myocardial necrosis and scarring (Rodriguez, Wolfe and Bergstrom, 1950; Keye, 1952). Similar lesions can be produced experimentally in the rat (Darrow and Miller, 1942). In the rat, myocardial necrosis becomes evident when myocardial potassium is severely reduced, usually to less than 72 mEq/kg wet weight (Nickerson, Karr and Dresel, 1961). Cannon, Frazier and Hughes (1953) showed, however, that lesions in the potassium-deficient rat myocardium do not occur if the animals are simultaneously deprived of sodium. Studies in isolated cat papillary muscle show that chronic intracellular potassium depletion leads to depressed velocity of contraction, reduction in the extent of shortening, and reduced maximal-developed isometric tension (Gunning, Harrison Coleman, 1972). The etiology of these changes is not clear but may be due in part to depressed mitochondrial function (Harrison et al., 1972). This conclusion is supported by earlier work which concluded that potassium regulates cellular respiration by a direct effect on mitochondria and secondarily through control of ADP production (Bland and Whittam, 1965). Not all species, however, respond to chronic potassium depletion with reduction in myocardial potassium content, cell necrosis, and cardiac failure. Potassium-depleted rabbits demonstrate a sharp decrease in skeletal muscle pH and potassium concentration but no such changes are noted in left ventricular muscle (Cameron and Hall, 1975). In the dog as well, chronic potassium depletion leads to a reduction in skeletal but not left ventricular muscle potassium concentration (Bahler and Rakita, 1971). Moreover, potassium-depleted dogs do not demonstrate cardiac histologic changes (Smith, Black-Schaffer and Lasater, 1950; Bahler and Rakita, 1971) and are more likely to die as the result of profound skeletal muscle weakness and anorexia or a ventricular arrhythmia (Bahler and Rakita, 1971). Potassium-depleted dogs have usually exhibited an increased cardiac output in the resting state (Bahler and Rakita, 1971; Bay et al., 1976; Knochel and Schlein, 1972), but a reduced cardiac output has also been reported (Abbrecht, 1972). Rather than a primary effect of potassium depletion, the rise in cardiac output usually seen may be a secondary response since total systemic resistance is reduced (Bahler and Rakita, 1971, Bay et al., 1976). It seems likely that those animals demonstrating an elevated cardiac output may have been masking a myocardium without reserve since potassium-depleted dogs fail to increase cardiac output in response to exercise and some developed overt pulmonary edema (Knochel and Schlein, 1972).

There is ample clinical and experimental evidence that potassium depletion results in increased cardiac sensitivity to digitalis preparations. Since the mechanism of action as well as toxicity of digitalis appears to depend on a reduction of fluxes of monovalent ions across cell membranes and in particular a decrease in the influx of potassium, the added burden of potassium depletion would accentuate digitalis toxicity. The mechanism of the apparent synergism between potassium depletion and digitalis on myocardial contractility is unknown but several pieces of data suggest that the link may be their similar effect on cell membrane Na-K-ATPase, the source of pump energy. Lowering in vitro potassium concentration reduces membrane Na-K-ATPase activity (Whittam and Ager, 1965) while conversely the rat kidney adapts to a chronic potassium load by increasing renal tissue concentrations of Na-K-ATPase, even in the absence of aldosterone (Silva, Hayslett and Epstein, 1973). Regarding digitalis, a growing body of experimental work suggests that the pharmacologic receptor for digitalis is the cell membrane Na-K-ATPase enzyme (Schwartz, 1976). Thus, digitalis binding to the receptor reduces sodium and potassium fluxes that provide additional calcium to the contractile proteins. There are several clinical consequences of these changes. Predictably, decreases in intracellular concentrations of potassium lower the membrane potential at the end of repolarization and could thereby lead to increased automaticity, decreased conduction velocity, and shortened refractory period. The electrocardiographic changes in potassium depletion are nonspecific and include prolongation of the Q-U interval, widening, flattening, or inverting of the T wave, S-T segment displacement, prominent U waves, diminished QRS voltage, and increased A-V conduction time (Blomberg and Lindqvist, 1954). A variety of atrial and ventricular arrhythmias may occur with potassium depletion, but the most common are premature atrial and ventricular beats. All of these tend to subside with correction of the hypokalemia. However, a reduction in ventricular arrhythmias in response to potassium administration is not necessarily proof of potassium depletion since of itself potassium has a nonspecific depressing effect on arrhythmias in the absence of potassium depletion or digitalis intoxication (Fisch et al., 1958).

In summary, chronic potassium depletion results in several derangements of the cardiovascular system, at least some of which are species specific. In man, the most important cardiac consequence of potassium depletion is an abnormality of cardiac rhythm, especially in the presence of digitalis. However, in severe cases myocardial failure apparently may occur.

On the vascular system

Potassium depletion may blunt circulatory reflexes as manifested by a modest postural fall in blood pressure, a blunted response to the cold pressor test, and a lack of hypertensive overshoot and bradycardia following the Valsalva maneuver (Biglieri and McIlroy, 1966). In addition, potassium depletion reduces blood pressure in hypertensive man (Perera, 1953), in normal rats (Freed and Friedman, 1950) in hypertensive rats (Friedman, Roseman and Freed, 1951), and in normotensive dogs (Bahler and Rakita, 1971, Abbrecht, 1972). Also, in the dog, chronic potassium depletion reduces heart rate (Bahler and Rakita, 1971) and systemic vascular resistance (Bahler and Rakita, 1971; Bay et al., 1976; Knochel and Schlein, 1972). One study of chronically potassium depleted dogs evaluated in the conscious state reported an increase in total peripheral resistance and a decrease in cardiac output (Abbrecht, 1972). With this exception, however, most studies show a depression of blood pressure and reduced peripheral resistance. Several factors may contribute to these circulatory changes. Hypokalemia appears to impair norepinephrine release from sympathetic neurons (Krakoff, 1972) and may thus account for the depression of circulatory reflexes. Potassium depletion may reduce systemic blood pressure by reducing secretion of aldosterone from the adrenal gland (Cannon et al., 1966). This control of aldosterone secretion by potassium is independent of the renin-angiotensin system since the effect persists in the presence of falling plasma renin activity (Brunner et al., 1970). Prostaglandins also may play a role in the vascular changes of potassium depletion. Bay and associates (1976) studied dogs made hypokalemic by chronic intermittent hemodialysis and noted a fall in total peripheral vascular resistance, an increase in the cardiac output, a rise in plasma renin activity, and a reduced sensitivity to the infusion of the angiotensin II. These findings were associated with a marked increase in urinary prostaglandin (PGE) excretion and were reversed upon administration of a prostaglandin synthesis inhibitor. Whether this relationship between hypokalemia and elevated urinary PGE levels was a cause and effect relationship or a nonspecific effect remains unclear.

While these preceding studies in general show that potassium depletion, either directly or indirectly, causes a relaxation of the vascular smooth muscle at the arteriolar level, in vitro data that have been examined on this point are conflicting. In vitro isolated vascular smooth muscle develops less tension when studied in a hypokalemic bath (Bohr, Brody and Cheu, 1958). In contrast, acute hypokalemic perfusion of normal canine gracilis muscle increases vascular resistance to blood flow (Anderson et al., 1972). A similar increase in coronary vascular resistance was found in the normal dog heart acutely perfused with a low potassium concentration (Brace et al., 1974). The presence of chronic potassium depletion in the former studies and its absence in the latter two studies may explain the apparently different effect of potassium depletion on vascular tone. However, during mild to moderate potassium depletion, the renal vasculature appears to respond differently when compared with the peripheral vasculature. Hollenberg and coworkers (1975) proposed that the state of potassium balance and its influence on the renin-angiotensin system was an important physiologic regulator of renal vascular tone. In this study, five to six days of moderate potassium restriction in normal man lead to increased circulating renin activity, increased angiotensin II concentration, and a reduction in basal renal blood flow without producing a change in mean arterial blood pressure. These authors also noted a reduced responsiveness of the renal vasculature to the infusion of intravenous angiotensin. Although the mechanism of this latter effect was not elucidated, the authors speculated that it was similar to the reduced responsiveness to angiotensin II infusions seen during sodium restriction (i.e., occupation of vascular receptors by endogenous intrarenal angiotensin II).

Potassium-depleted subjects are at greater risk of developing rhabdomyolysis following strenuous exercise (see above), and an alteration of muscle vasculature appears to be the cause. Exercise hyperemia correlates closely with interstitial potassium concentration (Kjellmer, 1965; Kjellmer, 1965). Knochel and Schlein (1972) demonstrated in the dog that chronic potassium depletion blunted exercise hyperemia of skeletal muscle following intense physical exercise. The mechanism proposed to explain this observation was the marked reduction of potassium release from muscle cells during exercise and the consequent failure to increase interstitial potassium concentration. Since the vascular smooth muscle in these potassium-depleted animals dilated when challenged with an exogenous potassium infusion, the defect appeared to be failure to release potassium from muscle cells per se. Thus the failure to increase muscle blood flow normally during exercise may lead to anoxic cell death.

In summary, vascular reflexes, systemic blood pressure, and renal hemodynamics are variably depressed during potassium depletion. These changes cause few clinically significant problems and may actually represent useful physiologic adaptations. In contrast, failure of the muscle vasculature to dilate in the postexercise period because of impaired muscle release of potassium may result in rhabdomyolysis and acute renal failure.

METABOLIC FACTORS IN POTASSIUM DEFICIENCY

Carbohydrate metabolism

Potassium depletion may disturb carbohydrate metabolism in several ways. The association of potassium deficiency with abnormal glucose handling in man was first made by McQuarrie, Zierler and Johnson (1943). Their report of atypical diabetes mellitus, hypertension, hypernatremia, and hypokalemia clearly resembled primary aldosteronism as defined later by Conn, Knopf and Nesbit (1964). Moreover, 50 percent of patients with primary hyperaldosteronism may have an abnormal glucose tolerance test that largely reverts to normal with correction of the potassium depletion (Conn, 1965). Potassium depletion in hepatic cirrhosis is also associated with a diabetic glucose tolerance curve and reduced output of insulin and growth hormone (Podolsky et al., 1973). In these patients, the etiology of the potassium deficiency may be multifactorial and includes poor intake, secondary hyperaldosteronism, gastrointestinal losses and diuretic drugs.

Following the introduction and clinical use of the benzothiadiazine diuretics, deficient glucose utilization and frank diabetes mellitus were observed in some patients (Shapiro, Benedek and Small, 1961; Saglid, Andersen and Andreasen, 1961; Rapoport and Hurd, 1964). This abnormality was corrected following potassium administration, suggesting that these diuretic agents caused the abnormality in carbohydrate handling through potassium depletion (Rapoport and Hurd, 1964). Conn et al. (1965) noted the association between potassium deficiency and impaired insulin release in cases of hyperaldosteronism in which potassium supplementation or surgery corrected both hypokalemia and the subnormal insulin production. Evidence from several investigators suggest that potassium depletion impairs glucose tolerance by impairing pancreatic release of insulin (Gordon, 1973; Howell and Taylor, 1968; Hiatt, Davidson and Bonorris, 1972). The cellular basis for these observations appears to be that potassium and calcium directly stimulate the pancreas to release insulin (Mondon, Burton and Grodsky, 1968; Grodsky and Bennett, 1966). Conversely, the beta cells of the pancreas appear to participate in feedback regulation of serum potassium concentration since insulin administration stimulates removal of potassium from the extracellular fluid compartment (Hiatt et al., 1973) and enhances muscle uptake of potassium (Santeusanio et al., 1973). The importance of the hormonal control of serum potassium level may be clearly appreciated in diabetic patients with the syndrome of hyporeninemic hypoaldosteronism in which both insulin and aldosterone are absent or reduced. In such patients, hyperkalemia may be life threatening (Goldfarb et al., 1976).

In addition to a quantitative effect on insulin release, potassium depletion also may have a qualitative effect since the proportion of the relatively inactive proinsulin fraction of total insulin released from the pancreas increased during potassium deficiency (Gorden, Sherman and Simopoulos, 1972). The impaired release of insulin in potassium depletion most likely represents a more widespread effect of potassium depletion on the synthesis and secretion of hormones in general. In this regard, elevated potassium concentration in vitro increases growth-hormone release from rat pituitary glands (MacLeod and Fontham, 1970), as well as stimulating release of luteinizing hormones (Samli and Gerschwind, 1968), thyrotropin (Vale and Guillemin, 1967) and ACTH (Kraicer et al., 1969).

Despite the foregoing evidence, alterations in blood glucose or glucose tolerance curves have not been a universal finding even with substantial diuretic-induced potassium deficits (Healy et al., 1970; Anderson et al.,

1971). In explaining this discrepancy, most authorities indicate that clinically significant carbohydrate intolerance following diuretic therapy is most likely to occur only in patients with frank or subclinical diabetes. Unfortunately, the serum potassium level in patients with thiazide-induced carbohydrate intolerance may be a poor index of total body potassium stores as equivalent reductions in total body potassium may be associated with either normakolemia or hypokalemia (Davidson et al., 1976). Despite reports that furosemide—a potent loop diuretic related to (but chemically distinct from) the thiazides—may not affect glucose tolerance in man (Anderson et al., 1971) or rat (Aynsley-Green and Alberti, 1973), any diuretic producing sufficient potassium depletion may bring on the abnormality. The apparent infrequency of this complication with furosemide as compared with thiazides probably reflects the comparatively shorter duration of action of furosemide and thus a reduced tendency for potassium wasting. Lastly, several reports suggest that thiazide diuretics impair carbohydrate tolerance by reducing peripheral uptake of glucose (Beardwood et al., 1965; Barnett and Whitney, 1966). The available evidence, however, strongly suggests that potassium depletion is the most important mechanism of the two.

Potassium is required for synthesis of hepatic and skeletal muscle glycogen (Gardner et al., 1950; Hastings, Renald and Teng, 1955) and in potassium depletion, both storage and synthesis of glycogen are seriously impaired (Blachley, Long and Knochel, 1974). The mechanism of this effect on glycogen storage and synthesis is not known but it may represent failure of a potassium-dependent release of insulin from the pancreas as well as a decline in glycogen synthetase (Blachley and Knochel, unpublished observations). The interrelationships between potassium homeostasis and glucoregulatory hormone has been reviewed recently (Knochel, 1977).

Nitrogen metabolism

Potassium deficiency is known to retard growth markedly in young animals (Brokaw, 1953; Kornberg and Endicott, 1946) as well as in man (Stefan, Helge, Merher, 1968). In the latter case, marked growth retardation associated with potassium depletion resulting from renal artery stenosis and secondary aldosteronism was corrected with potassium supplementation and spironolactone therapy. While the precise mechanism of impaired growth is not clear, potassium depletion has been associated with reversible impairment of growth-hormone release (Podolsky et al., 1973). In addition, insulin is known to stimulate protein synthesis in several tissues (Manchester, 1970). Presumably then, protein synthesis might be impaired by potassium depletion via a reduction of pancreatic insulin output.

CONCLUSIONS

The derangements described here are usually the consequence of serious and substantial potassium depletion as seen with the disorders listed in

Table 7.2. In terms of frequency, the disorders involving hyperaldosteronism are probably most common followed by chronic gastrointestinal loss of potassium. As a sole etiologic factor, diuretic use in a nonedematous patient generally does not result in serious total body potassium depletion (Kassirer and Harrington, 1977). When diuretics do participate, they do so usually by accelerating or unmasking some other potassium-losing process. Moreover, the risk of hypokalemia in the uncomplicated patient taking diuretics may sometimes be far less than the risk involved in treating or attempting to prevent hypokalemia using potassium chloride supplementation. Lawson and colleagues (1974) reported an incidence of hypokalemia of 4.9 percent in 1,294 patients receiving potassium-depleting diuretics without supplemental potassium chloride. None of this group developed a serious consequence as a result of the hypokalemia. In contrast, in a group of 4,900 patients receiving potassium chloride largely to prevent hypokalemia (86 percent), hyperkalemia represented a serious threat. Hyperkalemia developed in 176 patients and 7 of these patients died. For the present, the most prudent course to follow seems to be a careful analysis of serum potassium levels following institution of diuretic therapy followed by potassium chloride supplementation only in those patients demonstrating persistent serum potassium levels of 3.0–3.2 mEq/l or less.

Table 7.2 Etiology of potassium deficiency

I. Cellular dysfunction
II. Reduced intake
III. Gastrointestinal losses
   A. Protracted vomiting
   B. Malabsorption
   C. Ulcerative colitis
   D. Chronic diarrhea
   E. Laxative abuse
   F. Geophagia
   G. Ureterosigmoidostomy
   H. Villous adenoma
IV. Urinary losses
   A. Mineralocorticoid excess
     1. Primary hyperaldosteronism
     2. Secondary hyperaldosteronism
     3. Excess mineralocorticoid administration
     4. Glycyrrhezinic acid (contained in licorice, annisette, and carbenoxolone)
   B. Diuretic therapy
   C. Renal tubular acidosis
   D. Osmotic diuresis (e.g., diabetic ketoacidosis)
   E. Chronic respiratory acidosis

REFERENCES

Abbrecht, P. H. (1972). Cardiovascular effects of chronic potassium deficiency in the dog. *American Journal of Physiology*, 223, 555–560.

Adler, S. & Fraley, D. S. (1977). Potassium and intracellular pH. *Kidney International*, 11, 433–442.

Adler, S., Zett, B. & Anderson, B. (1972). The effect of acute potassium depletion on muscle cell pH in vitro. *Kidney International*, **2**, 159–163.
Adler, S., Zett, B. & Anderson, B. (1974). Renal citrate in the potassium-deficient rat: role of potassium and chloride ions. *Journal of Laboratory and Clinical Medicine*, **84**, 307–316.
Aitkins, E. L. & Schwartz, W. B. (1962). Factors governing correction of the alkalosis associated with potassium deficiency; the critical role of chloride in the recovery process. *Journal of Clinical Investigation*, **41**, 218–229.
Anderson, J., Godfrey, B. E., Hill, D. M., Munro-Favre, A. D. & Sheldon, J. (1971). A comparison of the effects of hydrochlorothiazide and of furosemide in the treatment of hypertensive patients. *Quarterly Journal of Medicine*, **40**, 541–560.
Anderson, D. K., Roth, S. A., Brace, R. A., Radawski, D., Haddy, F. J. & Scott, J. B. (1972). Effect of hypokalemia and hypomagnesemia produced by hemodialysis on skeletal muscle vascular resistance; role of potassium in active hyperemia. *Circulation Research*, **31**, 165–173.
Aynsley-Green, A. & Alberti, K. G. (1973). Diuretics and carbohydrate metabolism: the effects of furosemide and amiloride on blood glucose, plasma insulin and cations in the rat. *Diabetologia*, **9**, 34–42.
Baertl, J. M., Sancetta, S. M. & Gabuzda, G. J. (1963). Relation of acute potassium depletion to renal ammonium metabolism in patients with cirrhosis. *Journal of Clinical Investigation*, **42**, 696–706.
Bahler, R. C. & Rakita, L. (1971). Cardiovascular function in potassium-depleted dogs. *American Heart Journal*, **81**, 650–657.
Bank, N. & Aynedjian, H. S. (1964). A micropuncture study of the renal concentrating defect of potassium depletion. *American Journal of Physiology*, **206**, 1347–1354.
Barnett, C. A. & Whitney, J. E. (1966). The effect of diazoxide and chlorathiazide on glucose uptake in vivo. *Metabolism, Clinical and Experimental*, **15**, 88–93.
Bay, W. H., Galvez, O. G., Roberts, B. W. & Ferris, T. F. (1976). Hemodynamic changes with hypokalemia. *Kidney International*, **10**, 529A.
Beardwood, D. M., Alden, J. S., Graham, C. A., Beardwood, J. T. & Marble, A. (1965). Evidence for a peripheral action of chlorothiazide in normal man. *Metabolism, Clinical and Experimental*, **14**, 561–567.
Beck, N. & Davis, B. B. (1975). Impaired renal response to parathyroid hormone in potassium depletion. *American Journal of Physiology*, **228**, 179–183.
Beck, N. & Webster, S. K. (1976). Impaired urinary concentrating ability and cyclic AMP in K^+-depleted rat kidney. *American Journal of Physiology*, **231**, 1204–1208.
Berl, T., Anderson, R. J., Aisenbrey, G. A., Linas, S. L. & Schrier, R. W. (1977). On the mechanism of polyuria in potassium depletion: the role of polydipsia. *Journal of Clinical Investigation*, **60**, 620–625.
Berliner, R. W., Kennedy, T. J., Jr. & Orloff, J. (1951). Relationship between acidification of the urine and potassium metabolism. *American Journal of Medicine*, **11**, 274–282.
Beskind, H. & Mudge, G. H. (1959). Effect of potassium deficiency on renal tubular reabsorption and assimilation of glucose. *Bulletin of the Johns Hopkins Hospital*, **104**, 252–259.
Biglieri, E. G. & McIlroy, M. B. (1966). Abnormalities of renal function and circulatory reflexes in primary aldosteronism. *Circulation*, **33**, 78–86.
Bilbrey, G. L., Herbin, L., Carter, N. W. & Knochel, J. P. (1973). Skeletal muscle resting membrane potential in potassium deficiency. *Journal of Clinical Investigation*, **52**, 3011–3018.
Blachley, J., Long, J. & Knochel, J. P. (1974). Impaired muscle glycogen synthesis and prevention of muscle glycogen supercompensation by potassium deficiency. *Clinical Research*, **22**, (3), 517A.
Bland, W. H. & Bassett, S. H. (1953). Potassium deficiency in man. *Metabolism, Clinical and Experimental*, **2**, 218–224.
Bland, D. M. & Whittam, R. (1965). Effects of sodium and potassium ions on oxidative phosphorylation in relation to respiratory control by a cell-membrane adenosine triphosphate. *Biochemical Journal*, **97**, 523–531.
Blomberg, L. H. & Lindqvist, T. (1954). Electrocardiogram in paroxysmal essential hypopotassemia (periodic paralysis): report of two cases. *Acta Medica Scandinavica*, **147**, 437–446.

Boddy, K., Hume, R., White, C., Pack, A., King, P. C., Weyers, E., Rowan, T. & Mills, E. (1976). The relation between potassium in body fluids and total body potassium in healthy and diabetic subjects. *Clinical Science and Molecular Medicine*, 50, 455–461.

Bohr, D. F., Brody, D. C. & Cheu, D. H. (1958). Effect of eletrolytes on arterial muscle contraction. *Circulation*, 17, 746–749.

Brace, R. A., Anderson, D. K., Chen, W., Scott, J. B. & Haddy, F. J. (1974). Local effects of hypokalemia on coronary resistance and myocardial contractile force. *American Journal of Physiology*, 227, 590–597.

Brokaw, A. (1953). Renal hypertrophy and polydipsia in potassium deficient rats. *American Journal of Physiology*, 172, 333–346.

Brunner, H. R., Baer, L., Sealey, J. E., Ledingham, J. C. G. & Laragh, J. H. (1970). The influence of potassium administration and potassium deprivation on plasma renin in normal and hypertensive subjects. *Journal of Clinical Investigation*, 49, 2128–2138.

Burnell, J. M., Teubner, E. J. & Simpson, D. P. (1974). Metabolic acidosis accompanying potassium deprivation. *American Journal of Physiology*, 227, 329–333.

Cameron, I. R. & Hall, R. J. C. (1975). The effect of dietary K depletion and subsequent repletion on intracellular K concentration and pH of cardiac and skeletal muscle in rabbits. *Journal of Physiology* (London), 25, 70P–71P.

Campion, D. S., Arias, J. M. & Carter, N. W. (1972). Rhabdomyolysis and myoglobinuria association with hypokalemia of renal tubular acidosis. *Journal of American Medical Association*, 220, 967.

Cannon, P. R., Frazier, L. E. & Hughes, R. H. (1953). Sodium as a toxic ion in potassium deficiency. *Metabolism, Clinical and Experimental*, 2, 297–312.

Cannon, P. J., Ames, R. P. & Laragh, J. H. (1966). Relation between potassium balance and aldosterone secretion in normal subjects and in patients with hypertensive or renal tubular disease. *Journal of Clinical Investigation*, 45, 865–879.

Ching, S., Rogoff, T. M. & Gabuzda, G. J. (1973). Renal ammoniagenesis and tissue glutamine, glutamine synthetase, and glutaminase I levels in potassium-deficient rats. *Journal of Laboratory and Clinical Medicine*, 82, 208–214.

Clark, E., Evans, B. M., MacIntyre, I. & Milne, M. D. (1955). Acidosis in experimental electrolyte depletion. *Clinical Science*, 14, 421–440.

Cohen, T. & Hills, F. (1959). Hypokalemic muscular paralysis associated with administration of chlorthalidone. *Journal of American Medical Association*, 170, 2083–2085.

Conn, J. W., Knopf, R. F. & Nesbit, R. M. (1964). Clinical characteristics of primary aldosteronism from analysis of 145 cases. *American Journal of Surgery*, 107, 159–172.

Conn, J. W. (1965). Hypertension, the potassium ion and impaired carbohydrate tolerance. *New England Journal of Medicine*, 273, 1135–1143.

Conway, M. J., Seibel, J. A. & Eaton, R. P. (1974). Thyrotoxicosis and periodic paralysis. Improvements with beta blockade. *Annals of Internal Medicine*, 81, 332–336.

Coppen, A. J. & Reynolds, E. H. (1966). Electrolyte and water distribution in familial hypokalemic periodic paralysis. *Journal of Neurology, Neurosurgery, and Psychiatry*, 29, 107–112.

Cremer, W. & Bock, K. D. (1977). Symptoms and course of chronic hypokalemic nephropathy in man. *Clinical Nephrology*, 7, 112–119.

Creutzfeldt, O., Abbott, B. C., Fowler, W. M. & Pearson, C. M. (1963). Muscle membrane potentials in episodic adynamia. *Electroencephalography and Clinical Neurophysiology*, 15, 508–519.

Darrow, D. C. & Miller, H. C. (1942). The production of cardiac lesions by repeated injections of desoxycorticosterone acetate. *Journal of Clinical Investigation*, 21, 601–611.

Davidson, C., McLachlan, MSF, Burkinshaw, L. & Morgan, D. B. (1976). Effect of long-term diuretic treatment on body potassium in heart disease. *Lancet*, 2, 1044–1047.

Descamps, C., Vandenbroucke, J. M. & van Ypersele de Strihou, C. (1977). Rhabdomyolysis and acute tubular necrosis associated with carbenoxolone and diuretic treatment. *British Medical Journal*, 29, 272.

DiScala, V. A., Mautner, W., Cohen, A., Levitt, M. F., Churg, J. & Yunis, S. L. (1965). Tubular alterations produced by osmotic diuresis with mannitol. *Annals of Internal Medicine*, 63, 767–775.

Edmondson, R. P. S., Hilton, P. J., Thomas, R. D. & Patrick, J. (1974). Leukocyte electrolytes in cardiac and non-cardiac patients receiving diuretics. *Lancet*, 1, 12–14.

Fenn, W. O. & Cobb, D. M. (1934). The potassium equilibrium in muscle. *Journal of General Physiology*, 17, 629–645.

Fichman, M. P., Vorherr, H., Kleeman, C. R. & Telfer, N. (1971). Diuretic-induced hyponatremia. *Annals of Internal Medicine*, 75, 853–863.

Finkelstein, F. O. & Hayslett, J. P. (1975). Role of medullary Na-K-ATPase in renal potassium adaptation. *American Journal of Physiology*, 229, 524–528.

Finn, A. T., Handler, J. S. & Orloff, J. (1966). Relation between toad bladder potassium content and permeability response to vasopressin. *American Journal of Physiology*, 210, 1279–1284.

Fisch, C., Shields, J. P., Ridolfo, S. A. & Feigenbaum, H. (1958). Effect of potassium on conduction and ectopic rhythms in atrial fibrillation treated with digitalis. *Circulation*, 18, 98–106.

Fourman, P. S. & Hervey, G. R. (1955). An experimental study of oedema in potassium deficiency. *Clinical Science*, 14, 75–79.

Freed, S. C. & Friedman, M. (1950). Hypotension in the rat following limitation of potassium intake. *Science*, 112, 788–789.

Friedman, M., Rosenman, R. H. & Freed, S. C. (1951). The depressor effect of potassium deprivation on the blood pressure of hypertensive rats. *American Journal of Physiology*, 167, 457–461.

Fuller, G. R., MacLeod, M. G. & Pitts, R. F. (1955). Influence of administration of potassium salts on the renal tubular reabsorption of bicarbonate. *American Journal of Physiology*, 182, 111–118.

Gabuzda, G. J. & Hall, III, P. W. (1966). Relation of potassium depletion to renal ammonium metabolism and hepatic coma. *Medicine* (Baltimore), 45, 481–490.

Galvez, O. G., Roberts, B. W., Bay, W. H. & Ferris, T. F. (1976). Studies of the mechanism of polyuria with hypokalemia. *Kidney International*, 10, 583A.

Gardner, L. I., Talbot, N. B., Cook, C. D., Berman, H. & Uribe, C. (1950). The effect of potassium deficiency on carbohydrate metabolism. *Journal of Laboratory Clinical Medicine*, 35, 592–603.

Garella, S., Chazan, J. A. & Cohen, J. J. (1970). Saline-resistant metabolic alkalosis or "chloride-wasting nephropathy." *Annals of Internal Medicine*, 73, 31–38.

Gennari, F. J. & Cohen, J. J. (1975). Role of the kidney in potassium homeostasis: lessons from acid-base disturbances. *Kidney International*, 8, 1–5.

Goldfarb, S., Cox, M., Singer, I. & Goldberg, M. (1976). Acute hyperkalemia induced by hyperglycemia: hormonal mechanisms. *Annals of Internal Medicine*, 84, 426–432.

Gorden, P., Sherman, B. M. & Simopoulos, A. P. (1972). Glucose intolerance with hypokalemia: an increased proportion of circulating proinsulin-like component. *Journal of Clinical Endocrinology*, 34, 235–240.

Gordon, P. (1973). Glucose intolerance with hypokalemia: failure of short-term potassium depletion in normal subjects to reproduce the glucose and insulin abnormalities of clinical hypokalemia. *Diabetes*, 22, 544–551.

Graeff, J. de & Lameijer, L. D. (1965). Periodic paralysis. *American Journal of Medicine*, 39, 70–80.

Gregory, P. B., Broekelschen, P. H., Hill, M. D., Lipton, A. B., Knauer, C. M. Egger, M. & Miller, R. (1977). Complications of diuresis in the alcoholic patient with ascites: a controlled trial. *Gastroenterology*, 73, 534–538.

Grob, D., Johns, R. J. & Liljestrand, A. (1957). Potassium movement in patients with familial periodic paralysis: relationship to the defect in muscle function. *American Journal of Medicine*, 23, 356–375.

Grodsky, G. M. & Bennett, L. L. (1966). Cation requirement for insulin secretion in the isolated perfused pancreas. *Diabetes*, 15, 910–913.

Gross, E. G., Dexter, J. D. & Roth, R. G. (1966). Hypokalemic myopathy with myoglobinuria associated with licorice ingestion. *New England Journal of Medicine*, 274, 602–606.

Gunning, J. F., Harrison, C. E. & Coleman, H. N. III. (1972). The effects of chronic potassium deficiency on myocardial contractility and oxygen consumption. *Journal of Molecular and Cellular Cardiology*, 4, 139–153.

Harrison, C. E., Jr., Cooper, G., IV, Zujko, K. J. & Coleman, H. N., III. (1972). Myocardial and mitochondrial function in potassium depletion cardiomyopathy. *Journal of Cellular Cardiology*, 4, 633–649.

Harrison, T. R., Pilcher, C. & Ewing, G. (1930). Studies in congestive heart failure. IV The postassium content of skeletal and cardiac muscle. *Journal of Clinical Investigation*, **8**, 325-335.
Hastings, A. B., Renald, A. E. & Teng, C. T. (1955). Effects of ions and hormones on carbohydrate metabolism. *Recent Progress in Hormone Research*, **11**, 381–400.
Healy, J. J., McKenna, T. J., Canning, B., Brien, T. G., Duffy, G. J. & Muldowney, F. P. (1970). Body composition changes in hypertensive subjects on long term oral diuretic therapy. *British Medical Journal*, **1**, 716–719.
Hiatt, N., Davidson, M. B. & Bonorris, G. (1972). Effect of potassium chloride infusion on insulin secretion *in vivo*. *Hormone Metabolism Research*, **4**, 64–68.
Hiatt, N., Morgenstern, L., Davidson, M. B., Bonorris, G. & Miller, A. (1973). Role of insulin in the transfer of infused potassium to tissue. *Hormone Metabolism Research*, **5**, 84–88.
Hofmann, W. W. & Smith, R. A. (1970). Hypokalemic periodic paralysis studied *in vitro*. *Brain*, **93**, 445–474.
Hollenberg, N. K., Williams, G., Burger, B. & Hooshmand, I. (1975). The influence of potassium on the renal vasculature and the adrenal gland, and their responsiveness to angiotension II in normal man. *Clinical Science and Molecular Medicine*, **49**, 527–534.
Holliday, M. A., Winters, R. W., Welt, L. G., MacDowell, M. & Oliver, J. (1959). The renal lesions of electrolyte imbalance. II. The combined effect on renal architecture of phosphate loading and potassium depletion. *Journal of Experimental Medicine*, **110**, 161–168.
Howell, S. L. & Taylor, K. W. (1968). Potassium ions and the secretion of insulin by islets of Langerhans incubated *in vitro*. *Biochemical Journal*, **108**, 17–24.
Hulter, H. N., Ilnicki, L. P. & Sebastian, A. (1976). Pathogenic role of aldosterone deficiency in the metabolic acidosis resulting from dietary potassium restriction. *Kidney International*, **10**, 586A.
Kannegiesser, H. & Lee, J. B. (1971). Role of outer renal medullary metabolism in concentrating defect of potassium depletion. *American Journal of Physiology*, **220**, 1701–1707.
Kassirer, J. P. & Harrington, J. T. (1977). Diuretics and potassium metabolism: a reassessment of the need, effectiveness and safety of potassium therapy. *Kidney International*, **11**, 505–515.
Kassirer, J. P., Berkman, P. M., Lawrenz, D. R. & Schwartz, W. B. (1965). The critical role of chloride in the correction of hypokalemic alkalosis in man. *American Journal of Medicine*, **38**, 172–189.
Keye, J. D. (1952). Death in potassium deficiency. Report of a case including morphalogic findings. *Circulation*, **5**, 766–770.
Khuri, R. N., Wiederholt, M., Streider, N. & Giebisch, G. (1975). Effects of flow rate and potassium intake on distal tubular potassium transfer. *American Journal of Physiology*, **228**, 1249–1261.
Kjellmer, I. (1965). Studies on exercise hyperemia. *Acta Physiologica Scandinavica*. **244**, (suppl. 1).
Kjellmer, I. (1965). Potassium ion as a vasodilator during muscular exercise. *Acta Physiologica Scandinavica*, **63**, 460–468.
Knochel, J. P. & Carter, N. W. (1976). The role of muscle cell injury in the pathogenesis of acute renal failure after exercise. *Kidney International*, **10**, 558-64.
Knochel, J. P., Dotin, L. H. & Hamburger, R. J. (1972). Pathophysiology of intense physical conditioning in a hot climate. I. Mechanism of potassium depletion. *Journal of Clinical Investigation*, **51**, 242–255.
Knochel, J. P. & Schlein, E. M. (1972). On the mechanism of rhabdomyolysis in potassium depletion. *Journal of Clinical Investigation*, **51**, 1750–1758.
Knochel, J. P., Bilbrey, G. L., Fuller, T. J. & Carter, N. W. (1975). The muscle cell in chronic alcoholism: the possible role of phosphate depletion in alcoholic myopathy. *Annals of New York Academy of Science*, **252**, 274–286.
Knochel, J. P. (1977). Role of glucoregulatory hormones in potassium homeostasis. *Kidney International*, **11**, 443–452.
Kornberg, A. & Endicott, K. M. (1946). Potassium deficiency in the rat. *American Journal of Physiology*, **145**, 291–298.

Kraicer, J., Milligan, J. V., Gosbee, J. L., Conrad, R. G. & Branson, C. M. (1969). Potassium, corticosterone, and adrenocorticotropic hormone release in vitro. *Science*, **164**, 426.

Krakoff, L. R. (1972). Potassium deficiency and cardiac catecholamine metabolism in the rat. *Circulation Research*, **30**, 608–615.

Kurtzman, N. A., White, M. G. & Rogers, P. W. (1973). The effect of potassium and extracellular volume on renal bicarbonate reabsorption. *Metabolism*, **22**, 481–492.

Lawson, D. H., Boddy, K., Gray, M. G., Mahaffey, M. & Millis, E. (1976). Potassium supplements in patients receiving long-term diuretics for oedema. *Quarterly Journal of Medicine, Clinical and Experimental*, **45**, 469–478.

Loeb, R. F., Atchley, D. W. & Richards, D. W. (1932). On the mechanism of nephrotic edema. *Journal of Clinical Investigation*, **11**, 621–639.

Luke, R. G., Wright, F. S., Fowler, N. B. & Giebisch, G. (1976). Effect of K-depletion on segmental chloride transport in rat nephron. *Kidney International*, **10**, 591A.

McFarland, K. F. & Carr, A. A. (1977). Changes in the fasting blood sugar after hydrochlorothiazide and potassium supplementation. *Clinical Pharmacology*, **17**, 13–17.

MacLeod, R. M. & Fontham, E. H. (1970). Influence of ionic environment on the in vitro synthesis and release of pituitary hormones. *Endocrinology*, **86**, 863–869.

McQuarrie, I., Zierler, M. R. & Johnson, R. M. (1943). Plasma electrolyte disturbance in patient with hypercorticoadrenal syndrome contrasted with that found in Addison's disease. *Endocrinology*, **21**, 762–772.

Manchester, K. L. (1970). Insulin and protein synthesis. In *Biochemical Actions of Hormones*, ed. Litwack, G. Vol. I., pp. 267–320. New York: Academic Press.

Manitius, A., Levitin, H., Beck, D. & Epstein, F. H. (1960). On the mechanism of impairment of renal concentrating ability in potassium deficiency. *Journal of Clinical Investigation*, **39**, 684–697.

Martin, J. B., Craig, J. W., Eckel, R. E. & Munger, J. (1971). Hypokalemic myopathy in chronic alcoholism. *Neurology*, **21**, 1160–1168.

Mondon, C. E., Burton, S. D. & Grodsky, G. M. (1968). Glucose tolerance and insulin response of potassium-deficient rat and isolated liver. *American Journal of Physiology*, **215**, 779–787.

Muehrcke, R. C. & McMillan, J. C. (1963). The relationship of "Chronic pyelonephritis" to chronic potassium deficiency. *Annals of Internal Medicine*, **59**, 427–448.

Muehrcke, R. C. & Rosen, S. (1964). Hypokalemic nephropathy in rat and man: light and electron microscopic study. *Laboratory Investigation*, **13**, 1359–1373.

Muntwyler, E. & Griffin, G. E. (1953). Creatinine clearance in normal and potassium deficient rats. *American Journal of Physiology*, **173**, 145–150.

Nickerson, M., Karr, G. W. & Dresel, P. E. (1961). Pathogenesis of "electrolyte-steroid-cardiopathy." *Circulation Research*, **9**, 209–217.

Oh, S. H., Douglas, J. E. & Brown, R. A. (1971). Hypokalemic vacuolar myopathy associated with chlorthalidone treatment. *Journal of the American Medical Association*, **216**, 1858–1862.

Pagliara, A. S. & Goodman, A. D. (1970). Relation of renal cortical gluconeogenesis, glutamate content and production of ammonia. *Journal of Clinical Investigation*, **49**, 1967–1974.

Perera, G. W. (1953). Depressor effects of potassium deficient diets in hypertensive man. *Journal of Clinical Investigation*, **32**, 633–636.

Perkins, J. G., Petersen, A. B. & Riley, J. A. (1950). Renal and cardiac lesions in potassium deficiency due to chronic diarrhea. *American Journal of Medicine*, **8**, 115–123.

Podolsky, S., Zimmerman, H. J., Burrows, B. A., Cardarelli, H. A. & Pattavina, C. G. (1973). Potassium depletion in hepatic cirrhosis: a reversible cause of impaired growth-hormone and insulin response to stimulation. *New England Journal of Medicine*, **288**, 644–648.

Pollak, V. E., Mattenheimer, H., DeBruin, H. & Weinman, K. (1965). Experimental metabolic acidosis: the enzymatic basis of ammonia production by the dog kidney. *Journal of Clinical Investigation*, **44**, 169–181.

Rapoport, M. I. & Hurd, H. F. (1964). Thiazide-induced glucose intolerance treated with potassium. *Archives of Internal Medicine*, **113**, 405–408.

Rector, F. C., Jr., Buttram, H. & Seldin, D. W. (1962). An analysis of the mechanism of the inhibitory influence of K^+ on renal H^+ secretion. *Journal of Clinical Investigation*, 41, 611–617.

Rector, F. C., Jr., Bloomer, H. A. & Seldin, D. W. (1964). Effect of potassium deficiency on the reabsorption of bicarbonate in the proximal tubule of the rat kidney. *Journal of Clinical Investigation*, 43, 1976–1982.

Rector, F. C., Jr., Carter, N. W. & Seldin, D. W. (1967). The renal transport of hydrogen ion. In *Proceedings of the Third International Congress of Nephrology*. ed. Handler, J. S. Vol. 3, Physiology, pp. 76–85. Basel: S. Karger.

Relman, A. S. & Schwartz, W. B. (1956). Nephropathy of potassium depletion: a clinical and pathological entity. *New England Journal of Medicine*, 255, 195–203.

Relman, A. S. & Schwartz, W. B. (1958). The kidney in potassium depletion. *American Journal of Medicine*, 24, 764–773.

Roberts, K. E., Magida, M. G. & Pitts, R. F. (1953). Relationship between potassium and bicarbonate in blood and urine. *American Journal Physiology*, 172, 47–54.

Roberts, K. E., Randall, H. T., Sanders, H. L. & Hood, M. (1955). Effects of potassium on renal tubular reabsorption of bicarbonate. *Journal of Clinical Investigation*, 34, 666–672.

Rodriguez, C. E., Wolfe, A. L. & Bergstrom, V. W. (1950). Hypokalemic myocarditis. *American Journal of Clinical Pathology*, 20, 1050–1055.

Roystan, A. & Prout, B. J. (1976). Carbenoxolone-induced hypokalaemia simulating Guillain-Barré syndrome. *British Journal of Medicine*, 2, 150–151.

Sagild, U., Andersen, V. & Andreasen, P. B. (1961). Glucose tolerance and insulin responsiveness in experimental potassium depletion. *Acta Medica Scandinavica*, 169, 243–251.

Samli, M. H. & Geschwind, I. I. (1968). Some effects of energy-transfer inhibitors and of Ca^{++}-free or K^+-enhanced media on the release of luteinizing hormone (LH) from the rat pituitary gland *in vitro. Endocrinology*, 82, 225–231.

Santeusanio, F., Faloona, G. R., Knochel, J. P. & Unger, R. H. (1973). Evidence for a role of endogenous insulin and glucagon in the regulation of potassium homeostasis. *Journal of Laboratory Clinical Medicine*, 81, 809–817.

Schrier, R. W., Hano, J., Keller, H. I., Finkel, R. M., Gilliland, P. F., Criksena, W. J. & Teschan, P. E. (1970). Renal, metabolic and circulatory responses to heat and exercise. *Annals of Internal Medicine*, 73, 213–223.

Schwartz, A. (1976). Is the cell membrane Na^+, K^+ ATPase enzyme system the pharmacologic receptor for digitalis? *Circulation Research*, 39, 2–7.

Schwartz, W. B. & Relman, A. S. (1953). Metabolic and renal studies in chronic potassium depletion resulting from overuse of laxatives. *Journal of Clinical Investigation*, 32, 258–271.

Schwartz, W. B. & Relman. A. S. (1967). Effects of electrolyte disorders on renal structure and function. *New England Journal of Medicine*, 276, 383–389.

Schwartz, W. B., Van Ypersele de Strihou, C. & Kassirer, J. P. (1968). Role of anions in metabolic alkalosis and potassium deficiency. *New England Journal of Medicine*, 279, 630–639.

Sealey, J. E., & Laragh, J. H. (1974). A proposed cybernetic system for sodium and potassium homeostasis: coordination of aldosterone and intrarenal physical forces. *Kidney International*, 6, 281–290.

Seldin, D. W. & Rector, F. C., Jr. (1972). The generation and maintenance of metabolic alkalosis. *Kidney International*, 1, 306–321.

Shapiro, A. P., Benedek, T. G. & Small, J. L. (1961). Effect of thiazides on carbohydrate metabolism in patients with hypertension. *New England Journal of Medicine*, 265, 1028–1033.

Silva, P., Hayslett, J. P. & Epstein, F. H. (1973). The role of Na-K-activated adenosine triphosphate in potassium adaptation. Stimulation of enzymatic activity by potassium loading. *Journal of Clinical Investigation*, 52, 2665–2671.

Smith, S. G., Black-Schaffer, B. & Lasater, T. E. (1950). Potassium deficiency syndrome in the rat and the dog. *Archives of Pathology*, 49, 185–199.

Stanbury, S. W. & MaCaulay, D. (1957). Defects of renal tubular function in the nephrotic syndrome. *Quarterly Journal of Medicine*, 26, 7–30.

Stefan, H., Helge, H., Merher, H. J. et al. (1968). Nierenarterienstenose, renale Hypertonie und skundarer Hyperaldosteronismus bei einem 8 Jahre alten Knaben: Minderwuchs durch Kaliummangel *Helvetica Paediatrica Acta*, 23, 509–524.

Streeten, D. H. P. (1966). Periodic paralysis, in *The Metabolic Basis of Inherited Diseases*, 2/e, ed. Stanbury, J. B., Wyngaarden, J. B. & Fredrickson, D. S. Pp. 905–938. New York: McGraw-Hill.

Surawicz, B. (1964). Electrolytes and the electrocardiogram. *Modern Concepts of Cardiovascular Disease*, 33, 875–891.

Sweetin, J. C. & Thomson, W. H. S. (1973). Enzyme efflux and clearance. *Clinical Chimica Acta*, 48, 403–411.

Taher, S. M., Anderson, R. J., McCartney, R., Popovtzer, M. M. & Schrier, R. W. (1974). Renal tubular acidosis associated with toluene "sniffing." *New England Journal of Medicine*, 290, 765–768.

Talso, P. J., Glynn, M. F., Oester, Y. T. & Fudema, J. (1963). Body composition in hypokalemic familial periodic paralysis. *Annals of New York Academy of Science*, 110, 993–1008.

Tannen, R. L. (1970). The effect of uncomplicated potassium depletion on urine acidification. *Journal of Clinical Investigation*, 49, 813–827.

Tannen, R. L. & Terrien, T. (1975). Potassium-sparing effect of enhanced renal ammonia production. *American Journal of Physiology*, 228, 699–705.

Tannen, R. L. (1977). Relationships of renal ammonia production and potassium homeostasis. *Kidney International*, 11, 453–465.

Toback, F. G., Ordóñez, N. G., Bortz, S. L. & Spargo, B. H. (1976). Zonal changes in renal structure and phospholipid metabolism in potassium-deficient rats. *Laboratory Investigation*, 34, 115–124.

Vale, W. & Guillemin, R. (1967). Potassium-induced stimulation of thyrotropin release *in vitro*: requirements for presence of calcium and inhibition by thyroxine. *Experientia*, 23, 855–857.

van Ypersele de Strihou, C. & Dieu, J. (1977). Potassium deficiency acidosis in the dog: effect of sodium and potassium balance on renal response to a chronic acid load. *Kidney International*, 11, 335–347.

Wang, P. & Chausen, T. (1976). Treatment of attacks in hyperkalemic familial periodic paralysis by inhalation of salbutamol. *Lancet*, 1, 221–223.

Whittam, R. & Ager, M. E. (1965). The connection between active cation transport and metabolism in erythrocytes. *Biochemical Journal*, 97, 214–227.

Wilson, A. F. & Simmons, D. H. (1970). Relationships between potassium, chloride, intracellular and extracellular pH in dogs. *Clinical Science*, 39, 731–745.

Woods, J. W., Welt, L. G., Hollander, W., Jr. & Newton, M. (1959). Susceptibility of rats to experimental pyelonephritis following recovery from potassium depletion. *Journal of Clinical Investigation*, 39, 28–33.

Wrong, O. & Davies, H. E. F. (1959). The excretion of acid in renal disease. *Quarterly Journal of Medicine*, 28, 259–313.

Zierler, K. L. & Andres, R. (1957). Movement of potassium into skeletal muscle during spontaneous attack in family periodic paralysis. *Journal of Clinical Investigation*, 36, 730–737.

8
Mineralocorticoid excess and deficiency syndromes

MORRIS SCHAMBELAN
ANTHONY SEBASTIAN
HENRY N. HULTER

Physiologic effects of mineralocorticoid hormones
Syndromes of mineralocorticoid hormone excess
Experimental states of mineralocorticoid hormone excess
Primary aldosteronism
Adrenal enzyme defects
Cushing's syndrome
Syndromes possibly due to unknown mineralocorticoid hormones

Secondary hyperaldosteronism
Pseudohypermineralocorticoidism
Syndromes of mineralocorticoid hormone deficiency
Experimental states of mineralocorticoid hormone deficiency
Generalized adrenocortical insufficiency
Adrenal enzyme defects
Hyporeninemic hypoaldosteronism
Pseudohypomineralocorticoidism

Mineral homeostasis in man is dependent on the "mineralocorticoid activity" of steroid hormones secreted by the adrenal cortex. This activity is expressed physiologically in the regulation of effective arterial blood volume and of electrolyte and acid-base composition of extracellular fluid. Aldosterone is the only adrenal steroid known to participate in the physiologic feedback regulation of mineral homeostasis. Disorders in which the primary disturbance is either deficient or excessive production of aldosterone are characterized by abnormalities in blood pressure, in extracellular fluid volume, and in plasma electrolyte or acid-base composition. Changes in the rate of aldosterone production occur in patients with extraadrenal disorders that result in alterations in effective arterial blood volume or plasma electrolyte or acid-base composition, and, in turn, the changed mineralocorticoid production rate modifies the clinical expression of the primary disturbance. This chapter will consider the pathogenic and pathophysiologic features of clinical and experimental syndromes of mineralocorticoid excess and deficiency in the context of our current understanding of the physiologic effects of mineralocorticoid hormones (MCH).

PHYSIOLOGIC EFFECTS OF MINERALOCORTICOID HORMONES

Mineralocorticoid hormones modulate transepithelial transport of Na^+, K^+, H^+ (or HCO_3^-), Cl^-, Ca^{++}, Mg^{++}, and H_2O. Target epithelia are present in kidney, large bowel, sweat glands, and salivary ducts. In normal man not losing copious amounts of sweat or liquid stool, the renal effects of MCH provide the critical regulation of mineral homeostasis. The renal response to MCH reflects specific effects on the tubules and can occur independently of measured changes in glomerular filtration rate or renal hemodynamics (Garrod, Davies and Cahill, 1955; Yunis, et al., 1964). Functional responses attributed to differences in aldosterone concentration have been reported to occur in the distal nephron (dogs) (Vander et al., 1958), the proximal and distal convoluted tubules (rats) (Hierholzer, Wiederholt and Stolte, 1966; Wiederholt et al., 1972), the cortical collecting tubules (rabbits) (Gross, Imai and Kokko, 1975), and the medullary collecting ducts (rats) (Uhlich, Baldamus and Ullrich, 1969; Uhlich, Halbach, and Ullrich, 1970). It has been questioned whether the proximal tubular effect of aldosterone has physiologic significance. Recent studies indicate that isolated segments of the cortical collecting tubule in the rabbit are responsive to MCH but that early segments of the distal convoluted tubule are unresponsive (Gross et al., 1975). Earlier studies that revealed an effect of MCH in the distal convoluted tubule may have reflected changes occurring in the "late" segments, where cell morphology more closely resembles cortical collecting tubule. At least two cell types ("light" and "dark" cells) have been identified in the collecting tubules and ducts (Tisher, 1976), but it is not known if both cell types respond to MCH.

The urinary bladder of the toad (Lipton and Edelman, 1971; Ludens and Fanestil, 1974) and fresh water turtle (Al-Awqati et al., 1976) also have been found to respond to MCH. These epithelia have morphologic and functional characteristics that are similar to mammalian distal tubular epithelia, and they have been used extensively as a model for studies of MCH action. The specific effects of MCH on ion and water transport have been studied in such epithelial membranes and in isolated segments of tubule. Sodium reabsorption increases despite opposing transepithelial electrical (lumen-negative) and chemical (concentration) gradients; that is, active transport of Na^+ is stimulated (Crabbe, 1961). The stimulation of active Na^+ transport by MCH increases the lumen-negative transtubular electrical potential difference; that is, the transport process is electrogenic (Gross et al., 1975). This increase in potential difference may account in part for the stimulation of H^+ and K^+ secretion by MCH, but aldosterone stimulates secretion of these ions even when the effect of the hormone on Na^+ transport is prevented (Lifschitz, Schrier and Edelman, 1973). Aldosterone also stimulates net Cl^- reabsorption in isolated, perfused collecting tubules of rabbits, (Hanley and Kokko, 1977). Whether or not MCH stimulates Cl^- reabsorption independently of Na^+ transport is not known. Aldosterone has also been shown to enhance the effect of vasopressin to increase transepithelial water movement in the toad bladder (Handler, Preston and Orloff, 1969).

According to current concepts, the alterations in transport induced by MCH are mediated by molecular-ionic events at the luminal and peritubular cell membranes. Transepithelial transport of Na^+ in MCH-responsive epithelia appears to result from passive movement of Na⁺ from lumen to cell along a favorable electrical and chemical (concentration) gradient, followed by active extrusion of Na⁺ into peritubular fluid by an Na⁺-K⁺ exchange pump on the peritubular membrane. A considerable body of evidence suggests that MCH increases the permeability of the luminal cell membrane to Na⁺ and K⁺ (permease hypothesis) and concurrently increases the rate of peritubular Na⁺-K⁺ exchange (Hierholzer and Wiederholt, 1976) (Fig. 8.1). The mechanism whereby aldosterone increases the rate of transport of Na⁺ across the peritubular membrane has not been elucidated. Mineralocorticoid hormone increases the rate of pumping at least in part by increasing cellular entry of Na⁺ from the lumen. The transport process entails hydrolysis of ATP by an enzyme activated by increasing concentrations of Na⁺ (in cell) or K⁺ (in peritubular fluid): Na⁺-K⁺-ATPase (Katz and Epstein, 1967). This enzyme is present in highest concentration in the peritubular cell membrane and has been identified as the Na⁺-K⁺ exchange

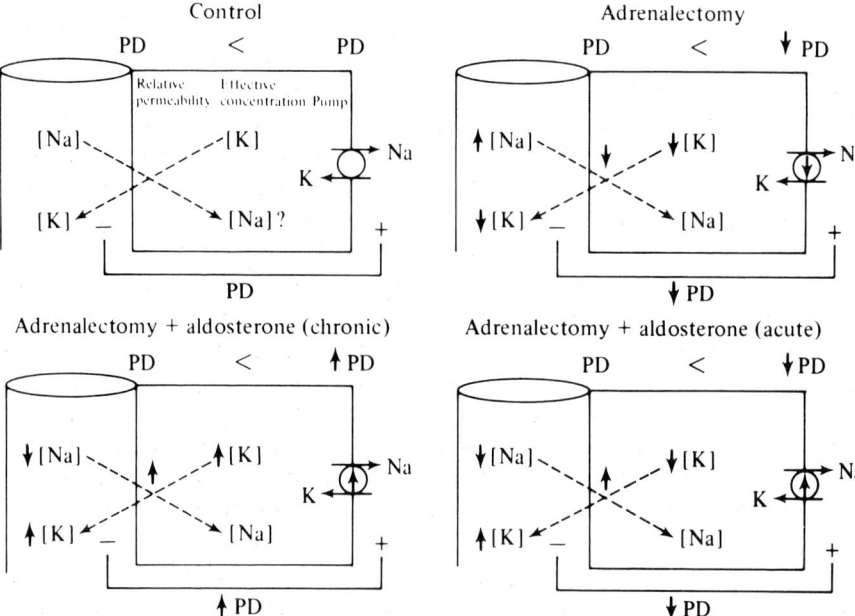

Figure 8.1 *Schematic representation of the effect of aldosterone on electrolyte transport and potential difference (PD) in the distal tubule of the rat.* Control (intact adrenal) conditions are depicted in the left upper panel. Adrenalectomy (upper right) reduced intracellular K⁺ concentration (K⁺-selective electrode), presumably because of a greater dampening effect on the peritubular Na⁺:K⁺ exchange pump than the simultaneous reduction in luminal membrane K⁺ permeability. The acute administration of aldosterone (lower right) did not result in restoration of a normal intracellular K⁺ concentration, presumably because the magnitude of the acute increase in activity of the peritubular Na⁺:K⁺ exchange pump was insufficient to counterbalance the aldosterone-induced acute increase in luminal membrane K⁺ permeability. In contrast, chronic administration of aldosterone to adrenalectomized rats (lower left) normalizes the PD at both membranes and returns intracellular K⁺ concentration to the normal range. Presumably, several days of hormone administration are required to repair the large deficit in the pool of intracellular K⁺. (Reprinted from Hierholtzer and Wiederholt, 1976, *Kidney International*, with permission.)

pump. It has been suggested that MCH increases the activity or concentration of Na^+ exchange pumps (pump hypothesis) (Goodman, Allen and Rasmussen, 1969), but MCH stimulates transepithelial Na^+ transport acutely before any discernible increase occurs in ATP hydrolytic activity of the pump (Crabbe, 1961; Lemann, Piering and Lennon, 1970; Ross et al., 1959). Evidence has been adduced that MCH increases the supply of ATP (metabolic hypothesis) and alters the coupling between ATP utilization and the transport process in such a way as to increase the "efficiency" of energy used for transport (Edelman and Fanestil, 1970). Electrophysiologic studies of the mechanism of aldosterone action on transepithelial Na^+ transport have yielded divergent results: certain studies (Saito and Essig, 1973; Siegal and Civan, 1976) have suggested that the "strength" of the pump (electromotive force) is not influenced by MCH but that the "ease" (conductance) with which Na^+ flows through the active pathway is enhanced. Studies by Spooner and Edelman (1975) suggest that the effect of aldosterone is greater than that predicted on the basis of increased conductance alone.

The specific effects of aldosterone on transepithelial H^+ transport have been studied in the isolated turtle bladder. In this epithelium, a component of the H^+ secretory mechanism is electrogenic in nature, (i.e., is capable of transporting a positive charge into the lumen) and is not dependent on Na^+ reabsorption. Secretion of H^+ occurs even under conditions in which Na^+ reabsorption is prevented and no favorable transepithelial electrical potential difference is present or is permitted to develop (Steinmetz, 1974). Under these conditions, the rate of H^+ secretion decreases linearly as the transepithelial H^+ concentration gradient against which H^+ is secreted increases; when the H^+ concentration in the mucosal fluid is approximately 1,000 times greater than that in the serosal fluid, the net rate of H^+ secretion is zero (Fig. 8.2) (Al-Awqati et al., 1976; Al-Awqati, 1977). The magnitude of the transepithelial H^+ concentration gradient required to nullify net H^+ transport is an index of the "driving force" of the H^+ secretory pump. Aldosterone augments the rate of H^+ secretion when the transepithelial H^+ concentration gradient is nonlimiting without apparently increasing the driving force of the pump (Fig. 8.2). It has been inferred that aldosterone increases the "conductance" of H^+ through the active transport pathway in the cell membrane (Al-Awqati et al., 1976).* One possible explanation for this apparent increase in H^+ conductance is an increase in the number of pumping sites.

---

* Although primary changes in the metabolic activity of the epithelial cells can influence H^+ conductance, studies of the effect of aldosterone on the coupling ratio between H^+ transport and glucose utilization suggest that the effect of the hormone is exerted primarily on the transport process rather than on metabolism: aldosterone initially stimulates H^+ secretion without affecting the "driving force" of the metabolic reaction coupled to the transport process. However, after prolonged stimulation by aldosterone, the transport system uses less glucose per unit of H^+ transport, and the driving force of the metabolic reaction linked to transport increases. This reduction in "glucose cost" of H^+ transport was considered to reflect an adaptive change to increased work load "reminiscent of the changes that occur in skeletal muscles of athletes undergoing physical conditioning." (Al-Awqati, 1977).

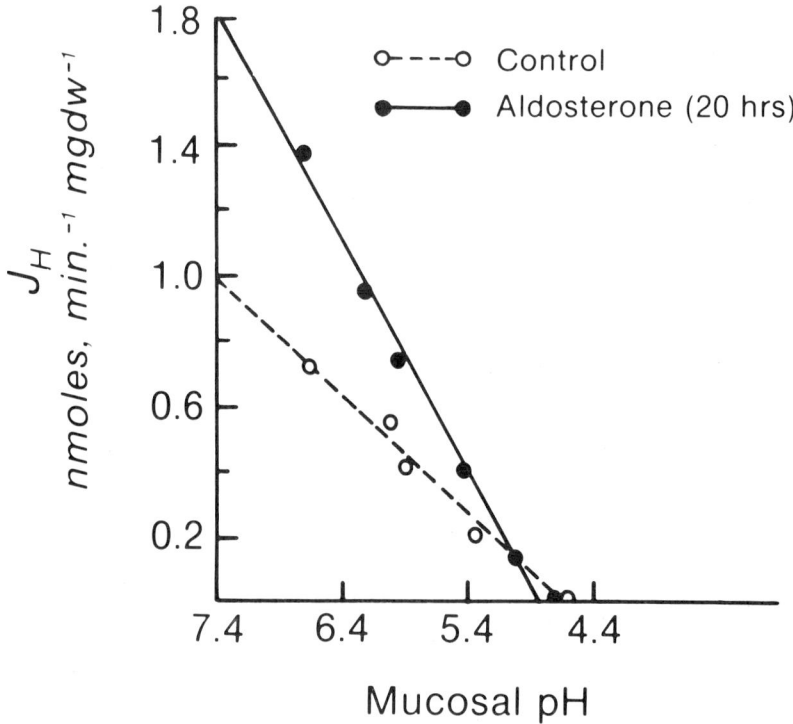

Figure 8.2 *The relationship of H^+ secretion (J_H) to the applied pH gradient (against which H^+ is secreted) in ouabain-treated fresh water turtle* (Pseudemys scripta) *bladders studied under short-circuit conditions in both control (○, freshly pithed) and aldosterone-treated (●, 20 hours) hemibladders. Aldosterone increased the rate of H^+ secretion at nonlimiting H^+ concentration gradients (increased slope) but did not alter the applied H^+ gradient at which net H^+ secretion ceases (the "proton motive force" of the pump).* (From Al-Awqati, 1977, *Journal of Clinical Investigation*, **60**, 1240)

The specific effects of MCH on transepithelial transport of K^+ have been studied less extensively than those on transport of Na^+ and H^+. An MCH-responsive K^+ transport mechanism has not been identified in the commonly studied vertebrate bladder preparations. In the mammalian distal nephron, transport effects of MCH have been demonstrated only in collecting tubules (Gross et al., 1975) and ducts (Uhlich et al., 1969; Uhlich et al., 1970); unresponsiveness to MCH has been demonstrated in early distal nephron segments (Gross et al., 1975). Peritubular transmembrane K^+ uptake can apparently occur against unfavorable electrochemical gradients (Wright, 1977), and the rate of K^+ secretion increases in response to MCH. Electrophysiologic data in collecting tubules have been interpreted to indicate that MCH does not increase the driving force of the K^+ secretory pumps but that it increases the conductance of K^+ through the epithelium. The possibility that MCH stimulates distal K^+ secretion by increasing the lumen-negative transepithelial potential difference in the distal nephron has been raised by renal clearance studies and is supported by the results of micropuncture experiments (Garcia, Malnic and Giebisch, 1976; Wiederholt et al., 1973). A determining role for potential differences in the K^+ secretory response to MCH has not been established; several instances of dissociation of changes in K^+ secretion and voltage have been reported (Good and Wright, 1977).

Potentiation by aldosterone of the hydroosmotic effect of vasopressin appears to be mediated by more than one mechanism: inhibition of cyclic nucleotide phosphodiesterase activity with resultant diminution of cyclic AMP degradation (Stoff et al., 1973) and inhibition of prostaglandin E synthesis, which ordinarily inhibits vasopressin-induced cyclic AMP (Zusman, Keiser and Handler, 1977). Greater potentiation occurs with adrenal steroids that have greater glucocorticoid potency than aldosterone, raising the possibility that this effect of aldosterone is mediated by its "glucocorticoid" activity.

A biochemical basis for the distinction between glucocorticoid and mineralocorticoid activity is provided by studies of the mechanism of steroid action at the subcellular level. The steroid crosses the cell membrane, presumably by simple diffusion, where it binds with high affinity receptor to proteins in the cytoplasm of the appropriate target tissue (Feldman, Funder and Edelman, 1972). Studies in rat kidney tissue have suggested that there are multiple steroid hormone-binding sites (Funder, Feldman and Edelman, 1973). Hormones that are classified as "mineralocorticoid" or "glucocorticoid" on the basis of their physiologic effects bind with different affinities to the various receptor sites. The mineralocorticoid and glucocorticoid potency of a particular steroid hormone correlates with its relative affinity for the so-called mineralocorticoid and glucocorticoid receptors (Baxter et al., 1976). Steroid action is dependent on the formation of an active steroid-receptor complex for subsequent initiation of genetic action (Samuels and Tomkins, 1970). Steroids with stimulatory (agonist) properties in a given tissue generate a steroid-receptor complex that is transferred to the cell nucleus. Studies with inhibitors of DNA-dependent RNA synthesis and inhibitors of synthesis of ribosomal protein suggest that the formation of an aldosterone-induced protein plays a primary role in mediating the Na^+ transport effects of the hormone (Crabbe and De Weer, 1964; Edelman, Bogoroch and Porter, 1963; Fanestil and Edelman, 1966). Steroids that act as competitive antagonists (e.g., progesterone, 17α-hydroxyprogesterone, and spironolactone) bind with high affinity to the MCH receptor but fail to initiate the subsequent steps required for MCH action (Fanestil, 1968).

SYNDROMES OF MINERALOCORTICOID HORMONE EXCESS (TABLE 8.1)

Experimental states of mineralocorticoid hormone excess

Acute administration of large amounts of MCH results in a reduction in the urinary excretion rates of Na^+ and Cl^- and an increase in the excretion

Table 8.1 Clinical Syndromes of MCH Excess

I. **Primary MCH excess**
  A. Primary aldosteronism
    1. Aldosterone-producing adenoma
    2. Idiopathic hyperplasia
    3. Aldosterone-producing carcinoma
    4. Glucocorticoid-suppressible hyperaldosteronism
  B. Adrenal enzyme defects
    1. 11β-hydroxylase deficiency
    2. 17α-hydroxylase deficiency
  C. Cushing's syndrome
    1. Ectopic ACTH syndrome
    2. Adrenal carcinoma
    3. Adrenal adenoma
    4. Hypothalamic-pituitary (Cushing's disease)
  D. Other MCH hormones
    1. Licorice
    2. Carbenoxolone
    3. (?) Low-renin essential hypertension

II. **Secondary MCH excess**
  A. Physiologic
    1. Sodium depletion—vomiting, diarrhea
    2. Hemorrhage
    3. Pregnancy
  B. Pathophysiologic
    1. Nonhypertensive
      a) Edematous states—cirrhosis, congestive heart failure, nephrotic syndrome
      b) Bartter's syndrome
      c) Renal tubular acidosis
      d) Diuretic therapy
    2. Hypertensive
      a) Accelerated hypertension
      b) Renovascular hypertension
      c) Renin-secreting tumor
      d) End-stage renal failure (rarely)
      e) Post-transplant hypertension
      f) Estrogen therapy

III. **Pseudohypermineralocorticoidism**
  A. Liddle's syndrome

rates of K^+ and net acid (Lemann et al., 1970; Mills, Thomas and Williamson, 1960; Ross et al., 1959). With continued administration (for days or weeks), extracellular fluid volume and Na^+ concentration increase and K^+ depletion and hypokalemia occur (Kassirer et al., 1970; Relman and Schwartz, 1952). In subjects ingesting normal amounts of Na^+ and K^+, the plasma bicarbonate concentration is not significantly changed, however, despite prolonged (three months) administration of even 10 times the normal amount of aldosterone (Kassirer et al., 1970). In dogs the alkalosis-producing effect of MCH is potentiated by deprivation of dietary K^+ (Hulter, Sigala and Sebastian, 1977a; Kurtzman, White and Rogers, 1973). Similar potentiation may occur in man because in patients with primary aldosteronism the severity of metabolic alkalosis is proportional to the degree of potassium depletion as indicated by the serum K^+ concentration (Kassirer et al., 1970) (Fig. 4.7).

The reduction in urinary excretion of Na^+ and Cl^- that occurs after initiation of MCH administration does not persist with continued hormone administration. The expansion of extracellular fluid volume that occurs as NaCl is retained initiates physiologic events that increase the filtered load of Na^+ and decrease the fraction of the filtered load of Na^+ reabsorbed by the renal tubules. As a consequence, after an initial decrement, the urinary excretion rates of Na^+ and Cl^- return to base-line values. Prior restriction of the amount of NaCl in the diet diminishes the initial decrement and prolongs the period of Na^+ and Cl^- retention, perhaps because a critical level of volume expansion is required to reestablish the steady state (Relman and Schwartz, 1952). Alterations in renal hemodynamics, interstitial hydrostatic pressure, and plasma oncotic pressure and increased plasma concentrations of natriuretic substances have been implicated in mediating the reduction in fractional Na^+ reabsorption that permits establishment of the steady state (Earley and Daugharty, 1969). The precise intrarenal physiologic mechanisms that mediate the reestablishment of Na^+ and Cl^- balance have not been elucidated and conflicting observations remain to be resolved.

The magnitude of the kaliuretic response to MCH administration in normal subjects is influenced by the amount of NaCl in the diet (Relman and Schwartz, 1952). Prior restriction of Na^+ intake limits the kaliuretic (and acid excretory) response and prior supplementation with Na^+ and Cl^- exaggerates the response. With continued MCH administration, kaliuresis may persist beyond the transition period wherein Na^+ balance is reestablished. Chronic K^+ depletion stimulates thirst (Berl et al., 1977), decreases renal concentrating ability (Rubini, 1961; Berl et al., 1977), and leads to polyuria; it reversibly decreases renal blood flow and glomerular filtration rate (Abbrecht, 1969), increases renal ammonia production (Tannen, 1977), and impairs renal conservation of Cl^- (Luke and Levitan, 1967).

Primary aldosteronism

The syndrome of primary aldosteronism is recognized by the classic findings of hypertension, hypokalemia, metabolic alkalosis, hyperaldosteronism, and suppressed plasma renin activity (Conn, 1955; Conn, Cohen and Rovner, 1964a). Hypertension appears to occur in part as a consequence of an expanded extracellular fluid volume (Biglieri and Forsham, 1961). Administration of mineralocorticoid antagonists (e.g., spironolactone) and other natriuretic agents will generally ameliorate the hypertension (Spark and Melby, 1968). When spironolactone treatment is discontinued abruptly in patients with primary aldosteronism, hypertension recurs promptly because total body Na^+ stores and plasma volume again become supernormal (Fig. 8.3) (Wenting et al., 1977). With continued observation, plasma volume tends to return toward normal levels but hypertension per-

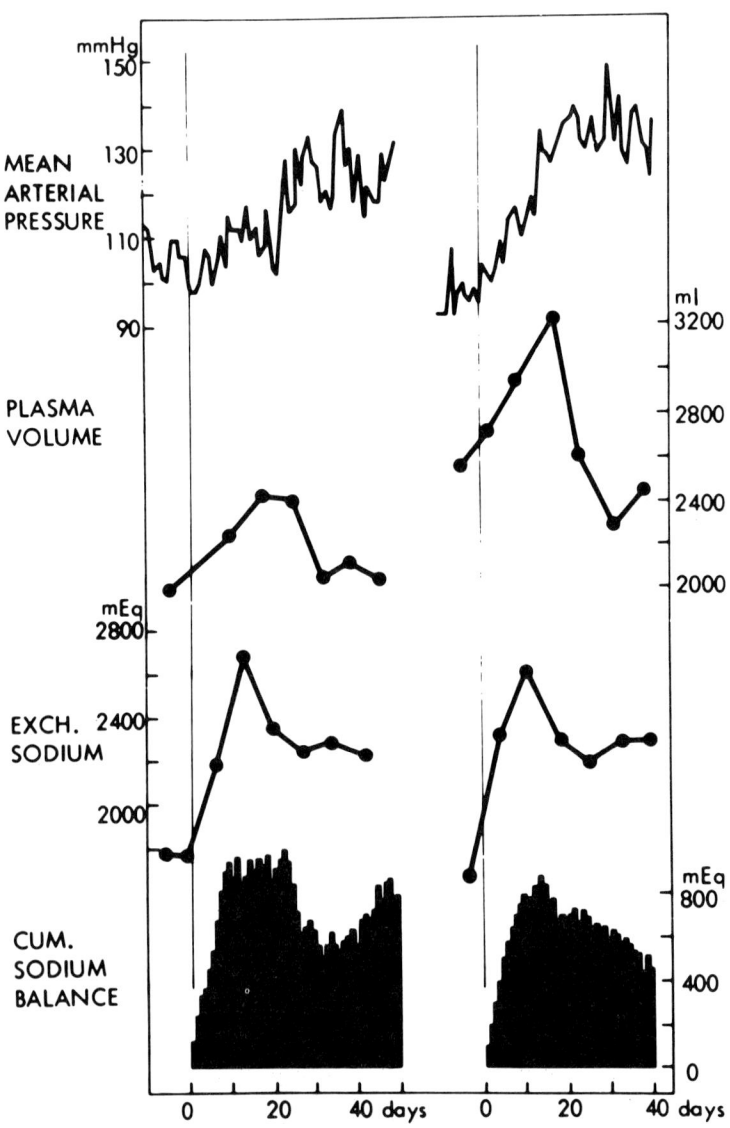

Figure 8.3 *Changes in cumulative Na^+ balance, exchangeable Na^+, and plasma volume during development of hypertension in two subjects with primary aldosteronism. Spironolactone treatment was stopped at day zero. Mean values of arterial pressure [diastolic pressure + 0.33 x (systolic pressure − diastolic pressure)] measured during the day, when the subject was recumbent, are presented. (From Wenting et al., 1977, Volume-pressure relationships during development of mineralocorticoid hypertension in man.* Circulation Research, **40** *[Suppl. I], I–163)*

sists. Whether or not factors other than plasma volume expansion contribute to hypertension in states of chronic MCH excess remains to be established.

Hypokalemia resulting from renal K⁺ wasting is characteristic of primary aldosteronism and is often the disturbance by which such patients are recognized clinically. Because the kaliuretic response to MCH is amplified by increased delivery of tubular fluid to the K⁺ secretory sites in the distal nephron (Wright, 1977) and because expansion of extracellular fluid volume fails to suppress aldosterone production normally in patients with primary aldosteronism (Biglieri et al., 1967; Kem et al., 1971), administration of large amounts of dietary NaCl tends to exacerbate renal K⁺ wasting and hypokalemia. Conversely, restriction of dietary NaCl reduces renal K⁺ wasting and ameliorates hypokalemia. Some patients with primary aldosteronism may not have frank hypokalemia despite normal amounts of dietary NaCl, but in contrast to normal subjects, such patients become hypokalemic when dietary Na⁺ intake is further increased.

In the majority of patients with primary aldosteronism, MCH excess is due to a unilateral adrenal adenoma (Conn, Knopf and Nesbit, 1964b), but a similar syndrome can occur in association with bilateral adrenal hyperplasia (Davis et al., 1967; Katz, 1967; Laragh, Ledingham and Sommers, 1967) or, rarely, adrenal carcinoma (Alterman et al., 1969). Preoperative distinction between adenoma and hyperplasia is necessary because surgical removal of the adenoma usually corrects the hypertension, whereas bilateral removal of hyperplastic glands usually does not. Adenoma and hyperplasia can be distinguished in greater than 90 percent of the cases by statistical analysis of selected biochemical and clinical parameters of MCH excess (Aitchison et al., 1971; Luetscher et al., 1974; Stockigt, Collins and Biglieri, 1971) and by differences in the circadian pattern of aldosterone secretion (Fig. 8.4) (Schambelan et al., 1976). In patients with hyperplasia, a directionally normal variation in plasma aldosterone concentration occurs in response to assumption of upright posture (Biglieri et al., 1974a; Ganguly et al., 1973). It is possible that the hyperaldosteronism in patients with hyperplasia may result from or be maintained by increased adrenal sensitivity to angiotensin because aldosterone secretion is stimulated by infusion of angiotensin at rates less than those required to stimulate aldosterone secretion in normal subjects (Wisgerhof, Carpenter and Brown, 1977). In patients with an adenoma, plasma aldosterone concentration fails to increase normally and frequently appears to decrease "paradoxically" with assumption of upright posture (Biglieri et al., 1974a; Ganguly et al., 1973). Because changes in plasma aldosterone concentration correlate positively with changes in plasma cortisol regardless of postural influences (Schambelan et al., 1976), it is possible that aldosterone secretion from the adenoma is controlled by ACTH production. Thus, determination of the effect of posturally mediated increases in renin activity on aldosterone secretion may be used to distinguish an adenoma from hyperplasia preoperatively.

Differences in the aldosterone level in the adrenal venous effluent can reliably distinguish between unilateral and bilateral hormonal overpro-

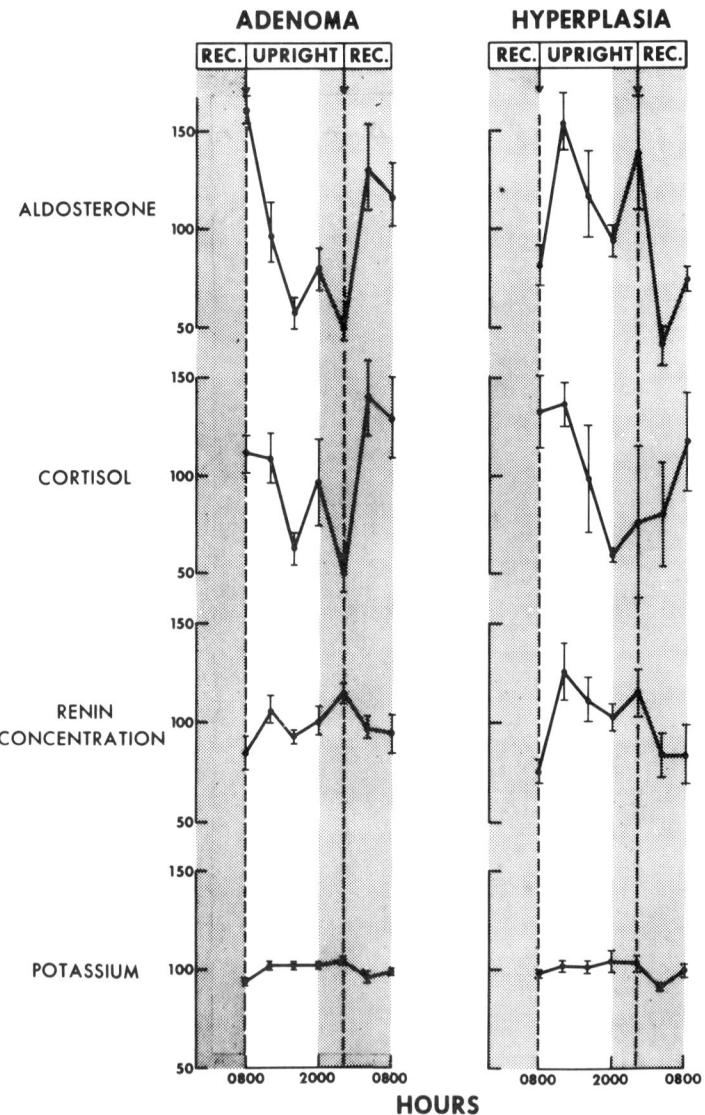

Figure 8.4 *Circadian patterns of plasma aldosterone, cortisol, renin, and K^+ concentrations on a day of upright posture in patients with primary aldosteronism due to an adenoma ($N = 6$) or idiopathic hyperplasia ($N = 5$). Each data point was normalized as the percentage of the average value for that parameter in the 24-hour period, and the mean ± SEM of the normalized data were plotted as the percentage of the average (100%) value for each parameter. In the patients with an adenoma, plasma aldosterone and plasma cortisol concentrations correlated significantly ($r = +0.68$, $P < 0.001$). In the patients with idiopathic hyperplasia, plasma aldosterone correlated significantly with plasma renin concentration ($r = +0.65$, $P < 0.001$) and plasma K^+ concentration ($r = +0.64$, $P < 0.001$). (From Schambelan et al., 1976, Journal of Clinical Endocrinology and Metabolism, 43, 115)*

duction (Melby et al., 1967), but technical difficulties have limited the widespread applicability of this procedure. Adrenal venography (Nicolis et al., 1972), iodocholesterol scanning (Hogan et al., 1976; Lieberman et al., 1971) and/or computerized axial tomography may also be of use in identifying the site of an anatomic abnormality.

Administration of small doses of dexamethasone to patients with primary aldosteronism usually results in a transient reduction in aldosterone secretion (Newton and Laragh, 1968; Slaton, Schambelan and Biglieri, 1969). Rarely, sustained suppression of aldosterone secretion and correction of the entire syndrome of primary aldosteronism can be effected with dexamethasone; this form is called glucocorticoid-suppressible hyperaldosteronism (Giebink et al., 1973; New, Siegal and Peterson, 1973; Sutherland, Ruse and Laidlaw, 1966). Such patients are generally young, and the disorder can be familial. An adrenal biosynthetic defect has been suspected but never established as the pathogenetic mechanism. Because surgery is unnecessary in these patients, a trial of dexamethasone therapy (two or three weeks) appears to be indicated in all young patients with primary aldosteronism.

Adrenal enzyme defects

Genetically transmitted deficiencies of the enzymes required for adrenal steroid biosynthesis lead to distinct clinical syndromes of disturbed MCH production. In two such enzyme deficiency states, 11β- and 17α-hydroxylase deficiency (Biglieri, Herron and Brust, 1966; Bongiovanni and Root, 1963; Gandy, Keutmann and Izzo, 1960; New and Seaman, 1970), impaired cortisol production results in increased ACTH production and secondarily in increased production of deoxycorticosterone (DOC), which produces a typical syndrome of mineralocorticoid excess. In 11β-hydroxylase deficiency, conversion of DOC to corticosterone and its derivatives is impaired, and thus corticosterone, 18-hydroxycorticosterone, and/or aldosterone may be deficient. In 17α-hydroxylase deficiency, conversion of DOC to its derivatives is not enzymatically blocked, hence corticosterone levels are markedly elevated and may contribute to the hypermineralocorticoid state. Despite the intact enzymatic pathway to aldosterone, aldosterone production is generally subnormal, presumably as a consequence of the reduced plasma renin activity and the hypokalemia that occur due to the hypermineralocorticoid state (Goldsmith, Solomon and Horton, 1967).

Although the MCH potency of DOC is less than that of aldosterone, the levels of DOC present in patients with these two disorders appear sufficiently great to account for the hypermineralocorticoid state. In studies using renal mineralocorticoid receptors incubated in the presence of normal plasma binding proteins, the estimated mineralocorticoid potency of DOC is approximately one-sixth that of aldosterone and accounts for little or no receptor occupancy at normal plasma concentrations (5 to 10 ng/dl) (Baxter et al., 1976). At plasma levels of DOC comparable to those observed in patients with these two enzyme deficiency states (200 to 500 ng/dl) (Biglieri, 1977), major receptor occupancy has been demonstrated. It has been hypothesized that the increased levels of corticosterone and 11-deoxycortisol may compete with DOC for available binding sites on corticosteroid-binding globulin and may therefore make more "free" DOC available for interaction with the mineralocorticoid receptor (Biglieri, 1977).

These two syndromes may be readily distinguished clinically by their different effects on sexual development. In the 17α-hydroxylase deficiency syndrome, biosynthesis of adrenal and gonadal androgen and estrogen is impaired, inhibiting normal sexual maturation. Genotypic female patients fail to undergo menarchy or to develop secondary sexual characteristics (Biglieri et al., 1966). Genotypic male patients become pseudohermaphrodites as a consequence of androgen deficiency in utero (Mantero et al., 1971; New, 1970). In the 11β-hydroxylase deficiency syndrome, the enzymatic steps required for the synthesis of adrenal sex steroids are unimpaired. Thus, the increased ACTH level stimulates increased production of adrenal androgen, frequently causing virilism (Bongiovanni and Root, 1963). In both syndromes, glucocorticoid hormone replacement therapy inhibits ACTH secretion and resultant excess MCH production and ameliorates the hypermineralocorticoid state. In patients with the 11β-hydroxylase deficiency syndrome, androgen production diminishes and signs of virilization tend to disappear. Such therapy does not correct the sexual abnormalities in patients with the 17α-hydroxylase deficiency syndrome. An apparent increased sensitivity to glucocorticoid hormone may occur in patients with 17α-hydroxylase deficiency and therefore replacement doses may have to be reduced accordingly.

Cushing's syndrome

Glucocorticoid hormone excess characteristically results in obesity, carbohydrate intolerance, muscle wasting, osteoporosis, and mild hirsutism. Hypertension and the electrolyte abnormalities typical of hypermineralocorticoidism may occur but are rarely severe. Because the severity of the hypokalemia and metabolic alkalosis in patients with Cushing's syndrome may be directly related to the degree of elevation of plasma cortisol (Christy and Laragh, 1961), it has been suggested that the electrolyte abnormalities reflect the "mineralocorticoid" activity of cortisol. Although cortisol has only 0.4 percent of the mineralocorticoid activity of aldosterone when measured in the presence of normal amounts of plasma-binding proteins (Baxter et al., 1976), the plasma concentrations may become sufficiently great to have a considerable mineralocorticoid effect. No systematic studies of the effects of chronic administration of large doses of glucocorticoid hormones on renal Na^+, K^+, and acid-base homeostasis have been reported.

When marked hypertension and hypokalemia are present, the presence of a more potent MCH than cortisol should be suspected. Elevated DOC levels,

as high as those observed in patients with the 17α-hydroxylase deficiency syndrome, have been reported in patients with Cushing's syndrome, usually in association with an underlying malignancy of either adrenal (Solomon et al., 1968) or extraadrenal (ectopic ACTH syndrome) (Schambelan, Slaton and Biglieri, 1971) origin. Less commonly, aldosterone can also be elevated in patients with the same underlying conditions and probably contributes to the severity of the hypermineralocorticoid state (Schambelan et al., 1971). One of the consequences of excess amounts of DOC and/or aldosterone is suppressed plasma renin activity, which is ordinarily normal or increased in patients with Cushing's syndrome due to benign hyperplasia.*
Thus, the finding of suppressed plasma renin activity in a patient with Cushing's syndrome should also suggest the presence of a potent MCH (Hogan, Schambelan and Biglieri, 1977).

Syndromes possibly due to unknown mineralocorticoid hormones

Administration of a potent MCH to normal subjects promptly results in two of the cardinal features of primary mineralocorticoid excess: hypokalemia and suppressed plasma renin activity (Shade and Grim, 1975). Urinary and plasma levels of aldosterone then decrease, presumably owing to the diminished trophic effect of K^+ and angiotensin on aldosterone production. Thus, when a patient with typical features of primary mineralocorticoid excess is found to have reduced levels of aldosterone, the presence of another potent MCH should be suspected. Such findings led to the recognition of the pathophysiologic role of DOC in the 17α-hydroxylase deficiency (Biglieri et al., 1966) and the ectopic ACTH syndromes (Schambelan et al., 1971) and have revealed the "mineralocorticoid" activity of licorice (Conn, Rovner and Cohen, 1968) and carbenoxolone (Royston and Prout, 1976). Recent isolated reports of patients with typical features of primary mineralocorticoid excess in whom levels of known steroid hormones with mineralocorticoid activity are normal or reduced have suggested that other as yet unidentified MCH may cause hypertensive syndromes in man (New et al., 1977).

The observation that 20 to 25 percent of patients with essential hypertension have reduced and subnormally responsive levels of plasma renin activity (Channick, Adlin and Marks, 1969; Jose and Kaplan, 1969), or so-called low-renin essential hypertension, has stimulated many investigators to consider the possibility of excess production of an MCH contributing to or directly causing the hypertension in this substantial population of patients. Because serum K^+ levels are not reduced and because aldosterone levels are

---

* Corrected values of plasma renin activity in patients with Cushing's syndrome remain within normal limits even when the effect of glucocorticoid-induced increases in renin substrate are taken into account (Krakoff, Nicolis and Amsel, 1975). Direct measurement of plasma renin concentration, a technique that is not influenced by endogenous levels of renin substrate, are usually normal in patients with Cushing's syndrome (Brown et al., 1965).

usually within the normal range, the evidence of MCH excess is somewhat circumstantial in these patients. Like patients with primary aldosteronism, hypertension appears to be "volume dependent." Administration of spironolactone as well as other natriuretic agents usually cures or substantially ameliorates the hypertension (Carey et al., 1972; Spark and Melby, 1971). Yet, unlike patients with primary aldosteronism, initial reports that claimed that total body Na^+ stores are increased in patients with low-renin essential hypertension (Woods et al., 1969) have not been confirmed (Lebel et al., 1974).

The search for a presumed MCH in patients with low-renin essential hypertension has been prompted further by the demonstration of a salt-retaining factor in extracts of urine obtained from these patients (Sennett et al., 1975). Evidence to support an adrenal source of the putative hormone has been derived from studies in which inhibitors of adrenal biosynthesis, such as aminoglutethamide, reduce blood pressure in patients with low-renin but not normal-renin hypertension (Woods et al., 1969). Furthermore, adrenalectomy has been reported to cure hypertension in such patients (Gunnells et al., 1970). The exact nature of the steroid or steroids presumed to be responsible for these findings has not been elucidated. Plasma DOC levels have been reported to be increased in some patients with low-renin hypertension (Brown et al., 1972), but other investigators have not found evidence for increased DOC secretion (Biglieri et al., 1968). The modest elevations in plasma DOC concentration reported would not be expected to have a major mineralocorticoid effect on the basis of quantitative considerations of plasma binding and the known affinity of DOC for the mineralocorticoid receptor (Baxter et al., 1976). Increased levels of 18-hydroxydeoxycorticosterone have also been described in some patients with low-renin essential hypertension (Melby, Dale and Wilson, 1971), but the mineralocorticoid potency of this steroid is minimal and, at the levels reported, would not be expected to contribute significantly to a hypermineralocorticoid state. Other steroids that have been reported to be present in increased amounts in patients with low-renin hypertension, such as 16β-hydroxydehydroepiandrosterone (Sennett et al., 1975) also appear to have minimal potency as mineralocorticoids (Baxter et al., 1976).

In contrast to these findings, recent studies using a rat kidney mineralocorticoid receptor assay system failed to demonstrate gross increases in plasma mineralocorticoid activity in patients with low-renin hypertension. Whereas the plasma of patients with primary aldosteronism resulted in displacement of labeled aldosterone from the receptor, there was no evidence of an increased competitor activity in the plasma of patients with low-renin hypertension compared with that from patients with normal-renin hypertension or from normal control subjects (Fig. 8.5). In this approach, the relative affinity for the mineralocorticoid receptor as well as the contribution of plasma binding to plasma mineralocorticoid activity is con-

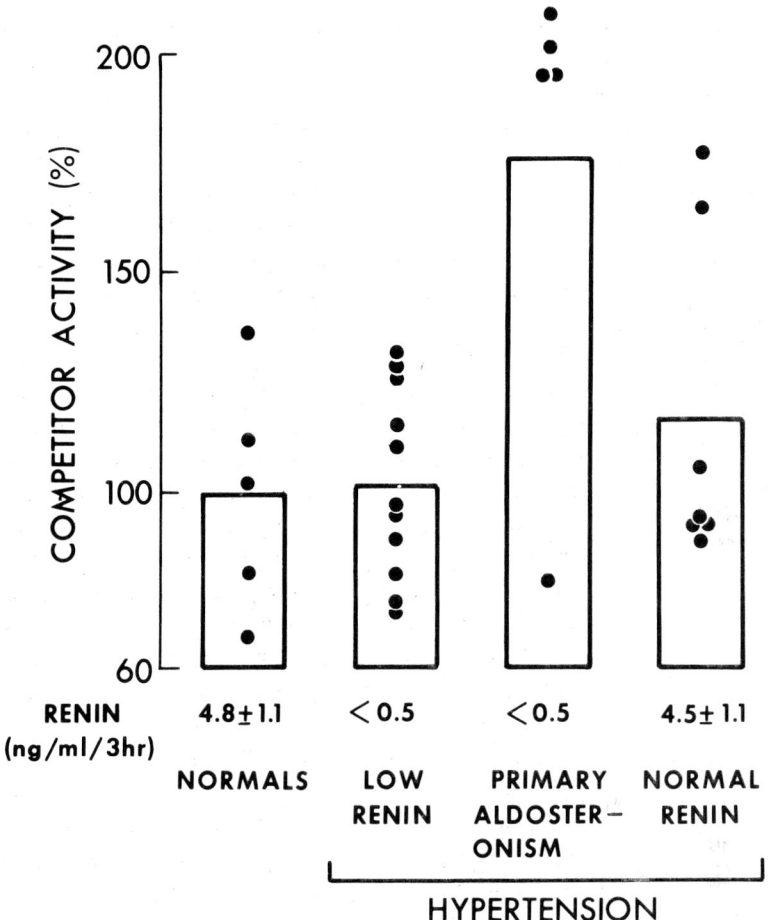

Figure 8.5 Comparison of plasma mineralocorticoid activity in normal control subjects ($N = 5$) with that in three groups of patients with hypertension: low-renin hypertension ($N = 11$), primary aldosteronism ($N = 5$), and normal-renin essential hypertension ($N = 7$). All plasma samples were obtained at 8 AM after overnight recumbency; the mean ± SEM plasma renin activity for each group is also shown. The mean of the values from normal subjects was assigned a value of 100 percent and the other determinations were expressed as a percentage of this value. The means for patients with low-renin and normal-renin hypertension were not significantly different from that of control subjects, whereas the mean value for patients with primary aldosteronism was significantly greater than those for the other groups ($P < 0.001$). (From Mitchell et al., 1977, Annals of Internal Medicine, 87, 591.)

sidered. The possibility that the effects of a putative mineralocorticoid might be mediated by another receptor system has not been excluded by these studies.

Several observations have suggested that abnormalities in aldosterone production and metabolism occur in patients with low-renin hypertension. Both a reduced aldosterone secretory response to Na⁺ restriction (Weinberger et al., 1968) and a subnormal suppression of aldosterone secretion with Na⁺ loading (Collins et al., 1970) have been reported. Reduced metabolic clearance of aldosterone has been reported to occur in patients with essential hypertension (Nowaczynski, Kuchel and Genest, 1971), but this finding has not been confirmed by other investigators (Brown, 1976). Of considerable interest is the recent observation that the aldosterone secretory response to infusion of angiotensin II may be increased in patients with low-renin hypertension (Wisgerhof and Brown, 1977). Thus, hypersensitivity to angiotensin at both the vascular and adrenal levels could conceivably account for the hypertension and maintenance of normal aldosterone secretion in patients with low-renin hypertension.

Secondary hyperaldosteronism

Hyperaldosteronism characteristically occurs in states of reduced effective arterial blood volume owing to activation of the renin-angiotensin system (Biglieri et al., 1974b). The resultant increases in plasma aldosterone concentration stimulate renal tubular reabsorption of Na⁺ and Cl⁻ and thereby tend to restore normal "effective" blood volume. When hypovolemia is of extrarenal origin (vomiting, diarrhea, and hemorrhage) and renal tubular reabsorption of Na⁺ and Cl⁻ is not specifically impaired, the Na⁺-retaining effect of hypermineralocorticoidism is manifested by reduced urinary excretion rates of Na⁺ that persist until effective arterial blood volume is restored. Provision of adequate amounts of dietary NaCl may effect normalization of effective arterial blood volume. When the reduction in effective arterial blood volume is due to congestive heart failure, cirrhosis, nephrotic syndrome, or other disorders, provision of dietary NaCl leads to progressive expansion of extracellular fluid volume and edema without restoration of effective blood volume. The Na⁺-retaining effect of hypermineralocorticoidism does not reduce excretion rates of Na⁺ when hypovolemia is caused predominantly by impaired renal tubular reabsorption of Na⁺ and Cl⁻ (e.g., diuretic administration, Bartter's syndrome, and some types of renal tubular acidosis).

Hyperaldosteronism also occurs in association with activation of the renin-angiotensin system in patients with accelerated hypertension, renovascular or segmental renal lesions, and rarely, renin-secreting neoplasms (Biglieri et al., 1974b). In such patients, the degree of hyperreninemia and/or hyperaldosteronism is too high for the effective blood volume and hypertension ensues. In many instances, hypertension occurs even when the levels of renin and/or aldosterone are within normal limits, presumably because of an inappropriate relationship between the degree of vasoconstriction and the state of the effective arterial blood volume.

The exact contribution of excess MCH production to the pathophysiology of states characterized by secondary hyperaldosteronism is not entirely clear. Hypertension persisted despite adrenalectomy in a patient with a renin-secreting tumor (Schambelan et al., 1973) and in animals with experimental renovascular hypertension (Blair-West et al., 1968). Reduction of Na^+ excretion and edema formation can occur in patients with cardiac, renal, or hepatic disease without increased aldosterone secretion (Chonko et al., 1977). Potassium wasting persists despite reduction of aldosterone secretion or administration of MCH antagonists in patients with Bartter's syndrome (Ch. 9) and may persist despite sustained correction of acidosis and secondary hyperaldosteronism in some patients with type 1 renal tubular acidosis (Sebastian, McSherry and Morris, 1971). In these disorders, however, it seems likely that the secondary aldosteronism serves to amplify the primary defect.

Pseudohypermineralocorticoidism

A rare familial syndrome characterized by hypertension, hypokalemia owing to renal K^+ wasting, and metabolic alkalosis was first described by Liddle, Bledsoe and Coppage (1963). Unlike patients with primary aldosteronism, the affected members of the family in this study had markedly reduced rates of aldosterone secretion. An assay for angiotensin in arterial blood was "negative." Whereas such findings might suggest the presence of another potent MCH, these patients differed from those with typical states of MCH excess in that Na^+ and K^+ excretion were not affected by agents that block the production of MCH (metyrapone) or the binding of MCH to the renal cytosol receptor (spironolactone). Furthermore, the $Na^+ : K^+$ ratio in saliva and sweat was high, in contrast to the low ratio usually seen in patients in states of MCH excess. Administration of triamterene, an agent that causes natriuresis and K^+ retention in the absence of MCH, resulted in correction of hypokalemia and hypertension. A primary renal transport defect was proposed to explain the findings in this family and in other such patients (Aarskog et al., 1967; Ross, 1959) but the exact nature of this defect has not been elucidated.

SYNDROMES OF MINERALOCORTICOID HORMONE DEFICIENCY (TABLE 8.2)

Experimental states of mineralocorticoid hormone deficiency

Few experimental studies of MCH deficiency in the absence of glucocorticoid hormone deficiency have been conducted. In dogs with isolated MCH deficiency, hyponatremia and hypovolemia due to Na^+ wasting, hyperkalemia due to K^+ retention and acidemia and metabolic acidosis due to reduced net acid excretion constitute the most prominent pathophysiologic features (Kurtzman, White and Rogers, 1971; Hulter et al., 1977b). The effect of aldosterone deficiency on the acidification process of the mammalian nephron has been studied in adrenalectomized dogs maintained postoperatively on

Table 8.2 Clinical Syndromes of MCH Deficiency

I. **Primary MCH deficiency**
   A. Addison's disease
      1. Tuberculosis, fungal disease
      2. Autoimmune ("idiopathic" adrenal atrophy)
      3. Metastatic carcinoma
      4. Bilateral adrenal hemorrhage
   B. Adrenal enzyme defects
      1. Desmolase deficiency
      2. 3β-ol-dehydrogenase deficiency
      3. 21-hydroxylase deficiency
      4. Corticosterone methyl oxidase deficiency, types 1 and 2
   C. Heparin
II. **Secondary MCH deficiency**
   A. Hyporeninemic hypoaldosteronism
      1. Diabetes mellitus
      2. Tubulointerstitial diseases
III. **Pseudohypomineralocorticoidism**
   A. Pseudohypoaldosteronism of infancy
   B. "Salt-wasting" nephropathy
   C. Miscellaneous case reports
      1. Arnold and Healy, 1969
      2. Gordon et al., 1970
      3. Spitzer et al., 1973
      4. Weinstein, Allan and Mendoza, 1974
      5. Schambelan, Sebastian and Rector, 1978

physiologic replacement doses of glucocorticoid and mineralocorticoid hormones (Hulter et al., 1977b). When administration of MCH is discontinued, net acid excretion decreases and hyperkalemic hyperchloremic acidosis develops and persists. The reduction in net acid excretion is due largely to a reduction in urinary excretion of ammonium, which in turn appears to be due to diminished renal production of ammonia, because it occurs in the absence of an increase in urinary pH or a decrease in urine flow; urinary pH remains constant or decreases (to values as low as 5.2). The reduction in ammonia production appears to be due in part to hyperkalemia, because the excretion rate of ammonium correlates inversely with the plasma K^+ concentration and does not decrease on discontinuation of MCH if hyperkalemia is prevented by restricting K^+ intake.

During the steady state of acidosis in these MCH-deficient dogs, urinary pH varies directly with the rate of urinary excretion of ammonium (Fig. 8.6). This finding indicates that the variation in urinary excretion of ammonium is not secondary to the variation in urinary pH but to differences in the rate of renal ammonia production and diffusion into the tubular lumen. The positive correlation between urinary pH and ammonium excretion reflects the buffer effect of diffusion of differing amounts of ammonia into the tubular lumen. A similar direct relationship between urinary pH and ammonium excretion is observed during the steady state of acidosis in MCH-replete dogs made acidotic experimentally by feeding them hydrochloric acid (Fig. 8.6). When the findings of the two groups are compared, however, a difference in the slope of the relationship is found: for a given increase in the amount of ammonia diffusing into the tubular lumen as indicated by a given increase in urinary ammonium excretion, the increase in urinary pH was greater in the MCH-deficient dogs than in the MCH-replete dogs, indicating that the secretion of H^+ increases at an abnormally low rate when luminal buffer increases. On the other hand, when the values of urinary pH are extrapolated to a zero value of urinary ammonium excretion, the values are less than 5.5 and are not significantly different in the two groups. Together these findings indicate that in the dog, aldosterone deficiency impairs the ability of the distal nephron to secrete H^+ at normal rates but does not impair its ability to generate normally steep lumen-to-blood H^+ concentration gradients. In addition, the findings indicate that aldosterone deficiency leads to a reduction in renal ammonia production, due in part to hyperkalemia.

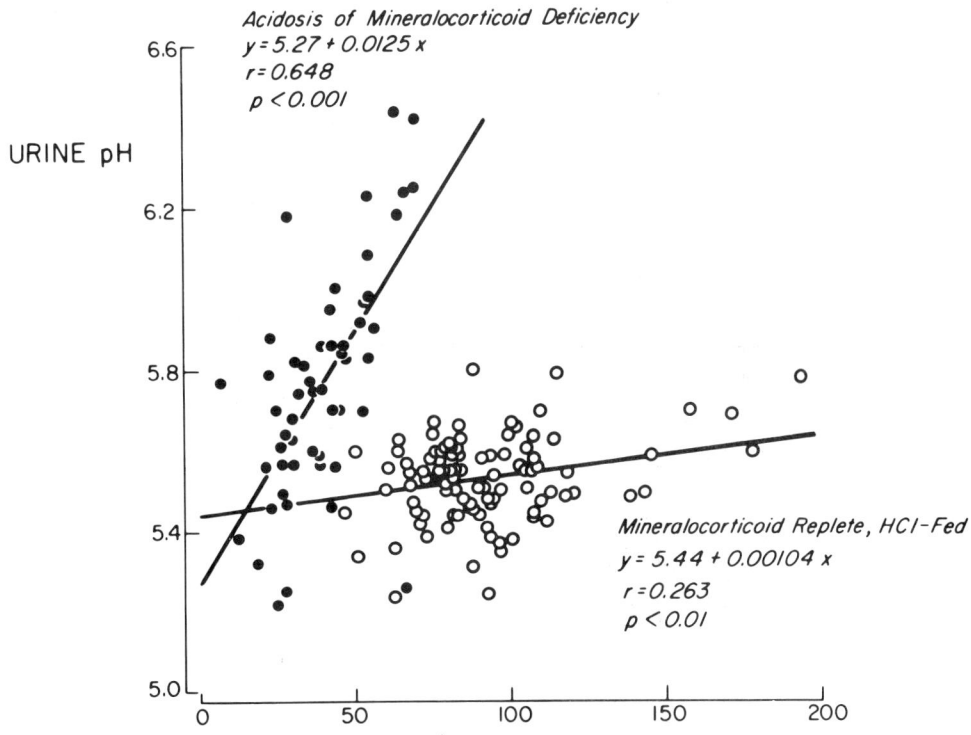

Figure 8.6 *Daily urine pH as a function of daily urinary ammonium excretion in two groups of dogs during steady-state metabolic acidosis: dogs with acidosis of MCH deficiency (●, N = 5) are compared with MCH-replete dogs fed hydrochloric acid (○, N = 8).* (From Hulter et al., 1977, *American Journal of Physiology,* **232**, F136)

Generalized adrenocortical insufficiency

Generalized adrenocortical insufficiency (Addison's disease) occurs when, as a consequence of destruction of adrenocortical tissue, secretory rates of both MCH and glucocorticoid hormone are inadequate for the physiologic needs of the organism. The combined deficiency of cortisol and aldosterone results in a progressive syndrome that is usually fatal if untreated.

Cortisol deficiency results in anorexia, weight loss, weakness, apathy, and a general inability to withstand "stress." Cortisol normally inhibits ACTH secretion so that markedly elevated levels of ACTH are characteristic of Addison's disease. Mineralocorticoid deficiency results in impaired renal Na^+ conservation. If the Na^+ intake is sufficiently large, extracellular fluid volume can be maintained at normal or near normal levels. However, if the Na^+ intake is low or if extrarenal losses of Na^+ occur, the inability to conserve Na^+ maximally results in marked Na^+ deficits, hyponatremia, hypovolemia, and increased plasma renin levels (Brown et al., 1968). Glucocorticoid deficiency may contribute further to the severity of the hypovolemia by an apparent effect of glucocorticoid to promote transfer of fluid to the intravascular compartment (Haack et al., 1977; Swingle et al., 1936).

The failure of the kidney to correct hyponatremia in patients with Addison's disease by the excretion of urine that is appropriately dilute with respect to the prevailing level of plasma hypoosmolality appears to be a consequence of the deficiency of both MCH and glucocorticoid hormone. The impairment of renal-diluting ability that occurs in MCH-deficient glucocorticoid-replete animals is prevented by administration of sufficient amounts of NaCl to minimize extracellular fluid volume depletion (Ufferman and Schrier, 1972). Extracellular fluid volume depletion is associated with enhanced proximal reabsorption of filtrate and thereby with a decreased delivery of tubular fluid to the diluting segments of the distal nephron, which limits the rate of excretion of solute-free water. Hypovolemia is also known to stimulate release of antidiuretic hormone (Robertson et al., 1973), which would further limit water excretion. Impaired diluting ability also occurs in patients who have glucocorticoid deficiency without MCH deficiency (panhypopituitarism and isolated ACTH deficiency), and this abnormality is reversible by administration of small doses of glucocorticoid hormones (Agus and Goldberg, 1971). Elevated levels of antidiuretic hormone have been observed in adrenalectomized dogs maintained only on MCH (Berns et al., 1977), and the associated defect in water excretion has been attributed to impaired suppression of secretion of antidiuretic hormone secondary to hemodynamic factors (e.g., decreased left ventricular stroke volume).

Hyperkalemia and metabolic acidosis might be expected to occur in patients with Addison's disease as a consequence of a diminished mineralocorticoid effect on K^+ and H^+ secretion in the distal nephron. However, in adult subjects with normal renal function, ingesting liberal amounts of dietary Na^+ and Cl^- and receiving glucocorticoid hormone therapy, isolated

MCH deficiency has not been observed to result in frank hyperkalemia or metabolic acidosis (Perez, Oster and Vaamonde, 1976). Hyperkalemia and acidosis have been reported more frequently in children with isolated MCH deficiency (18-hydroxylation deficiency, see below), but the need for replacement MCH therapy has been found to decrease with age.

A variety of pathologic processes involving the adrenal gland can result in generalized adrenocortical insufficiency: tuberculosis, histoplasmosis and other fungal diseases, metastatic carcinoma, amyloidosis, and bilateral adrenal hemorrhage. Presently, Addison's disease appears to occur most frequently as a component of an autoimmune process that results in selective atrophy of the adrenal cortex ("idiopathic" adrenal insufficiency), usually sparing the adrenal medulla (Blizzard and Kyle, 1963). Patients so affected often manifest evidence of other autoimmune processes affecting the thyroid, islet cells, gonads, and other tissues.

The diagnosis of adrenal insufficiency is relatively simple once the disorder is suspected. Plasma cortisol and urinary 17-hydroxycorticoid levels are subnormal and plasma ACTH concentration is elevated. Administration of ACTH, on either an acute or chronic basis, fails to increase cortisol production to normal levels. Treatment with hydrocortisone or an equivalent glucocorticoid is required on a lifelong basis. Because aldosterone production is also decreased, MCH replacement is frequently added, but many patients may be maintained on a high Na^+ intake. With proper therapy, patients with Addison's disease of benign cause should have a normal life expectancy.

Adrenal enzyme defects

Mineralocorticoid deficiency occurs in several of the syndromes associated with a deficiency of the enzymes required for adrenal hormone biosynthesis. If the deficiency occurs early in the biosynthetic pathway (desmolase step, 3β-ol-dehydrogenase step), production of nearly all adrenal secretory products is impaired and survival is rare. In the most common form of the adrenogenital syndrome, deficiency of 21-dehydroxylase results in impaired cortisol production and, in a significant percentage of patients, in impaired aldosterone production as well (Bongiovanni and Root, 1963). Infants fail to thrive and often present with salt wasting, hyperkalemia, and acidosis. As a consequence of impaired cortisol secretion, ACTH levels are increased, resulting in increased production of adrenal androgen. This can result in premature development of secondary sexual characteristics in male infants and ambiguous genitalia or signs of virilization in female infants. Administration of glucocorticoid hormone readily reverses the major abnormalities of carbohydrate metabolism and virilization. In those children affected with salt wasting, MCH therapy is also required. Monitoring plasma renin activity, which is markedly elevated in untreated patients, is useful in adjusting the dose of MCH to provide optimal physiologic replacement therapy.

A deficiency in the mixed function oxidase required for the final steps of aldosterone biosynthesis results in a syndrome of isolated mineralocorticoid deficiency characterized by salt wasting, hyperkalemia, and metabolic acidosis. Examples of impaired hydroxylation of corticosterone to 18-hydroxycorticosterone (Degenhart et al., 1966) and of subsequent dehydrogenation of 18-hydroxycorticosterone to aldosterone (Ulick et al., 1964) have been reported. The terms corticosterone methyl oxidase types 1 and 2 have been proposed to describe these enzyme deficiency states (Ulick, 1976). In patients with the type 2 defect, plasma levels of aldosterone may occasionally be in the "normal range" (Rösler et al., 1977a), but these values should be considered to be inappropriately low for the marked degree of hyperreninemia and hyperkalemia. The presence of the type 2 defect can be established by the demonstration of an increase in the excretory ratio of the major urinary metabolites of 18-hydroxycorticosterone and aldosterone (Ulick, 1976). Because these enzymes are not required for the biosynthesis of cortisol, there are no associated abnormalities in the levels of cortisol, ACTH, or adrenal androgen. The severity of the consequences of MCH deficiency in affected children can be minimized by the maintenance of a high salt intake or by MCH replacement therapy. Whereas the requirements for MCH and/or a high salt intake may decrease with age, the impaired ability to conserve Na^+ in response to a low salt intake persists.

Hyporeninemic hypoaldosteronism

Isolated deficiency of aldosterone production, once considered a rare condition (Hudson, Chobanian and Relman, 1957; Vagnucci, 1969), has been reported with increasing frequency in the past five years (Perez, Siegel and Schreiner, 1972; Schambelan, Stockigt and Biglieri, 1972; Weidmann et al., 1973). In affected patients adrenal production of glucocorticoid hormones is intact. In most adult patients the deficiency in aldosterone production appears to be a direct consequence of impaired renal secretion of renin and is called hyporeninemic hypoaldosteronism (Schambelan et al., 1972). As might be predicted from the known physiologic effect of aldosterone to augment renal excretion of K^+ and H^+, hyperkalemia and metabolic acidosis are major clinical manifestations in patients with this disorder. Chronic glomerular insufficiency is commonly present, which suggests the possibility that the impairment in renin secretion is the consequence of a primary pathologic process in the kidney. Renal disease may be necessary for the overt clinical manifestations of aldosterone deficiency. In some patients, physiologic or even superphysiologic doses of MCH fail to correct the K^+ and H^+ secretory defect completely, suggesting that some degree of resistance to the renal effects of aldosterone may result from the primary renal lesion (Sebastian et al., 1977).

In our experience and in the reported experience of other investigators, diabetes mellitus has been the most frequently recognized, associated disorder. Even before the recognition of the pathophysiologic role of impaired renin and aldosterone secretion, the frequent association of hyperkalemic hyperchloremic acidosis and "chronic pyelonephritis" was noted predominantly in patients with diabetes mellitus (Carroll and Farber, 1964). The occurrence of hyporeninemic hypoaldosteronism has also been observed in association with a variety of tubulointerstitial diseases of the kidney: gout, lead nephropathy, and hypercalcemic nephropathy. In many cases the cause of the underlying renal disease is unknown.

A recent study of 22 patients with hyporeninemic hypoaldosteronism provides additional evidence to support the hypothesis that the occurrence of hypoaldosteronism is a consequence of hyporeninemia (Schambelan, Sebastian and Biglieri, 1977). Plasma aldosterone concentration increased normally in response to the increase in plasma renin activity induced by salt restriction as predicted from the observed relationship between these variables in normal subjects so studied; plasma aldosterone concentration was a positive linear function of plasma renin activity in both groups, and the slopes in the two groups were not significantly different. Furthermore, plasma aldosterone increased in response to administration of exogenous angiotensin II. That hyperkalemia stimulates aldosterone secretion in these patients is suggested by the observation that the concentration of plasma aldosterone for any level of plasma renin activity was significantly greater in the patients than in the normal subjects. In addition, changes in serum K^+ during manipulations of K^+ intake and during therapy with furosemide produced directionally similar changes in plasma aldosterone when plasma renin activity was constant. Similarly, administration of ACTH was found to increase aldosterone secretion. Thus, aldosterone production in patients with hyporeninemic hypoaldosteronism is blunted by the low levels of plasma renin activity but increases appropriately in response to normal stimulatory factors. The magnitude of the reduction in plasma renin activity appears to be quantitatively sufficient to account for the hypoaldosteronism.

The pathogenesis of the primary defect in renin secretion has not been elucidated. Several possible mechanisms, particularly in patients with diabetes mellitus, must be considered as possible causes of the hyporeninemia.

1. Hyalinization of the afferent arteriole might result in destruction of the juxtaglomerular cells and thereby reduce renin secretory capacity. This lesion (Sparagana, 1974), as well as arteriolar sclerosis that results in an abnormal separation of the juxtaglomerular cells and macula densa cells, occurs commonly in diabetic patients with nephropathy (Schindler and Sommers, 1966).

2. The autonomic neuropathy that occurs commonly in diabetic patients may be causal or contribute to the reduction of the renin secretory capacity (Tuck et al., 1977).

3. Hyporeninemia and hypoaldosteronism may result from a volume-mediated suppression of renin secretion. In studies of four patients with hypoaldosteronism, extracellular fluid volume and total exchangeable Na^+ were noted to be increased (Oh et al., 1974). A primary reduction in renal clearance of Na^+ was proposed as the cause of the hypervolemia.

4. The demonstration of an inactive form of renin (i.e., "big renin") in the plasma of diabetic patients (deLeiva et al., 1976) suggests that a failure to convert normally a presumed renin "precursor" to the smaller active form of renin may underlie the hyporeninemia in some patients.

It is possible that the causes of hyporeninemia will be found to be multifactorial.

In patients with hyporeninemic hypoaldosteronism associated with diabetes mellitus, aldosterone secretion may be reduced in part because of an associated primary defect in adrenal hormonal biosynthesis (deLeiva et al., 1976). In one such patient studied in our laboratory, the magnitude of the increase in aldosterone secretion for a given increase in plasma renin activity was markedly blunted in comparison with that observed in patients with hyporeninemic hypoaldosteronism as a group. The aldosterone secretory response to oral K^+ loading and to infusion of ACTH was also markedly blunted in this patient. In such patients a primary deficiency in both renin and aldosterone secretion may contribute to the pathogenesis of the MCH deficiency state.

In diabetic patients with hyporeninemic hypoaldosteronism, the severity of the hyperkalemia may be contributed to by the associated defect in insulin secretion. In two insulin-dependent diabetic patients with hyporeninemic hypoaldosteronism, paradoxic hyperkalemia occurred in response to glucose administration and during periods of spontaneous hyperglycemia (Goldfarb et al., 1976). Insulin appears to be an important regulatory factor in the maintenance of K^+ homeostasis by increasing in response to administration of K^+ and by promoting cellular uptake of K^+ (Santeusanio et al., 1973).

The severity of the hyperchloremic metabolic acidosis in patients with hyporeninemic hypoaldosteronism correlates with the degree of glomerular insufficiency, but frank acidosis occurs in patients with a glomerular filtration rate greater than 25 ml/minute (Perez, Oster and Vaamonde, 1974; Sebastian et al., 1973). The acidosis appears to result from an impairment in renal acidification that is pathophysiologically distinct from other forms of renal tubular acidosis (Sebastian et al., 1973). In contrast to type 1 renal tubular acidosis, the urine is acidic and bicarbonate-free during spontaneously occurring acidosis. In similarity with type 2 renal tubular acidosis, the reabsorption of filtered bicarbonate is subnormal at normal plasma bicarbonate concentrations, but the magnitude of the reduction is not sufficiently great to implicate an impairment of H^+ secretion in the proximal convoluted tubule. Proximal tubular dysfunction indicated by hyperaminoaciduria, glucosuria, or increased renal clearance of phosphate is not present. The urinary excretion rate of ammonium is greatly reduced even when the urine is highly acidic. These findings, together with the finding of a reduced renal clearance of K^+, constitute a pathophysiologically separable form of renal tubular dysfunction that is designated as type 4 renal tubular acidosis.

Administration of exogenous MCH in patients with hyporeninemic hypoaldosteronism can augment both renal H^+ secretion and renal production of ammonia and thereby ameliorate or correct the metabolic acidosis (Fig. 8.7) (Sebastian et al., 1977). In four such patients with a glomerular filtration rate ranging from 13 to 44 ml/min, prolonged administration of MCH increased urinary K^+ and net acid excretion, corrected hyperkalemia, and ameliorated acidosis substantially. Except in the patient with the lowest glomerular filtration rate, the increased net acid excretion was due largely to increased excretion of ammonium. Urinary pH decreased initially in each patient, but in the three patients with the highest glomerular filtration rates, it increased subsequently as ammonium excretion increased, indicating that after stimulation of renal H^+ secretion, renal production of ammonia increased (Fig. 8.8a).

The enhancement of renal production of ammonia during administration of fludrocortisone might have been causally related to the correction of hyperkalemia that occurred. In a patient with hyporeninemic hypoaldosteronism studied by Szylman et al. (1976), the urinary excretion rate of ammonium varied inversely with the serum K^+ concentration when hyperkalemia was corrected by oral administration of a K^+-binding resin. A similar inverse relationship between these two variables is also evident in patients with hyporeninemic hypoaldosteronism treated with MCH (Fig. 8.8b). When MCH is discontinued in patients with hyporeninemic hypoaldosteronism, excretion of ammonium decreases unless hyperkalemia is prevented by a concomitant decrease in K^+ intake (Sebastian et al., 1977). Acidosis occurs despite maintenance of normokalemia largely because of an increase in urinary pH and consequent reduction in excretion of titratable acid. Thus, isolated aldosterone deficiency in man, like experimental isolated MCH deficiency in dogs, reduces net acid excretion and causes hyperkalemia by two mechanisms: reduction in renal H^+ secretory capacity independent of K^+ retention and reduction of renal production of ammonia secondary to hyperkalemia.

Whereas treatment of patients with hyporeninemic hypoaldosteronism with MCH will correct hyperkalemia and hyperchloremic metabolic acidosis, such therapy may not always be indicated. Some patients are hypertensive and have measured increases in extracellular fluid volume and total exchangeable Na^+ content (Oh et al., 1974). By increasing renal tubular reabsorption of NaCl, this treatment might exacerbate the degree of hypertension and might produce congestive heart failure in patients with limited cardiac reserve. Administration of furosemide has recently been demonstrated to increase K^+ and net acid excretion and to ameliorate substantially hyperchloremia and metabolic acidosis in such patients (Sebastian and

258 ACID BASE AND POTASSIUM HOMEOSTASIS

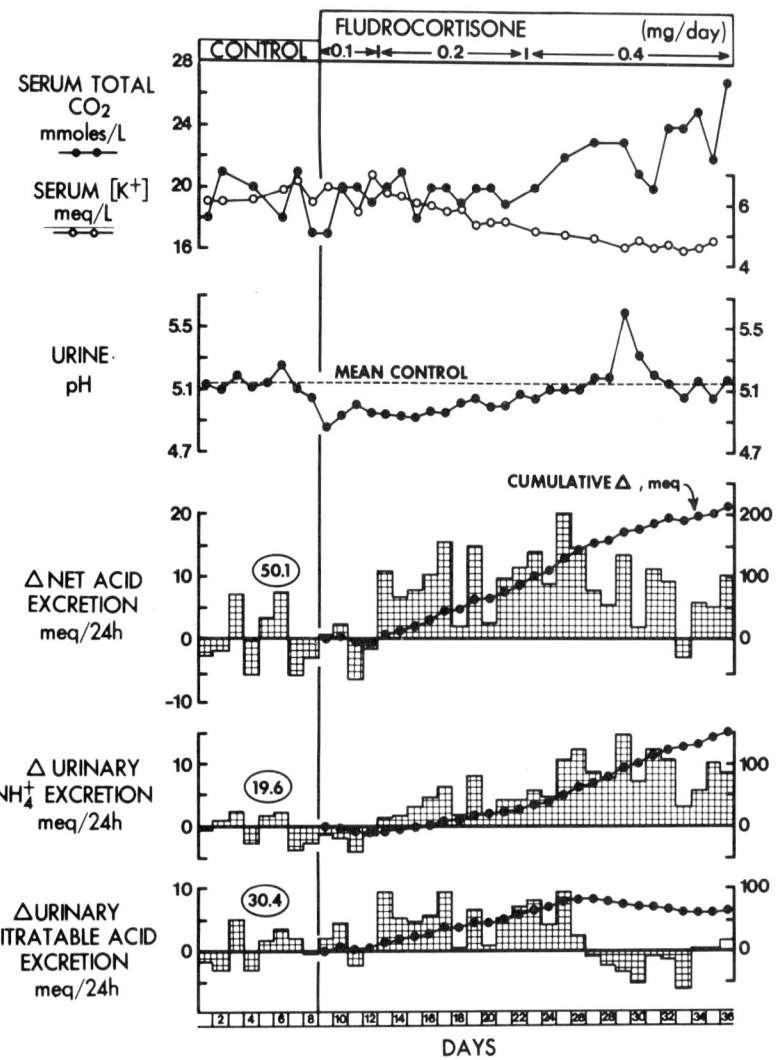

Figure 8.7 *Effect of fludrocortisone on serum CO_2 and K^+ concentrations, urine pH, and urinary titratable acid, ammonium, and net acid excretion in a patient with hyporeninemic hypoaldosteronism and chronic renal insufficiency (creatinine clearance 35 ml/min/1.73 m^2).* In the three bottom panels, the hatched bars represent the difference between the measured value of acid excretion and the mean value before therapy; the magnitude of these differences is indicated by the scale on the left-hand side of the panel. For reference, the mean control value is designated by the numerals within the ellipses. The accumulated values of the daily differences are shown by the solid circles and are indicated by the scale on the right-hand side of the panel. (Reprinted by permission from Sebastian et al., 1977, *New England Journal of Medicine*, 297, 576)

Schambelan, 1977). The studies indicate that furosemide increases net acid excretion both by stimulating H^+ secretion and by increasing the availability of ammonia for titration of secreted H^+. In the patients studied, the magni-

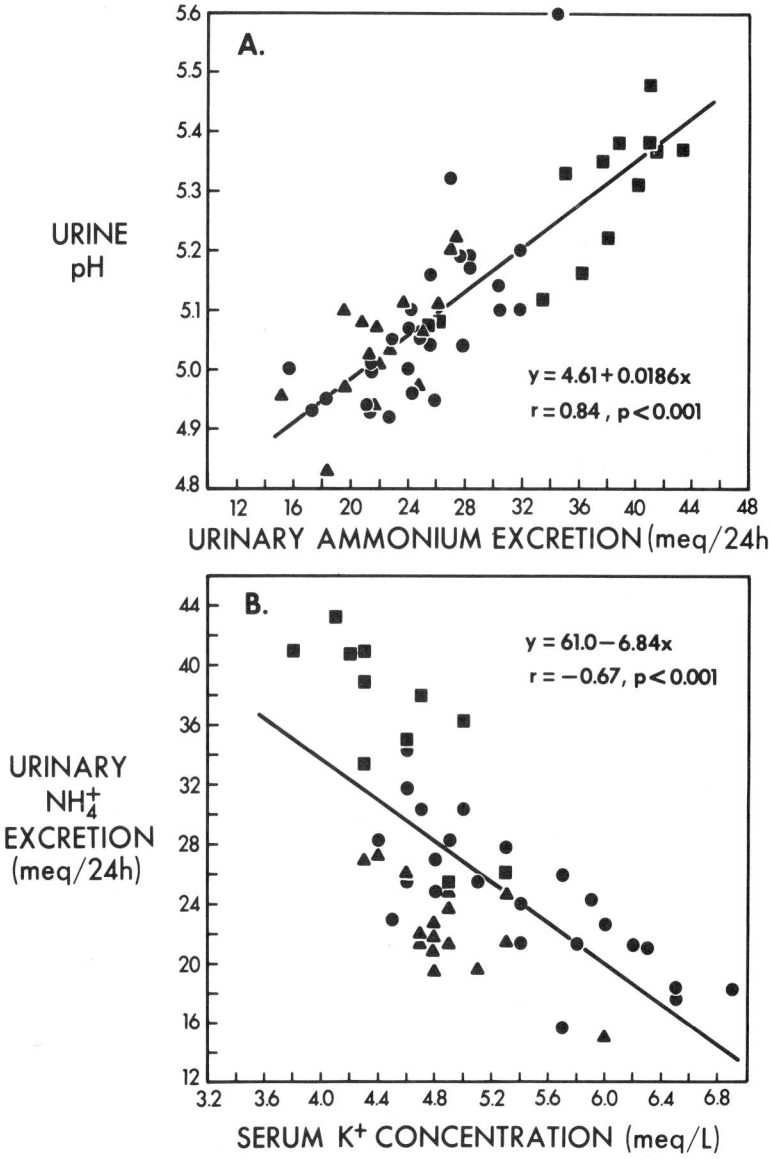

Figure 8.8 *Relation between the rate of urinary ammonium excretion and serum K+ concentration (a) and between urine pH and urinary ammonium excretion (b) during treatment with fludrocortisone in three patients with hyporeninemic hypoaldosteronism.* (Reprinted by permission from Sebastian et al., 1977, *New England Journal of Medicine*, **297**, 576)

tude of the acid excretory and kaliuretic effect of furosemide was directly correlated with the level of endogenous aldosterone. In those patients with the most severe degree of hypoaldosteronism, the ameliorative effect of furosemide was greatly enhanced by pretreatment with small doses of MCH.

Pseudohypomineralocorticoidism

A syndrome similar to MCH deficiency can occur despite elevated aldosterone levels, apparently as a consequence of renal tubular "resistance" to the action of MCH. In addition to hyperkalemia and metabolic acidosis, affected patients have extracellular fluid volume depletion and hypotension due to renal Na^+ wasting. Plasma renin levels are characteristically elevated. Administration of MCH in superphysiologic amounts fails to ameliorate this disorder. Rare examples of such a syndrome have been reported in infants without renal parenchymal disease (pseudohypoaldosteronism of infancy) (Cheek and Perry, 1958; Donnell, Litman, and Roldan, 1959) and in children and adults with azotemia that is typically secondary to chronic tubulointerstitial damage ("salt-wasting nephropathy") (Stanbury and Mahler, 1959; Thorn, Koepf, and Clinton, 1944; Walker et al., 1965).

In contrast to infants with an enzymatic defect in aldosterone biosynthesis, in both subgroups of patients with pseudohypomineralocorticoidism aldosterone levels are markedly elevated as a consequence of hyperreninemia and hyperkalemia, and they increase further in response to more severe degrees of salt depletion and to direct infusion of angiotensin II, K^+, and ACTH (Rösler et al., 1977b). The markedly elevated levels of plasma aldosterone reported in patients with pseudohypomineralocorticoidism would be predicted to be sufficient to saturate the mineralocorticoid receptor completely. Despite these high endogenous levels of aldosterone and/or the additional administration of large amounts of exogenous MCH, acid-base and electrolyte abnormalities persist. Treatment with large quantities of NaCl and sodium bicarbonate is usually required to maintain normal extracellular fluid volume; with adequate replacement, plasma renin and aldosterone levels return toward normal values. In infants with pseudohypoaldosteronism, as in infants with enzymatic defects, the electrolyte and acid-base abnormalities and the severity of the salt wasting tends to decrease with age.

Renal tubular responsiveness to the action of MCH also appears to be impaired in a rare syndrome of "MCH deficiency" that is characterized by hyperkalemia, hyperchloremic metabolic acidosis, hypertension, hyporeninemia, and abnormally reduced aldosterone production (Arnold and Healy, 1969; Gordon et al., 1970; Schambelan, Sebastian and Rector, 1978; Spitzer et al., 1973; Weinstein, Allan and Mendoza, 1974). Glomerular filtration rate is in the normal range. Mineralocorticoid resistance is apparent by persistence of hyperkalemia and a markedly reduced kaliuretic response to large amounts of exogenously administered MCH. The antinatriuretic and antichloriuretic responses to MCH may be intact. A primary defect in renal K^+ secretion has been proposed. In one such patient, during normal NaCl intake, fractional renal K^+ excretion was subnormal and increased only minimally during administration of large amounts of MCH. However, distal tubule K^+ secretion increased greatly when distal Na^+ delivery was increased with anions other than Cl^- (sulfate or bicarbonate). Thus, hyperkalemia and MCH resistance did not appear to be due to an intrinsic defect in renal K^+ secretion. The finding of normal salivary and fecal secretion of K^+ further mitigated against a generalized defect in transepithelial transport of K^+. In this patient it was proposed that the primary disturbance was abnormally increased distal reabsorption of Cl^-, which would effectively shunt the Na^+- and MCH-dependent driving forces, for K^+ secretion resulting in hyperkalemia, and by augmenting distal Na^+ reabsorption, would result in volume expansion, hypertension, and hyporeninemic hypoaldosteronism (Schambelan et al., 1978).

REFERENCES

Aarskog, D., Stoa, K. F., Thorsen, T. & Wefring, K. W. (1967). Hypertension and hypokalemic alkalosis associated with underproduction of aldosterone. *Pediatrics*, 39, 884.

Abbrecht, P. H. (1969). Effects of potassium deficiency on renal function in the dog. *Journal of Clinical Investigation*, 48, 432.

Agus, Z. S. & Goldberg, M. (1971). Role of antidiuretic hormone in the abnormal water diuresis of anterior hypopituitarism in man. *Journal of Clinical Investigation*, 50, 1478.

Aitchison, J., Brown, J. J., Ferriss, J. B., Fraser, R., Kay, A. W., Lever, A. F., Neville, A. M., Symington, T. & Robertson J. I. S. (1971). Quadric analysis in the preoperative distinction between patients with and without adrenocortical tumors in hypertension with aldosterone excess and low plasma renin. *American Heart Journal*, 82, 660.

Al-Awqati, Q., Norby, L. H., Mueller, A. & Steinmetz, P. R. (1976). Characteristics of stimulation of H^+ transport by aldosterone in turtle urinary bladder. *Journal of Clinical Investigation*, 58, 351.

Al-Awqati, Q. (1977). Effect of aldosterone on the coupling between H^+ transport and glucose oxidation. *Journal of Clinical Investigation*, 60, 1240.

Alterman, S. L., Dominguez, C., Lopez-Gomez, A. & Lieber, A. L. (1969). Primary adrenocortical carcinoma causing aldosteronism. *Cancer*, 24, 602.

Arnold, J. E. & Healy, J. K. (1969). Hyperkalemia, hypertension and systemic acidosis without renal failure associated with a tubular defect in potassium excretion. *American Journal of Medicine*, 47, 461.

Baxter, J. D., Schambelan, M., Matulich, D. T., Spindler, B. J., Taylor, A. A. & Bartter, F. C. (1976). Aldosterone receptors and the evaluation of plasma mineralocorticoid activity in normal and hypertensive states. *Journal of Clinical Investigation*, 58, 579.

Berl, T., Anderson, R. J., Aisenbrey, G. A., Linas, S. L. & Schrier, R. W. (1977). On the mechanism of polyuria in potassium (K) depletion: the role of polydipsia. *Clinical Research*, 25, 136A.

Berns, A. S., Pluss, R. G., Erickson, A. L., Anderson, R. J., McDonald, K. M. & Schrier, R. W. (1977). Renin-angiotensin system and cardiovascular homeostasis in adrenal insufficiency. *American Journal of Physiology*, 233, F509.

Biglieri, E. G. & Forsham, P. H. (1961). Studies on expanded extracellular fluid and responses to various stimuli in primary aldosteronism. *American Journal of Medicine*, 30, 564.

Biglieri, E. G., Herron, M. A. & Brust, N. (1966). 17-hydroxylation deficiency in man. *Journal of Clinical Investigation*, 45, 1946.

Biglieri, E. G., Slaton, P. E., Jr., Kronfield, S. J. & Schambelan, M. (1967). Diagnosis of an aldosterone-producing adenoma in primary aldosteronism: an evaluative maneuver. *Journal of the American Medical Association*, 201, 510.

Biglieri, E. G., Slaton, P. E., Schambelan, M. & Kronfield, S. J. (1968). Hypermineralocorticoidism. *American Journal of Medicine*, 45, 170.

Biglieri, E. G., Schambelan, M., Brust, N., Chang, B. & Hogan, M. (1974a). Plasma aldosterone concentration. Further characterization of aldosterone-producing adenomas. *Circulation Research*, 34 & 35 (suppl. I), I–183.

Biglieri, E. G., Stockigt, J. R., Schambelan, M. & Collins, R. D. (1974b). Secondary hyperaldosteronism. In *Angiotensin*, ed. Page, I. K. & Bumpus, F. M. Berlin: Springer-Verlag.

Biglieri, E. G. (1977). Plasma deoxycorticosterone concentrations in the adrenal enzymatic deficiencies causing hypertension. In *Juvenile Hypertension*, ed. New, M. I. & Levine, L. S. New York: Raven Press.

Blair-West, J. R., Coghlan, J. P., Denton, D. A., Funder, J. W., Scoggins, B. A. & Wright, R. D. (1968). Effects of adrenal steroid withdrawal on chronic renovascular hypertension in adrenalectomized sheep. *Circulation Research*, 23, 803.

Blizzard, R. M. & Kyle, M. (1963). Studies of the adrenal antigens and antibodies in Addison's disease. *Journal of Clinical Investigation*, 42, 1653.

Bongiovanni, A. M. & Root, A. W. (1963). The adrenogenital syndrome. *New England Journal of Medicine*, 268, 1283; 1342; 1391.

Brown, J. J., Davies, D. L., Lever, A. F. & Robertson, J. I. S. (1965). Plasma renin concentration in human hypertension. II. Renin in relation to aetiology. *British Medical Journal*, 2, 1215.

Brown, J. J., Fraser, R., Lever, A. F., Robertson, J. I. S., James, V. H. T., McCusker, J. & Wynn, V. (1968). Renin, angiotensin, corticosteroids, and electrolyte balance in Addison's disease. *Quarterly Journal of Medicine*, 37, 97.

Brown, J. J., Ferriss, J. B., Fraser, R., Lever, A. F., Love, D. R., Robertson, J. I. S. & Wilson, A. (1972). Apparently isolated excess deoxycorticosterone in hypertension. A variant of the mineralocorticoid-excess syndrome. *Lancet*, 2, 243.

Brown, R. D. (1976). Aldosterone metabolic clearance is normal in low-renin essential hypertension. *Journal of Clinical Endocrinology and Metabolism*, 42, 661.

Carey, R. M., Douglas, J. G., Schweikert, J. R. & Liddle, G. W. (1972). The syndrome of essential hypertension and suppressed plasma renin activity. Normalization of blood pressure with spironolactone. *Archives of Internal Medicine*, 130, 849.

Carroll, H. J. & Farber, S. J. (1964). Hyperkalemia and hyperchloremic acidosis in chronic pyelonephritis. *Metabolism: Clinical and Experimental*, 13, 808.

Channick, B. J., Adlin, E. V. & Marks, A. D. (1969). Suppressed plasma renin activity in hypertension. *Archives of Internal Medicine*, 123, 131.

Cheek, D. B. & Perry, J. W. (1958). A salt-wasting syndrome in infancy. *Archives of Disease in Childhood*, 33, 252 .

Chonko, A. M., Bay, W. H., Stein, J. H. & Ferris, T. F. (1977). The role of renin and aldosterone in the salt retention of edema. *American Journal of Medicine*, 63, 881.

Christy, N. P. & Laragh, J. H. (1961). Pathogenesis of hypokalemic alkalosis in Cushing's syndrome. *New England Journal of Medicine*, 265, 1083.

Collins, R. D., Weinberger, M. H., Dowdy, A. J., Nokes, G. W., Gonzales, C. M. & Luetscher, J. A. (1970). Abnormally sustained aldosterone secretion during salt loading in patients with various forms of benign hypertension; relation to plasma renin activity. *Journal of Clinical Investigation*, 49, 1415.

Conn, J. W. (1955). Presidential address. II. Primary aldosteronism, a new clinical syndrome. *Journal of Laboratory and Clinical Medicine*, 45, 6.

Conn, J. W., Cohen, E. L. & Rovner, D. R. (1964a). Suppression of plasma renin activity in primary aldosteronism. *Journal of the American Medical Association*, 190, 213.

Conn, J. W., Knopf, R. F. & Nesbit, R. M. (1964b). Clinical characteristics of primary aldosteronism from an analysis of 145 cases. *American Journal of Surgery*, 107, 159.

Conn, J. W., Rovner, D. R. & Cohen, E. L. (1968). Licorice-induced pseudoaldosteronism. Hypertension, hypokalemia, aldosteronopenia, and suppressed plasma renin activity. *Journal of the American Medical Association*, 205, 80.

Crabbe, J. (1961). Stimulation of active sodium transport by the isolated toad bladder with aldosterone *in vitro*. *Journal of Clinical Investigation*, 40, 2103.

Crabbe, J. & De Weer, P. (1964). Action of aldosterone on the bladder and skin of the toad. *Nature* (London), 202, 298.

Davis, W. W., Newsome, H. H., Wright, L. D., Hammond, W. G., Easton, J. & Bartter, F. C. (1967). Bilateral adrenal hyperplasia as a cause of primary aldosteronism with hypertension, hypokalemia, and suppressed renin activity. *American Journal of Medicine*, 42, 642.

Degenhart, H. J., Frankena, L., Visser, J. K. A., Cost, W. S. and van Seters, A. P. (1966). Further investigation of a new hereditary defect in the biosynthesis of aldosterone: evidence for a defect in 18-hydroxylation of corticosterone. *Acta Physiologica et Pharmacologica*, **14**, 89.

deLeiva, A., Christlieb, A. R., Melby, J. C., Graham, C. A., Day, R. P., Luetscher, J. A. & Zager, P. G. (1976). Big renin and biosynthetic defect of aldosterone in diabetes mellitus. *New England Journal of Medicine*, **295**, 639.

Donnell, G. N., Litman, N. & Roldan, M. (1959). Pseudohypo-adrenalocorticism. Renal sodium loss, hyponatremia, and hyperkalemia due to a renal tubular insensitivity to mineralocorticoids. *American Journal of Diseases of Childhood*, **97**, 813.

Earley, L. E. & Daugharty, T. M. (1969). Sodium metabolism. *New England Journal of Medicine*, **281**, 72.

Edelman, I. S., Bogoroch, R. & Porter, G. A. (1963). On the mechanism of action of aldosterone on sodium transport: the role of protein synthesis. *Proceedings of the National Academy of Sciences of the United States of America*, **50**, 1169.

Edelman, I. S. & Fanestil, D. D. (1970). Mineralocorticoids. In *Biochemical Actions of Hormones*, ed. Litwack, G. Ch. 8. New York: Academic Press.

Fanestil, D. D. & Edelman, I. S. (1966). On the mechanism of action of aldosterone on sodium transport: effects of inhibitors of RNA and of protein synthesis. *Federation Proceedings; Federation of American Societies for Experimental Biology*, **25**, 912.

Fanestil, D. D. (1968). Mode of spironolactone action: competitive inhibition of aldosterone binding to mineralocorticoid receptors. *Biochemical Pharmacology*, **17**, 2240.

Feldman, D., Funder, J. W. & Edelman, I. S. (1972). Subcellular mechanisms in the action of adrenal steroids. *American Journal of Medicine*, **53**, 545.

Funder, J. W., Feldman, D. & Edelman, I. S. (1973). The roles of plasma binding and receptor specificity in the mineralocorticoid action of aldosterone. *Endocrinology*, **92**, 994.

Gandy, H. M., Keutmann, E. H. & Izzo, A. J. (1960). Characterization of urinary steroids in adrenal hyperplasia: isolation of metabolites of cortisol, compound S, and desoxycorticosterone from normotensive patient with adrenogenital syndrome. *Journal of Clinical Investigation*, **39**, 364.

Ganguly, A., Dowdy, A. J., Luetscher, J. A. & Melada, G. A. (1973). Anomalous postural response of plasma aldosterone concentration in patients with aldosterone-producing adrenal adenoma. *Journal of Clinical Endocrinology and Metabolism*, **36**, 401.

Garcia, E., Malnic, G. & Giebisch, G. (1976). Effects of changes in electrical potential difference upon distal tubular potassium concentrations. *Proceedings of the American Society of Nephrology*, **9**, 98A.

Garrod, O., Davies, S. A. & Cahill, G., Jr. (1955). The action of cortisone and desoxycorticosterone acetate on glomerular filtration rate and sodium and water exchange in the adrenalectomized dog. *Journal of Clinical Investigation*, **34**, 761.

Giebink, G. S., Gotlin, R. W., Biglieri, E. G. & Katz, F. H. (1973). A kindred with familial glucocorticoid-suppressible aldosteronism. *Journal of Clinical Endocrinology and Metabolism*, **36**, 715.

Goldfarb, S., Cox, M., Singer, I. & Goldberg, M. (1976). Acute hyperkalemia induced by hyperglycemia: hormonal mechanisms. *Annals of Internal Medicine*, **84**, 426.

Goldsmith, O., Solomon, D. H. & Horton, R. (1967). Hypogonadism and mineralocorticoid excess: the 17-hydroxylase deficiency syndrome. *New England Journal of Medicine*, **277**, 673.

Good, D. W. & Wright, F. S. (1977). Effects of sodium reabsorption, transepithelial voltage and fluid flow rate on potassium secretion by renal distal tubule. *Proceedings of the American Society of Nephrology*, **10**, 106A.

Goodman, D. D., Allen, J. E. & Rasmussen, H. (1969). On the mechanism of action of aldosterone. *Proceedings of the National Academy of Sciences of the United States of America*, **64**, 330.

Gordon, R. D., Geddes, R. A., Pawsey, C. G. K. & O'Halloran, M. W. (1970). Hypertension and severe hyperkalaemia associated with suppression of renin and aldosterone and completely reversed by dietary sodium restriction. *Australasian Annals of Medicine*, **4**, 287.

Gross, J. B., Imai, M. & Kokko, J. P. (1975). A functional comparison of the cortical collecting tubule and the distal convoluted tubule. *Journal of Clinical Investigation,* **55,** 1284.

Gunnells, J. C., Jr., McGuffin, W. L., Jr., Robinson, R. R., Grim, C. E., Wells, S., Silver, D. & Glenn, J. F. (1970). Hypertension, adrenal abnormalities, and alterations in plasma renin activity. *Annals of Internal Medicine,* **73,** 901.

Haack, D., Möhring, J., Möhring, B., Petric, M. & Hachenthal, E. (1977). Comparative study on development of corticosterone and DOCA hypertension in rats. *American Journal of Physiology,* **233,** F403.

Handler, J. S., Preston, A. S. & Orloff, J. (1969). Effect of adrenal steroid hormones on the response of the toad's urinary bladder to vasopressin. *Journal of Clinical Investigation,* **48,** 823.

Hanley, M. J. & Kokko, J. P. (1977). Characteristics of chloride transport across the rabbit cortical collecting tubule: response to desoxycorticosterone. *Clinical Research,* **25,** 506A.

Hierholzer, K., Wiederholt, M. & Stolte, H. (1966). Hemmung der natriumresorption im proximalen und distalen konvolut adrenalektomierter ratten. *Pfluegers Archiv,* **291,** 43.

Hierholzer, K. & Wiederholt, M. (1976). Some aspects of distal tubular solute and water transport. *Kidney International,* **9,** 198.

Hogan, M. J., McRae, J., Schambelan, M. & Biglieri, E. G. (1976). Location of aldosterone-producing adenomas with ^{131}I-19-iodocholesterol. *New England Journal of Medicine,* **294,** 410.

Hogan, M. J., Schambelan, M. & Biglieri, E. G. (1977). Concurrent hypercortisolism and hypermineralocorticoidism. *American Journal of Medicine,* **62,** 777.

Hudson, J. B., Chobanian, A. V. & Relman, A. S. (1957). Hypoaldosteronism. A clinical study of a patient with an isolated adrenal mineralocorticoid deficiency, resulting in hyperkaliemia and Stokes-Adams attacks. *New England Journal of Medicine,* **257,** 529.

Hulter, H. N., Sigala, J. F. & Sebastian, A. (1977a). Effect of pre-existing dietary K+ restriction on the renal action of mineralocorticoid hormone (MH). *Clinical Research,* **25,** 436A.

Hulter, H. N., Ilnicki, L. P., Harbottle, J. A. & Sebastian, A. (1977b). Impaired renal H+ secretion and NH_3 production in mineralocorticoid-deficient glucocorticoid-replete dogs. *American Journal of Physiology,* **232,** F136.

Jose, A. & Kaplan, N. M. (1969). Plasma renin activity in the diagnosis of primary aldosteronism. Failure to distinguish primary aldosteronism from essential hypertension. *Archives of Internal Medicine,* **123,** 141.

Kassirer, J. P., London, A. M., Goldman, D. M. & Schwartz, W. B. (1970). On the pathogenesis of metabolic alkalosis in hyperaldosteronism. *American Journal of Medicine,* **49,** 306.

Katz, A. I. & Epstein, F. H. (1967). The role of sodium-potassium activated adenosine triphosphatase in the reabsorption of sodium by the kidney. *Journal of Clinical Investigation,* **46,** 1999.

Katz, F. H. (1967). Primary aldosteronism with suppressed plasma renin activity due to bilateral nodular adrenocortical hyperplasia. *Annals of Internal Medicine,* **67,** 1035.

Kem, D. C., Weinberger, M. H., Mayes, D. M. & Nugent, C. A. (1971). Saline suppression of plasma aldosterone in hypertension. *Archives of Internal Medicine,* **128,** 380.

Krakoff, L., Nicolis, G. & Amsel, B. (1975). Pathogenesis of hypertension in Cushing's syndrome. *American Journal of Medicine,* **58,** 216.

Kurtzman, N. A., White, M. G. & Rogers, P. W. (1971). Aldosterone deficiency and renal bicarbonate reabsorption. *Journal of Laboratory and Clinical Medicine,* **77,** 931.

Kurtzman, N. A., White, M. G. & Rogers, P. W. (1973). Pathophysiology of metabolic alkalosis. *Archives of Internal Medicine,* **131,** 702.

Laragh, J. H., Ledingham, J. G. G. & Sommers, S. C. (1967). Secondary aldosteronism and reduced plasma renin in hypertensive disease. *Transactions of the Association of American Physicians,* **80,** 168.

Lebel, M., Schalekamp, M. A., Beevers, D. G., Brown, J. J., Davies, D. L., Fraser, R., Kremer, D., Lever, A. F., Morton, J. J., Robertson, J. I. S., Tree, M. & Wilson, A. (1974). Sodium and the renin-angiotensin system in essential hypertension and mineralocorticoid excess. *Lancet,* **2,** 308.

Lemann, J., Jr., Piering, W. F. & Lennon, E. J. (1970). Studies of the acute effects of aldosterone and cortisol on the interrelationship between renal sodium, calcium and magnesium excretion in normal man. *Nephron*, 7, 117.

Liddle, G. W., Bledsoe, T. & Coppage, W. S., Jr. (1963). A familial renal disorder simulating primary aldosteronism but with negligible aldosterone secretion. *Transactions of the Association of American Physicians*, 76, 199.

Lieberman, L. M., Beierwaltes, W. H., Conn, J. W., Ansari, A. N. & Nishiyama, H. (1971). Diagnosis of adrenal disease by visualization of human adrenal glands with ^{131}I-19-iodocholesterol. *New England Journal of Medicine*, 285, 1387.

Lifschitz, M. D., Schrier, R. W. & Edelman, I. S. (1973). Effect of actinomycin D on aldosterone-mediated changes in electrolyte excretion. *American Journal of Physiology*, 224, 376.

Lipton, P. & Edelman, I. S. (1971). Effects of aldosterone and vasopressin on electrolytes of toad bladder epithelial cells. *American Journal of Physiology*, 221, 733.

Ludens, J. H. & Fanestil, D. D. (1974). Aldosterone stimulation of acidification of urine by isolated urinary bladder of the Colombian toad. *American Journal of Physiology*, 226, 1321.

Luetscher, J. A., Ganguly, A., Melada, G. A. & Dowdy, A. J. (1974). Preoperative differentiation of adrenal adenoma from idiopathic adrenal hyperplasia in primary aldosteronism. *Circulation Research*, 34 & 35 (suppl. I), I-175.

Luke, R. G. & Levitan, H. (1967). Impaired renal conservation of chloride and the acid-base changes associated with potassium depletion in the rat. *Clinical Science*, 32, 511.

Mantero, F., Busnardo, B., Riondel, A., Veyrat, R. & Austoni, M. (1971). Hypertension artérielle, alcalose hypokaliémque et pseudohermaphrodisme mâle par déficit en 17-alpha-hydroxylase. *Schweizerische Medizinische Wochenschrift*, 101, 38.

Melby, J. C., Spark, R. F., Dale, S. L., Egdahl, R. H. & Kahn, P. C. (1967). Diagnosis and localization of aldosterone-producing adenomas by adrenal-vein catheterization. *New England Journal of Medicine*, 277, 1050.

Melby, J. C., Dale, S. L. & Wilson, T. E. (1971). 18-hydroxy-deoxycorticosterone in human hypertension. *Circulation Research*, 28 (suppl. II), II-143.

Mills, J. N., Thomas, S. & Williamson, K. S. (1960). The acute effect of hydrocortisone, deoxycorticosterone and aldosterone upon the excretion of sodium, potassium and acid by the human kidney. *Journal of Physiology* (London), 151, 312.

New, M. I. (1970). Male pseudohermaphroditism due to 17α-hydroxylase deficiency. *Journal of Clinical Investigation*, 49, 1930.

New, M. I. & Seaman, M. P. (1970). Secretion rates of cortisol and aldosterone precursors in various forms of congenital adrenal hyperplasia. *Journal of Clinical Endocrinology and Metabolism*, 30, 361.

New, M. I., Siegal, E. J. & Peterson, R. E. (1973). Dexamethasone-suppressible hyperaldosteronism. *Journal of Clinical Endocrinology and Metabolism*, 37, 93.

New, M. I., Levine, L. S., Biglieri, E. G., Pareira, J. & Ulick, S. (1977). Evidence for an unidentified steroid in a child with apparent mineralocorticoid hypertension. *Journal of Clinical Endocrinology and Metabolism*, 44, 924.

Newton, A. & Laragh, J. H. (1968). Effects of glucocorticoid administration on aldosterone excretion and plasma renin in normal subjects, in essential hypertension and in primary aldosteronism. *Journal of Clinical Endocrinology and Metabolism*, 28, 1914.

Nicolis, G. L., Mitty, H. A., Modlinger, R. S. & Gabrilove, J. L. (1972). Percutaneous adrenal venography: a clinical study of 50 patients. *Annals of Internal Medicine*, 76, 899.

Nowaczynski, W., Kuchel, O. & Genest, J. (1971). A decreased metabolic clearance rate of aldosterone in benign essential hypertension. *Journal of Clinical Investigation*, 50, 2184.

Oh, M. S., Carroll, H. J., Clemmons, J. E., Vagnucci, A. H., Levison, S. P. & Whang, E. S. M. (1974). A mechanism for hyporeninemic hypoaldosteronism in chronic renal disease. *Metabolism: Clinical and Experimental*, 23, 1157.

Perez, G., Siegel, L. & Schreiner, G. E. (1972). Selective hypoaldosteronism with hyperkalemia. *Annals of Internal Medicine*, 76, 757.

Perez, G. O., Oster, J. R. & Vaamonde, C. A. (1974). Renal acidosis and renal potassium handling in selective hypoaldosteronism. *American Journal of Medicine*, 57, 809.

Perez, G. A., Oster, J. R. & Vaamonde, C. A. (1976). Renal acidification in patients with mineralocorticoid deficiency. *Nephron*, 17, 461.

Relman, A. S. & Schwartz, W. B. (1952). The effect of DOCA on electrolyte balance in normal man and its relation to sodium chloride intake. *Yale Journal of Biology and Medicine*, 24, 540.

Robertson, G. L., Mahr, E. A., Athar, S. & Sinha, T. (1973). Development and clinical application of a new method for the radioimmunoassay of arginine vasopressin in human plasma. *Journal of Clinical Investigation*, 52, 2340.

Rösler, A., Rabinowitz, D., Theodor, R., Ramirez, L. C. & Ulick, S. (1977a). The nature of the defect in a salt-wasting disorder in Jews of Iran. *Journal of Clinical Endocrinology and Metabolism*, 44, 279.

Rösler, A., Theodor, R., Boichis, H., Gerty, R., Ulick, S., Alagem, M., Tabachnik, E., Cohen, B. & Rabinowitz D. (1977b). Metabolic responses to the administration of angiotensin II, K and ACTH in two salt-wasting syndromes. *Journal of Clinical Endocrinology and Metabolism*, 44, 292.

Ross, E. J. (1959). Hypertension and hypokalemia associated with hypoaldosteronism. *Proceedings of the Royal Society of Medicine*, 52, 1056.

Ross, E. J., Reddy, W. J., Rivera, A. & Thorn, G. W. (1959). Effects of intravenous infusions of *dl*-aldosterone acetate on sodium and potassium excretion in man. *Journal of Clinical Endocrinology and Metabolism*, 19, 389.

Royston, A. & Prout, B. J. (1976). Carbenoxolone-induced hypokalaemia simulating Guillain-Barré syndrome. *British Medical Journal*, 2, 150.

Rubini, J. E. (1961). Water excretion in potassium-deficient man. *Journal of Clinical Investigation*, 40, 2215.

Saito, T. & Essig, A. (1973). Effect of aldosterone on active and passive conductance and E_{Na} in the toad bladder. *Journal of Membrane Biology*, 13, 1.

Samuels, H. H. & Tomkins, G. M. (1970). Relation of steroid structure to enzyme induction in hepatoma tissue culture cells. *Journal of Molecular Biology*, 52, 57.

Santeusanio, F., Faloona, G. R., Knochel, J. P. & Unger, R. H. (1973). Evidence for a role of endogenous insulin and glucagon in the regulation of potassium homeostasis. *Journal of Laboratory and Clinical Medicine*, 81, 809.

Schambelan, M., Slaton, P. E., Jr. & Biglieri, E. G. (1971). Mineralocorticoid production in hyperadrenocorticism. Role in pathogenesis of hypokalemic alkalosis. *American Journal of Medicine*, 51, 299.

Schambelan, M., Stockigt, J. R. & Biglieri, E. G. (1972). Isolated hypoaldosteronism in adults. A renin-deficiency syndrome. *New England Journal of Medicine*, 287, 573.

Schambelan, M., Howes, E. L., Jr., Stockigt, J. R., Noakes, C. A. & Biglieri, E. G. (1973). Role of renin and aldosterone in hypertension due to a renin-secreting tumor. *American Journal of Medicine*, 55, 86.

Schambelan, M., Brust, N. L., Chang, B. C. F., Slater, K. L. & Biglieri, E. G. (1976). Circadian rhythm and effect of posture on plasma aldosterone concentration in primary aldosteronism. *Journal of Clinical Endocrinology and Metabolism*, 43, 115.

Schambelan, M., Sebastian, A. & Biglieri, E. G. (1977). Control of aldosterone secretion in hyporeninemic hypoaldosteronism. *Clinical Research*, 25, 466A.

Schambelan, M., Sebastian, A. & Rector, F. C., Jr. (1978). Mineralocorticoid (MC) resistant renal potassium (K^+) secretory defect: proposed distal tubule chloride shunt. *Clinical Research*, 26, 545A.

Schindler, A. M. & Sommers, S. C. (1966). Diabetic sclerosis of the renal juxtaglomerular apparatus. *Laboratory Investigation*, 15, 877.

Sebastian, A., McSherry, E. & Morris, R. C., Jr. (1971). Renal potassium wasting in renal tubular acidosis (RTA). Its occurrence in types 1 and 2 RTA despite sustained correction of systemic acidosis. *Journal of Clinical Investigation*, 50, 667.

Sebastian, A., McSherry, E., Schambelan, M., Connor, D., Biglieri, E. & Morris, R. C., Jr. (1973). Renal tubular acidosis (RTA) in patients with hypoaldosteronism caused by renin deficiency. *Clinical Research*, 21, 706A.

Sebastian, A., Schambelan, M., Lindenfeld, S. & Morris, R. C., Jr. (1977). Amelioration of metabolic acidosis with fludrocortisone therapy in hyporeninemic hypoaldosteronism. *New England Journal of Medicine*, 297, 576.

Sebastian, A. & Schambelan, M. (1977). Amelioration of type 4 renal tubular acidosis (RTA) in chronic renal failure (CRF) with furosemide (F). *Proceedings of the American Society of Nephrology*, 10, 82A.

Sennett, J. A., Brown, R. D., Island, D. P., Yarbro, L. R., Watson, J. T., Slaton, P. E., Hollifield, J. W. & Liddle, G. W. (1975). Evidence for a new mineralocorticoid in patients with low-renin essential hypertension. *Circulation Research*, 36 & 37 (suppl. I), I–2.

Shade, R. E. & Grim, C. E. (1975). Suppression of renin and aldosterone by small amounts of DOCA in normal man. *Journal of Clinical Endocrinology and Metabolism*, 40, 652.

Siegel, B. & Civan, M. M. (1976). Aldosterone and insulin effects on driving force of Na+ pump in toad bladder. *American Journal of Physiology*, 230, 1603.

Slaton, P. E., Jr., Schambelan, M. & Biglieri, E. G. (1969). Stimulation and suppression of aldosterone secretion in patients with an aldosterone-producing adenoma. *Journal of Clinical Endocrinology and Metabolism*, 29, 239.

Solomon, S. S., Swersie, S. P., Paulsen, C. A. & Biglieri, E. G. (1968). Feminizing adrenocortical carcinoma with hypertension. *Journal of Clinical Endocrinology and Metabolism*, 28, 608.

Sparagana, M. (1974). Hyporeninemic hypoaldosteronism associated with diabetic glomerulosclerosis. *Journal of Steroid Biochemistry*, 5, 369.

Spark, R. F. & Melby, J. C. (1968). Aldosteronism in hypertension. The spironolactone response test. *Annals of Internal Medicine*, 69, 685.

Spark, R. F. & Melby, J. C. (1971). Hypertension and low plasma renin activity: presumptive evidence for mineralocorticoid excess. *Annals of Internal Medicine*, 75, 831.

Spitzer, A., Edelmann, C. M., Jr., Goldberg, L. D. & Henneman, P. H. (1973). Short stature, hyperkalemia and acidosis: a defect in renal transport of potassium. *Kidney International*, 3, 251.

Spooner, P. M. & Edelman, I. S. (1975). Further studies on the effect of aldosterone on electrical resistance of toad bladder. *Biochimica et Biophysica Acta*, 406, 304.

Stanbury, S. W. & Mahler, R. F. (1959). Salt-wasting renal disease. Metabolic observations on a patient with "salt-losing nephritis." *Quarterly Journal of Medicine*, 28, 425.

Steinmetz, P. R. (1974). Cellular mechanisms of urinary acidification. *Physiological Reviews*, 54, 890.

Stockigt, J. R., Collins, R. D. & Biglieri, E. G. (1971). Determination of plasma renin concentration by angiotensin I immunoassay. Diagnostic import of precise measurement of subnormal renin in hyperaldosteronism. *Circulation Research*, 28 & 29 (suppl. II), II-175.

Stoff, J. S., Handler, J. S., Preston, A. S. & Orloff, J. (1973). The effect of aldosterone on cyclic nucleotide phosphodiesterase activity in toad urinary bladder. *Life Sciences*, 13, 545.

Sutherland, D. J. A., Ruse, J. L. & Laidlaw, J. C. (1966). Hypertension, increased aldosterone secretion and low plasma renin activity relieved by dexamethasone. *Canadian Medical Association Journal*, 95, 1109.

Swingle, W. W., Parkins, W. M., Taylor, A. R. & Hays, H. W. (1936). Relation of serum sodium and chloride levels to alterations of body water in the intact and adrenalectomized dog, and the influence of adrenal cortical hormone upon fluid distribution. *American Journal of Physiology*, 116, 438.

Szylman, P., Better, O. S., Chaimowitz, C. & Rosler, A. (1976). Role of hyperkalemia in the metabolic acidosis of isolated hypoaldosteronism. *New England Journal of Medicine*, 294, 361.

Tannen, R. L. (1977). Relationship of renal ammonia production and potassium homeostasis. *Kidney International*, 11, 453.

Thorn, G. W., Koepf, G. F. & Clinton, M., Jr. (1944). Renal failure simulating adrenocortical insufficiency. *New England Journal of Medicine*, 231, 76.

Tisher, C. C. (1976). Anatomy of the kidney. In *The Kidney*, ed. Brenner, B. M. & Rector, F. C., Jr., Ch. 1. Philadelphia: Saunders.

Tuck, M. L., Barrett, J. D., Eggena, P., Mayes, D. M. & Sambhi, M. P. (1977). Mineralocorticoid biosynthesis and neural control of renin release in selective hypoaldosteronism. *Clinical Research*, 25, 140A.

Ufferman, R. C. & Schrier, R. W. (1972). Importance of sodium intake and mineralocorticoid hormone in the impaired water excretion in adrenal insufficiency. *Journal of Clinical Investigation*, 51, 1639.

Uhlich, E., Baldamus, C. A. & Ullrich, K. J. (1969). Einfluβ von aldosteron auf den natriumtransport in den sammelrohren der säugetierniere. *Pfluegers Archiv,* **308**, 111.

Uhlich, E., Halbach, R. & Ullrich, K. J. (1970). Einfluβ von aldosteron auf donausstrom markierten natriums aus den sammelrohren der ratte. *Pfluegers Archiv,* **320**, 261.

Ulick, S., Gautier, E., Vetter, K. K., Markello, J. R., Yaffe, S. & Lowe, C. U. (1964). An aldosterone biosynthetic defect in a salt-losing disorder. *Journal of Clinical Endocrinology and Metabolism,* **24**, 669.

Ulick, S. (1976). Diagnosis and nomenclature of the disorders of the terminal portion of the aldosterone biosynthetic pathway. *Journal of Clinical Endocrinology and Metabolism,* **43**, 92.

Vagnucci, A. H. (1969). Selective aldosterone deficiency. *Journal of Clinical Endocrinology and Metabolism,* **29**, 279.

Vander, A. J., Malvin, R. L., Wilde, W. S., Lapides, J., Sullivan, L. P. & McMurray, V. M. (1958). Effects of adrenalectomy and aldosterone on proximal and distal tubular sodium reabsorption. *Proceedings of the Society for Experimental Biology and Medicine,* **99**, 323.

Walker, W. G., Jost, L. J., Johnson, J. R. & Kowarski, A. (1965). Metabolic observations on salt wasting in a patient with renal disease. *American Journal of Medicine,* **39**, 505.

Weidmann, P., Reinhart, R., Maxwell, M. H., Rowe, P., Coburn, J. W. & Massry, S. G. (1973). Syndrome of hyporeninemic hypoaldosteronism and hyperkalemia in renal disease. *Journal of Clinical Endocrinology and Metabolism,* **36**, 965.

Weinberger, M. H., Dowdy, A. J., Nokes, G. W. & Luetscher, J. A. (1968). Plasma renin activity and aldosterone secretion in hypertensive patients during high and low sodium intake and administration of diuretic. *Journal of Clinical Endocrinology and Metabolism,* **28**, 359.

Weinstein, S. F., Allan, D. M. E. & Mendoza, S. A. (1974). Hyperkalemia, acidosis, and short stature associated with a defect in renal potassium excretion. *Journal of Pediatrics,* **85**, 355.

Wenting, G. J., Man in 't Veld, A. J., Verhoeven, R. P., Derkx, F. H. M. & Schalekamp, M. A. D. H. (1977). Volume-pressure relationships during development of mineralocorticoid hypertension in man. *Circulation Research,* **40** (suppl. I), I–163.

Wiederholt, M., Behn, C., Schoormans, W. & Hansen, L. (1972). Effect of aldosterone on sodium and potassium transport in the kidney. *Journal of Steroid Biochemistry,* **3**, 151.

Wiederholt, M., Schoormans, W., Fischer, F. & Behn, C. (1973). Mechanism of action of aldosterone on potassium transfer in the rat kidney. *Pfleugers Archiv,* **345**, 159.

Wisgerhof, M. & Brown, R. D. (1977). Aldosterone-renin dissociation and increased adrenal sensitivity in essential hypertension. *Clinical Research,* **25**, 16A.

Wisgerhof, M., Carpenter, P. C. & Brown, R. D. (1977). Increased adrenal sensitivity to angiotensin II in idiopathic hyperaldosteronism. *Clinical Research,* **25**, 568A.

Woods, J. W., Liddle, G. W., Stant, E. G., Jr., Michelakis, A. M. & Brill, A. B. (1969). Effect of an adrenal inhibitor in hypertensive patients with suppressed renin. *Archives of Internal Medicine,* **123**, 366.

Wright, F. S. (1977). Sites and mechanisms of potassium transport along the renal tubule. *Kidney International,* **11**, 415.

Yunis, S. L., Bercovitch, D. D., Stein, R. M., Levitt, M. F. & Goldstein, M. H. (1964). Renal tubular effects of hydrocortisone and aldosterone in normal hydropenic man: comment on sites of action. *Journal of Clinical Investigation,* **43**, 1668.

Zusman, R. M., Keiser, H. R. & Handler, J. S. (1977). Adrenal steroids enhance vasopressin-stimulated water flow in the toad bladder by inhibiting prostaglandin E biosynthesis. *Proceedings of the American Society of Nephrology,* **10**, 127A.

9 | Pathophysiology of Bartter's syndrome

JOHN J. BARDGETTE
JAY H. STEIN

Introduction
Symptoms
Pathogenesis
Laboratory and radiographic findings
Histologic changes
Pathophysiology
Hypokalemia
Metabolic alkalosis

Renin and aldosterone hypersecretion
Renal sodium and chloride conservation
Prostaglandins
Vascular insensitivity to angiotensin II
Theories of pathogenesis
Differential diagnosis
Treatment

INTRODUCTION

Bartter's syndrome is an uncommon disorder characterized by distinctive metabolic abnormalities. The hallmarks of this entity are hypokalemia, metabolic alkalosis, normal blood pressure in the face of increased secretion of renin and aldosterone, resistance to the pressor effects of angiotensin II, and hyperplasia of the juxtaglomerular apparatus. The etiology and pathogenesis of this syndrome have provoked imaginative speculation but remain enigmatic.

Recognition of the syndrome occurred in 1960 when Pronove, MacCardle and Bartter delineated the characteristic abnormalities in a 5-year-old boy. Subsequently, these same workers amplified this case report with an extensive clinical description and metabolic evaluation of the original subject and of an additional patient (Bartter et al., 1962). More than 50 case reports of similar patients have appeared in the English literature.

The disorder has been diagnosed in subjects ranging from one month (Wald, Perrin and Bolande, 1971) to 52 years of age (Tomko, Yeh and Falls, 1976). Despite the wide range of ages at the time of diagnosis, symptoms become manifest during childhood in most patients. Indeed, the appearance of the initial symptoms during adulthood is unusual. Of the more than 50 cases reviewed, only two developed their symptoms after age

25. Another eight individuals noted the appearance of symptoms between the ages of 15 and 25. Thus, more than 80 percent of patients became symptomatic in childhood.

Not all authors distinguish between age of onset and age at the time of diagnosis, but from those reports in which the distinction is made, it is customary for diagnosis to ensue shortly after the development of symptoms. The interval between onset and diagnosis is commonly one to two years. In a few extraordinary individuals, diagnosis has been delayed for several decades. The patient described by Tomko et al. (1976) became symptomatic at age 13 but a diagnosis was not made until the age of 52. Two factors would seem to explain diagnostic delay: the uncommonness of this disorder and the variable severity of symptoms.

Males and females are equally affected. Age at onset, age at diagnosis, and symptoms are similar for men and women.

Although some have considered blacks to be more commonly affected (Hall, 1971), not all reports remark on the ethnic origins of patients and no strict conclusions can be reached. Of those articles that describe patients from the United States and comment on race, however, 15 of 18 individuals have been black. Reports from northern and southern Europe as well as Israel and the Orient have appeared.

SYMPTOMS

The clinical manifestations of Bartter's syndrome are few in number and their enumeration in individual cases is strikingly similar (Table 9.1).

Table 9.1 Clinical findings in Bartter's syndrome

Symptoms	Signs
Muscle weakness	Proximal muscle weakness
Muscle cramps	Truncal muscle weakness
Enuresis	Failure to thrive
Nocturia	Short stature
Polyuria	Convulsions
Anorexia	Tetany
Salt craving	Chvostek sign
Vomiting	Trousseau sign
Constipation	Ileus
	Hypoventilation
	Gout
	Mental retardation

Symptoms and signs include abnormalities of the neuromuscular, urinary, and gastrointestinal systems as well as developmental retardation. The first category includes muscle weakness and cramps, convulsions, tetany, carpopedal spasm, and positive Chvostek and Trousseau signs. Urinary symptoms include enuresis, nocturia, and polyuria. Salt craving, vomiting, and constipation are the frequently mentioned gastrointestinal symptoms. Mental·retardation as well as growth impairment may be found.

Failure to thrive and muscle weakness are the symptoms that most commonly bring patients to medical attention. Among affected adults proximal muscle weakness stands out as the most frequent and prominent symptom. It should be emphasized that the muscle weakness seen in adults may be of recent onset. The patient of White (1972) became symptomatic and was diagnosed at age 22, and Solomon and Brown (1975) similarly reported a one-year duration of symptoms in a patient diagnosed as having Bartter's syndrome at age 41.

There are single reports of rickets (Arant et al., 1970), hypoventilatory attacks (Dehart et al., 1974), ileus and weakness involving truncal as well as proximal muscles (Tomko et al., 1976).

Only two asymptomatic patients have been reported. In one instance, evaluation of the sibling of a symptomatic individual led to detection (White, 1972) and in the other instance, hypokalemia was found in the course of evaluation of proteinuria (Dehart et al., 1974).

The association of gout and Bartter's syndrome has been found in three individuals (Meyer, Gill and Bartter, 1975; Modlinger et al., 1973). Although the creatinine clearance was diminished in two of the three (50 and 61 cc/minute), the reduction was not seemingly severe enough to account for the degree of hyperuricemia noted. In four of eight affected individuals, who were not gouty, hyperuricemia was found. In two of these four, however, creatinine clearances were less than 18 cc/min. The pathophysiologic connection between hyperuricemia and Bartter's syndrome is unknown.

PATHOGENESIS

The majority of symptoms are attributable to the consequences of hypokalemia. Hypokalemia is the likely cause of proximal muscle weakness as replacement therapy results in disappearance of this symptom in most cases.

Several authors have examined the impaired renal concentrating ability in these patients (Bartter et al., 1962; Cannon et al., 1968). The patient described by Cannon et al. excreted from four to seven liters of urine daily and was unable to form concentrated urine in response to either dehydration or pitressin infusion. Even with normalization of the serum potassium for two years their patient remained polyuric and unable to concentrate his urine. Yet, it still seems likely that the isosthenuria and antidiuretic hormone resistance may be ascribed in most patients to the consequences of chronic hypokalemia (Relman and Schwartz, 1958; Rubini, 1961). Enuresis, nocturia, polyuria, and polydipsia are the consequences of the concentrating defect.

Although potassium deficiency uncommonly affects gut motility in the normal adult, children seem more susceptible. This may explain the frequency of constipation among children with the syndrome and the rarity of constipation among affected adults.

The pathogenesis of mental retardation and growth impairment is unknown. The growth abnormalities have been examined most closely by

Simopoulos and Bartter (1972). Growth retardation is most prominent in infancy and early childhood. The children studied all exhibited preadolescent growth failure but experienced normal adolescent maturation. As all patients eventually reached reasonable adult height and weight, they concluded that true dwarfism did not occur in Bartter's syndrome. Of interest is the further observation that the normal growth spurt occurred in the absence of treatment in one individual and in the face of unsuccessful treatment of hypokalemia in another patient. These observations suggest that normal body potassium stores are not critical for growth in the adolescent period. These findings are supported by the case report of Modlinger et al. (1973). This patient demonstrated symptoms in infancy but remained undiagnosed until age 35. Growth impairment was present in childhood but showed spontaneous improvement at age 17. In contrast, Cannon et al. believed that potassium supplementation was responsible for the increase in height seen in their patient. Thus the precise role played by potassium in the growth retardation seen in infancy and early childhood remains unclear.

LABORATORY AND RADIOGRAPHIC FINDINGS

Hormonal and electrolyte abnormalities may be classified as those that are obligatory for the diagnosis and those that are occasionally but not commonly found (Table 9.2). Those in the former group include elevated

Table 9.2 Laboratory findings in Bartter's syndrome

Common Laboratory Findings	Less Common Laboratory Findings
Hypokalemia	Hypomagnesemia
Metabolic alkalosis	Hypercalcemia
Hyperreninemia	Hyperuricemia
Hyperaldosteronism	Glucose intolerance
Angiotensin II insensitivity	Polycythemia
Vasopressin resistance (isosthenuria)	Erythrocyte sodium abnormalities
	EEG abnormalities

secretion and excretion of aldosterone, increased plasma renin activity, hypokalemia, and alkalosis.

Except for abnormal aldosterone secretion, adrenocortical function is normal. Secretion of compound S, cortisol, and desoxycorticosterone have been found to be normal (Goodman, Vagnucci and Hartroft, 1969). Excretion of 17-OH and 17-ketosteroids is unaffected and responds with an appropriate rise when ACTH is infused (Bartter et al., 1962).

The renal concentrating defect was alluded to in the previous section.

Hypomagnesemia is not uncommon. It has been suggested in at least one case to be responsible for some of the symptomatology (Mace et al., 1973). An 11-week-old child with marked weakness and hypoactive stretch reflexes failed to improve when serum potassium rose to normal with supplementation while dramatic improvement occurred within 48 hours after magnesium administration and correction of hypomagnesemia.

Asymptomatic mild hypercalcemia has been found (Trygstad et al., 1969). In no instance has hypocalcemia been noted. As mentioned above, Meyer et al. (1975) have called attention to the occurrence of hyperuricemia in patients with this syndrome.

Glucose intolerance, probably attributable to hypokalemia, occurs but its frequency is unknown. Of the six patients seen by Simopoulos and Bartter (1972) in whom glucose tolerance was examined, only two exhibited abnormalities. With potassium repletion was performed, glucose tolerance became normal. Despite the presence of glucose intolerance, fasting hyperglycemia is not present.

Erkelens and Statius van Eps (1973) described the occurrence of increased red cell volume and increased erythropoietin levels in a 33-year-old man with Bartter's syndrome. They suggested that this finding further supported the view that erythropoietin or its precursor was released from the juxtaglomerular apparatus.

Alterations of erythrocyte sodium transport are present in some but not all patients (Gall et al., 1971) (Gardner et al., 1972). Diminished red cell sodium efflux and increased red cell sodium concentrations were seen in six of eight patients studied by Gardner et al. Red cells from relatives of individuals with abnormal sodium transport were similarly affected.

Electroencephalography has shown dysrrhythmias in some children (Bartter et al., 1962).

Three radiographic findings, all nonspecific, may be seen. The most common finding is dilatation of the ureters and hydronephrosis, presumably related to the polyuric state. Bartter et al. (1962) mention retarded bone age. Barium studies of a patient whose symptoms were predominantly gastrointestinal showed "hypomotility" and dilatation of the duodenum (Tarm et al., 1973).

HISTOLOGIC CHANGES

There are few extensive descriptions of the histologic changes in the kidney in Bartter's syndrome. The most constant histologic feature is hyperplasia of the juxtaglomerular apparatus (Fig. 9.1). With the use of the Bowie stain, which detects renin containing granules, it has been possible to show in the majority of instances that the hyperplastic cells produce renin. Examination of the kidney in some patients, however, while revealing hyperplasia of the juxtaglomerular cell, has not shown increased numbers of renin-containing granules (Bartter et al., 1962; Brackett et al., 1968; Wald et al., 1971). Wald found no Bowie positive granules but observed with electron microscopy prominent protogranular forms in juxtaglomerular cells. In an effort to reconcile these seemingly contrary findings, it has been speculated that the Bowie positive granules are a mature storage organelle whereas the protogranular forms are the actual sites of renin synthesis.

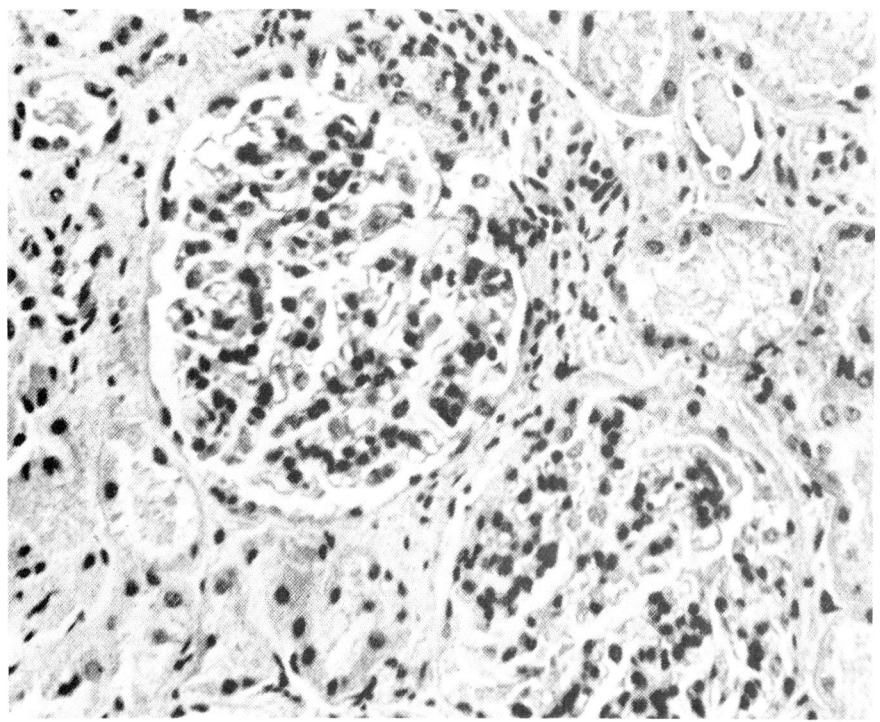

Figure 9.1 Renal biopsy from a patient with Bartter's syndrome. The juxtaglomerular apparatus is hyperplastic. (Reproduced from Bartter et al., 1962, with permission of the publisher)

As a result of the hyperplasia, there is replacement of vascular muscle cells with juxtaglomerular cells causing thickening and elongation of afferent arterioles.

The macula densa also participates in the hyperplasia of the juxtaglomerular apparatus. Although the cells of the macula densa are increased in number, their nuclei may be flattened and the cells may appear atrophic and at times pyknotic. Cannon et al. (1968) observed desquamation of macula densa cells and found loss of the normal continuity between macula densa and juxtaglomerular cells. Some investigators have found no changes in the histology of the individual macula densa cells (Schwartz and Cornfield, 1975).

It should also be emphasized that juxtaglomerular hyperplasia is not pathognomonic of Bartter's syndrome and has been seen in other conditions associated with long-standing hyperreninemia: familial chloridorrhea (Pasternack et al., 1967), Addison's disease (Cannon et al., 1968), and laxative-induced hyperaldosteronism (Fleischer et al., 1969), to name a few of these conditions.

Glomerular changes are also prominent but variable. The most consistent finding is glomerular hypercellularity, in large measure due to mesangial hyperplasia. The glomeruli are usually otherwise normal. Crescent formation and capillary and epithelial hyperplasia have been described by a few investigators but probably are indicative of a concomitant primary glomerular disorder.

The proximal tubule commonly shows vacuole formation. The remainder of the nephron has not been well characterized histologically.

France, Shelley and Gray (1974) and Verberckmoes et al. (1976) made the intriguing observation of interstitial cell hyperplasia of the medulla (Fig. 9.2). The latter group observed that the histochemical properties of

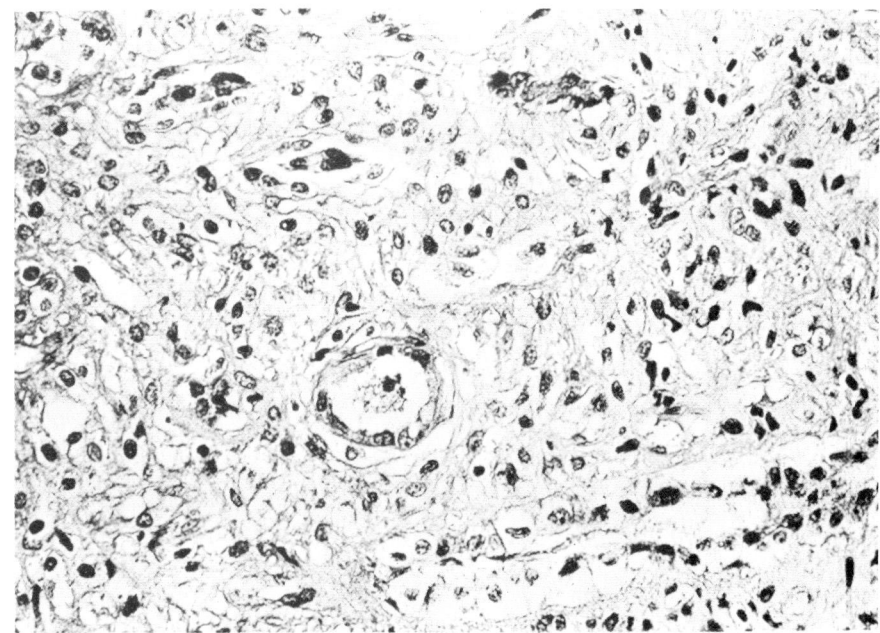

Figure 9.2 Renal biopsy from a patient with Bartter's syndrome. The medullary interstitial cells are hyperplastic. (Reproduced from Verberckmoes et al., 1976, with permission of the publisher)

these cells are identical to those of cells capable of synthesizing prostaglandins. The implications of this finding and the possible role of prostaglandins in the pathogenesis of Bartter's syndrome are discussed below.

PATHOPHYSIOLOGY

In the following sections, we will discuss in detail the major abnormalities which have been identified in this syndrome.

Hypokalemia

Hypokalemia is an invariable feature of the disorder and must be considered essential to the diagnosis. The potassium deficit results from excessive urinary loss of potassium which may exceed 300 mEq/day. Although increased aldosterone secretion may play a role in the potassium wasting seen in Bartter's syndrome, urinary losses of the cation persist even in the absence of aldosterone excess (Greenberg et al., 1966). When aldosterone is suppressed to the normal range by feedback inhibition with dexamethasone, by synthetic inhibition with aminoglutethimide, or by administration of inhibitors of distal tubular potassium secretion, serum potassium remains low (Goodman et al., 1969). As secretion of cortisol, corticosterone, and desoxycorticosterone are normal, there is no evidence that other adrenocorticoid hormones are responsible for the potassium wasting. Finally, total adrenalectomy performed on one patient with Bartter's syndrome had no significant effect on potassium losses (Trygstad et al., 1969).

Alterations in sodium balance have a variable effect on the magnitude of urinary potassium loss in patients with the syndrome. Thus, Tomko et al. (1976), Cannon et al. (1968), and Schwartz and Cornfeld (1975) found that sodium restriction led to a decrease in potassium supplementation. Others have found the converse (Solomon and Brown, 1975), and some workers have observed no effect of alterations in sodium intake on urinary excretion of potassium (Bartter et al., 1962; Goodman et al., 1969).

Thus, a variety of interventions known to decrease potassium excretion have little if any effect in patients with Bartter's syndrome.

Metabolic alkalosis

It seems likely that the combination of hypokalemia and mineralocorticoid excess are responsible for the metabolic alkalosis. Although modest potassium deficiency per se probably does not cause metabolic alkalosis in man, the potassium deficit in Bartter's patients must be considered severe as witnessed by the failure of enormous amounts of potassium to consistently restore the serum potassium concentration to normal. Potassium deficiency may, therefore, be of primary importance. In addition, Cannon et al. (1968) noted that withdrawal of supplemental potassium in the amount of 350 mEq/day resulted in minimal increases of urinary titratable acid but marked increases in urinary ammonium and total acid excretion. These alterations in acid excretion were associated with the development of metabolic alkalosis. This increase in ammonium and net acid excretion is presumably due to the reciprocal relationship between the serum potassium concentration and urinary ammonium excretion (Tannen, 1970). It should be emphasized once again, however, that enhanced aldosterone secretion may also be of major importance in the development of metabolic alkalosis in these patients.

Renin and aldosterone hypersecretion

It seems established that enhanced renin secretion is responsible for the hyperaldosteronism in these patients. Numerous studies provide evidence that renin and aldosterone may respond appropriately to well-characterized physiologic maneuvers. When sodium intake is restricted, both plasma renin activity and aldosterone secretion rise although to a higher level than is usually seen in normal subjects (Cannon et al., 1968; Goodman et al., 1969; Solomon and Brown, 1975). The majority of studies have shown that oral or intravenous salt loading or albumin infusion causes a major reduction in plasma renin activity and aldosterone levels (Goodman et al., 1969; White, 1972; Modlinger et al., 1973; Solomon and Brown, 1975). Several reports, however, describe a paradoxical rise in plasma renin activity and aldosterone secretion in response to salt loading (Bartter et al., 1962; Cannon et al., 1968; Imai et al., 1974). The failure of plasma renin activity to normalize with volume expansion suggests that factors other than the state of the extracellular fluid volume are operative in controlling renin production.

As in normal subjects, hypokalemia will tend to suppress aldosterone secretion in these patients. Thus, potassium supplementation causes enhanced aldosterone production and omission of these supplements causes a reduction in aldosterone levels. The inhibitory effect of hypokalemia explains the observation that plasma renin activity is often elevated proportionately more than aldosterone in the untreated patient with the syndrome (Goodman et al., 1969). It also explains the occasional finding of normal or even decreased aldosterone levels in patients with Bartter's syndrome (Greenberg et al., 1966; Brackett et al., 1968).

Renal sodium and chloride conservation

The response to sodium restriction is variable. The patient of Cannon et al. (1968) during a 15-day period of restricted sodium intake persistently excreted amounts of sodium greater than his intake and developed azotemia. The patient of Schwartz and Cornfeld (1975) became dehydrated while the subject of Tomko et al. (1976) lost four pounds and developed postural hypotension when salt restricted. In contrast, most patients have come into sodium balance rapidly when salt restricted (Bartter et al., 1962; Goodman et al., 1969; Norby, Mark and Kaloyanides, 1976).

Whether these discrepant observations mean that Bartter's syndrome is a condition with multiple etiologies or whether this reflects a spectrum of the same pathogenetic entity is not clear. It is worthy of note that patients with this syndrome, when treated with large potassium supplements and sodium restriction, may be excreting sodium because of the natriuretic effect of potassium administration (Vander, 1970) with consequent mobilization of intracellular sodium.

Other abnormalities of renal sodium handling have been observed. White (1972) showed that two patients given three liters of saline excreted the salt load more rapidly than normal subjects. Chaimovitz et al. (1973) found that fractional distal delivery of sodium was normal in a patient with Bartter's syndrome and inferred that there was no abnormality of proximal sodium reabsorption. They also noted impaired free-water clearance at any given level of distal sodium delivery and suggested that sodium transport in the ascending limb of Henle's loop was impaired. Tomko et al. (1976) found increased free-water clearance and increased distal delivery of sodium. Their findings suggested a proximal sodium leak. There is, however, no evidence of generalized proximal tubular dysfunction. Glycosuria is almost never found and minimal elevation of amino acid excretion has been noted by only one observer (Tomko et al., 1976). Bicarbonate reabsorption is increased (White, 1972). Phosphate handling has not been examined. Thus, we would conclude that the evidence for a primary sodium leak is much less persuasive than the findings previously discussed in regard to potassium.

Kurtzman and Gutierrez (1975) first suggested that Bartter's syndrome may be caused by a deficit in chloride transport in the ascending limb of Henle's loop. Thus, there would be an increased delivery of both sodium and potassium to the distal tubular exchange site. In the presence of high levels of aldosterone, substantial reabsorption of the increased sodium load might occur in exchange for potassium and thus further enhance potassium wasting. Yet, it is difficult to understand why an individual with a defect in ascending limb chloride transport would readily come into sodium balance when placed on a low sodium diet, as is usually the case in patients with Bartter's syndrome. In an analogous sense, it seems equally unlikely that a patient treated with small doses of furosemide (which also inhibits chloride transport in the ascending limb) could rapidly come into sodium balance when sodium was removed from the diet. It has been shown that potassium wasting persists in patients with bilateral adrenalectomy (Trygstad et al., 1969).

Prostaglandins

Several investigators have studied the role of prostaglandins in Bartter's syndrome. Using a sensitive mass spectrometric technique, Gill et al. (1976) found that excretion of urinary PGE was notably higher in four patients with Bartter's syndrome than in normal individuals. In further exploring the role of prostaglandins, they administered indomethacin, an inhibitor of prostaglandin synthesis, to ascertain its effect on prostaglandin excretion and on the other metabolic abnormalities (Fig. 9.3). Indomethacin treatment resulted in reductions in prostaglandin excretion. Concomitantly, plasma renin activity and aldosterone secretion fell. Associated with these changes were a diminution in sodium excretion, development of positive sodium balance, and an increase in body weight. Similarly, potassium balance became less negative and the serum potassium rose but still remained in the hypokalemic range. These findings have been found by several groups (Verberckmoes et al., 1976; Fichman, Telfer and Zia, 1976).

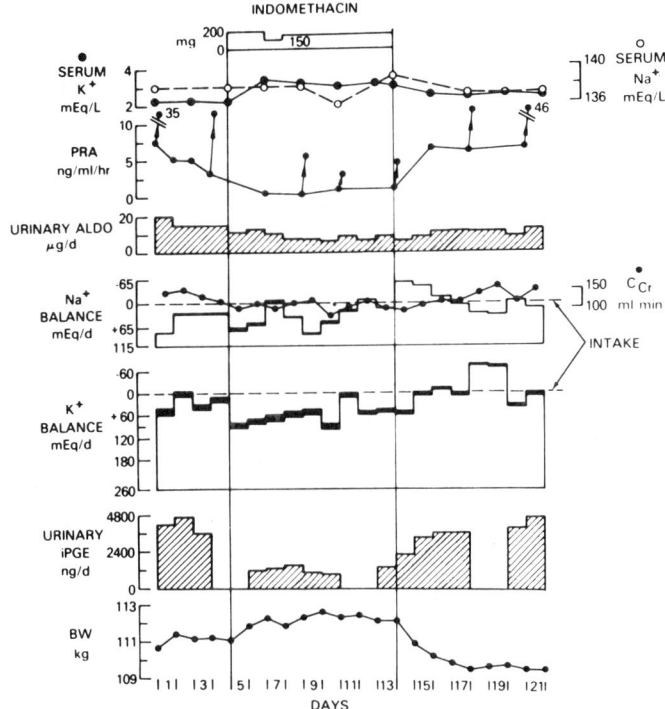

Figure 9.3 The effect of indomethacin on prostaglandin excretion and on the metabolic abnormalities of Bartter's syndrome. (Reproduced from Gill et al., 1976, with permission of the publisher)

Vascular insensitivity to angiotensin II

Infusions of angiotensin II fail to elicit the expected blood pressure elevations in patients with Bartter's syndrome. The insensitivity appears to be specific for angiotensin II as the response to norepinephrine is usually normal. Responsiveness to angiotensin II may return to normal with saline infusion (White, 1972; Goodman et al., 1969), but this is unusual. On the other hand, as shown in Figure 9.4, indomethacin promptly restores the vascular sensitivity to angiotensin II (Halushka et al., 1977).

THEORIES OF PATHOGENESIS

There are few, if any, conditions whose pathogenesis is so mystifying. Bartter and associates (1962) originally proposed that the basic defect in the condition was a resistance to the vasoconstrictor action of angiotensin II. Several observations are against this hypothesis. First, studies by Kaplan

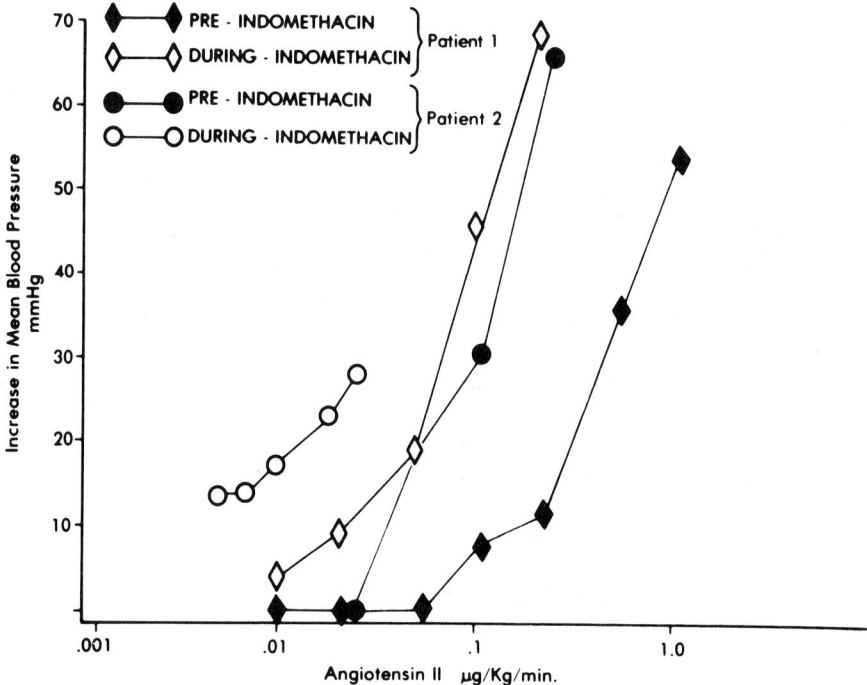

Figure 9.4 The effect of indomethacin on the vascular sensitivity to angiotensin II infusion in Bartter's syndrome. Note how indomethacin shifts the dose response curve to the left. (From Halushka et al., 1977, with permission of the publisher)

and Silah (1964) showed similar resistance to exogenous angiotensin II in many clinical disorders with elevated endogenous levels of renin and presumably angiotensin II. More importantly, saralasin infusion produces hypotensive responses in these patients (Fig. 9.5) (Kono, Oseko and Shimbo, 1976; Sasaki et al., 1976). Saralasin, a competitive antagonist of angiotensin II, causes a fall in blood pressure in those pathophysiologic states in which angiotensin contributes to blood pressure maintenance (Hollenberg et al., 1974; Brunner et al., 1971). These findings suggest that the blood pressure in patients with Bartter's syndrome is maintained, at least in part, by the high circulating levels of angiotensin II and that angiotensin excess acts to compensate for some alteration in the systemic circulation.

Some investigators have advanced the notion that renal salt wasting is the fundamental abnormality in Bartter's syndrome (Cannon et al., 1968; White, 1972). It is suggested that there is a primary defect in transport that causes a decrease in effective blood volume with resultant hyperreninemia and hyperaldosteronism. The evidence in favor of renal sodium wasting, as

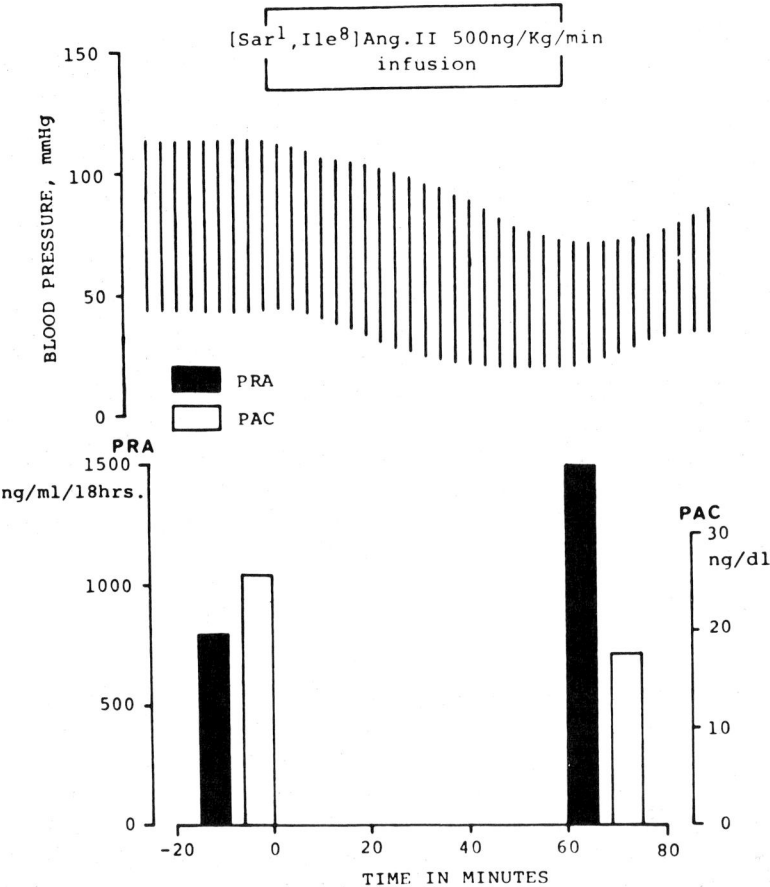

Figure 9.5 The effect of angiotensin II blockade on blood pressure in Bartter's syndrome. (From Sasaki et al., 1976, with permission of the publisher)

discussed previously, includes the demonstration of negative sodium balance in some patients on a low sodium diet and an exaggerated natriuresis after acute volume expansion. In several careful studies, however, evidence for negative sodium balance is lacking. Thus, the evidence is not persuasive that renal salt wasting is a major contributor to the abnormalities of Bartter's syndrome. We have also previously discussed the issue of a primary defect in chloride transport.

It is possible that the basic defect in Bartter's syndrome is renal potassium wasting. This abnormality is prominent in every patient and is not due to hyperaldosteronism (Trygstad et al., 1969). How can potassium wasting explain the other manifestations of the syndrome? One possible scheme is shown in Figure 9.6.

Galvez et al. (1977) have recently found that urinary PGE excretion rapidly and markedly increased when potassium depletion was induced in

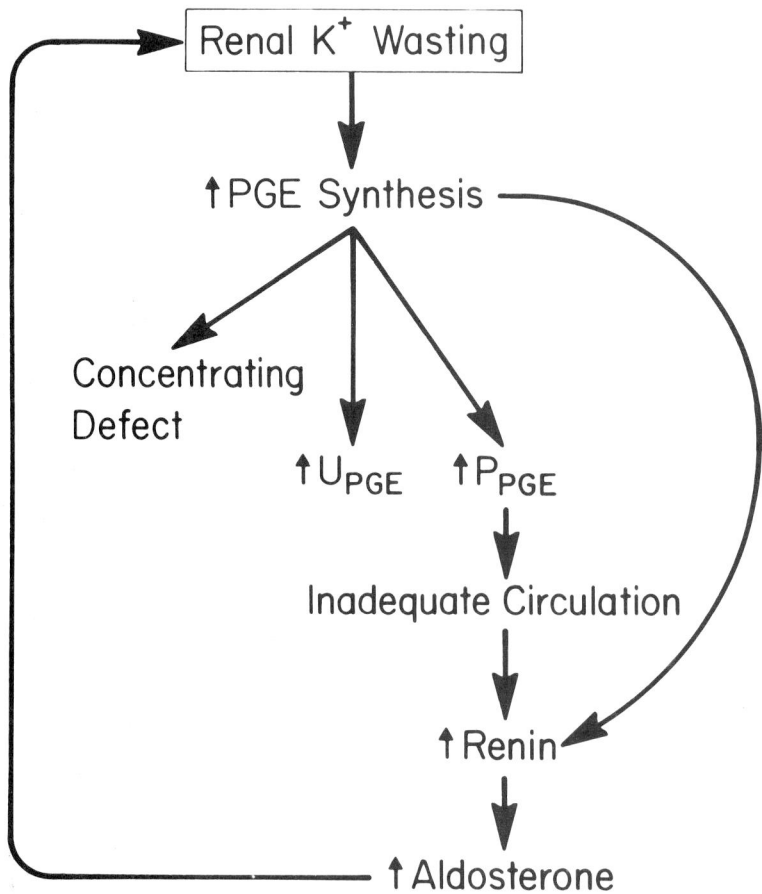

Figure 9.6 The proposed pathophysiology of Bartter's syndrome. The relationship of urinary potassium wasting to the other metabolic derangements is emphasized.

the dog (Fig. 9.7). Thus, it may be that the marked increase in urinary PGE in patients with Bartter's syndrome is secondary to the renal potassium loss. Systemic infusions of prostaglandins may alter the circulation and lead to renin release (Fichman et al., 1972). Although PGE produced in the kidney is metabolized substantially in the lung, recent studies by Golub et al. (1975) have shown that this degradation is incomplete. Therefore, a marked increase in renal synthesis of PGE may alter systemic hemodynamics, leading to enhanced renin and angiotensin II and subsequently to aldosterone production. In addition, there is evidence that enhanced prostaglandin production may have a direct stimulating effect on renin release independent of systemic hemodynamic changes (Weber et al., 1976). The resistance to exogenous angiotensin II would then be due to the high endogenous level of angiotensin II. This pathogenic sequence of

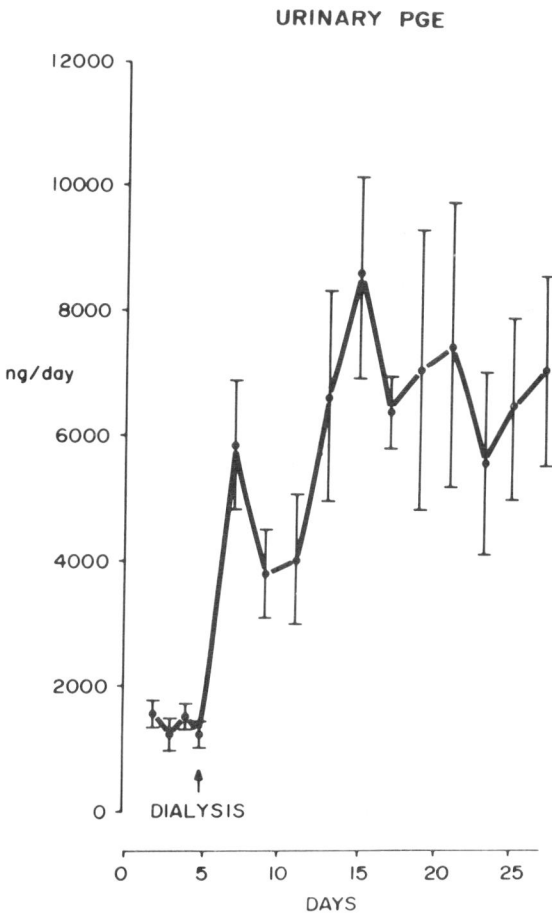

Figure 9.7 The effect of dialysis induced hypokalemia on urinary excretion of potassium in the dog. (From Galvez et al., 1977, with permission of the publisher)

events would also explain the reversal of angiotensin insensitivity with indomethacin (Halushka et al., 1977). Enhanced release of kinins (Lecchi et al., 1976) may also play a role in this phenomenon.

Excessive PGE production may also be responsible for the concentrating defect seen in Bartter's syndrome. Several in vitro studies have suggested that PGE may antagonize the hydro-osmotic action of vasopressin (Grantham and Orloff, 1968; Orloff, Handler and Bergstrom, 1965). Anderson et al. (1975) have also convincingly demonstrated a similar interrelationship between PGE and vasopressin in clearance studies in the dog. Thus, the increased renal production of PGE may, in some manner, interfere with the local action of vasopressin leading to a concentrating defect in Bartter's syndrome.

DIFFERENTIAL DIAGNOSIS

Unexplained hypokalemia should lead to a consideration of this disorder. Many of the disorders that cause hypokalemia will be readily excluded by the history and physical examination. The presence of hypertension in association with hypokalemia will suggest primary aldosteronism, a renin-producing tumor or renal artery stenosis. Normal glucocorticoid homeostasis excludes other hyperfunctioning adrenocorticoid states that produce hypokalemia. A history of diarrhea should be sought, but sleuthlike persistence is necessary to unmask the laxative abuser who may mimic in all respects the patient with Bartter's syndrome. Fleischer et al. (1969) reported a patient with all of the characteristic abnormalities of the syndrome who eluded correct diagnosis until the laxative phenolphthalein was discovered in the urine. In like fashion, surreptitious diuretic administration should be excluded.

TREATMENT

In most instances, initial attention should be directed to repletion of body potassium stores. A variety of therapeutic maneuvers, including potassium supplementation, potassium-sparing diuretics, propranolol, indomethacin, aspirin, and adrenalectomy, have been tried. Many of these interventions produce improvement in the serum potassium concentration but only rarely have they resulted in long-term normalization of this parameter.

Potassium-sparing diuretics have rarely been effective when used alone, but when coupled with potassium supplementation, are modestly useful. Bartter's original two patients demonstrate the variable response to spironolactone. While on a 10 mEq sodium, 160 mEq potassium diet, the serum potassium rose from 2.6 to 6.8 mEq/l in one patient, whereas the other patient showed a rise from 2.2 to only 2.6 mEq/l. The patient of White (1972) showed a remarkable refractoriness to the regimen of potassium supplementation and spironolactone. When given a daily dosage of 460 mEq of potassium chloride and 200 mg of spironolactone, this patient raised his potassium concentration to only 3.1 mEq/l. Although potassium-sparing diuretics may be effective initially, long-term results are less gratifying. An "escape" from the effects of spironolactone commonly occurs, and hypokalemia again ensues over a period of months (Brackett et al., 1968; Modlinger et al., 1973; Solomon and Brown, 1975). When potassium supplements are used alone, monumental amounts are usually required to correct the potassium deficit. The patient of Cannon et al. (1968) was able to maintain a normal serum potassium concentration only with the daily provision of 500 mEq of potassium.

Propranolol has been completely ineffective in restoring potassium balance. Modlinger et al. (1973) found that propranolol (360 mg/day) caused a reduction in plasma renin activity and plasma aldosterone but had little effect on serum potassium. Moreover, propranolol in high doses has not prevented the rise in renin and aldosterone secretion which accompanies spironolactone therapy.

Numerous investigators have documented the short term effectiveness of indomethacin in correcting some of the metabolic abnormalities of Bartter's syndrome (Verberckmoes et al., 1976; Fichman et al., 1976; Gill et al., 1976). With the administration of indomethacin, urinary excretion of potassium may diminish somewhat in association with a rise in the serum potassium concentration (Fig. 9.3). This is presumably due to the inhibitory effect of prostaglandin inhibition on the component of potassium excretion, which is aldosterone dependent. As noted previously, however, indomethacin does not correct the potassium-wasting state seen in patients with this syndrome.

Other inhibitors of prostaglandin synthesis, such as aspirin (Norby et al., 1976) and ibuprofen (Gill et al., 1976) have been found to cause changes similar to those seen with indomethacin. It may be of interest that aspirin seemed to readily increase the serum potassium concentration in one instance, but the patient under study had a relatively mild degree of potassium wasting (Norby et al., 1976).

Magnesium supplementation for hypomagnesemia may rarely have dramatic symptomatic effects. Although supplementation seems advised when hypomagnesemia is present, magnesium deficit is uncommonly a cause of symptoms. As discussed previously, adrenalectomy has neither a marked nor a lasting beneficial effect.

ACKNOWLEDGMENT

Portions of this work were supported by NIH Program Grant AM-17387 and Training Grant AM-07103.

REFERENCES

Anderson, R. J., Berl, T., McDonald, K. M. & Schrier, R. W. (1975). Evidence for an in-vivo antagonism between vasopressin and prostaglandin in the mammalian kidney. *Journal of Clinical Investigation*, 56, 420–426.
Arant, B. S., Brackett, N. C., Young, R. B. & Still, W. J. S. (1970). Case studies of siblings with juxtaglomerular hyperplasia and secondary aldosteronism associated with severe azotemia and renal rickets. Bartter's syndrome or disease? *Pediatrics*, 46, 344–361.
Bartter, F. C., Pronove, P., Gill, J. R., Jr. & MacCardle, R. C. (1962). Hyperplasia of the juxtaglomerular complex with hyperaldosteronism and hypokalemic alkalosis: a new syndrome. *American Journal of Medicine*, 33, 811–828.
Brackett, N. C., Jr., Koppel, M., Randall, R. E. & Nixon, W. P. (1968). Hyperplasia of the juxtaglomerular complex with secondary aldosteronism without hypertension (Bartter's syndrome). *American Journal of Medicine*, 44, 803–819.
Brunner, H. R., Kirshman, J. D., Sealey, J. D. & Laragh, J. H. (1971). Hypertension of renal origin: evidence for two different mechanisms. *Science*, 174, 1344–1346.
Bryan, G. T., MacCardle, R. C. & Bartter, F. C. (1966). Hyperaldosteronism, hyper-

plasia of the juxtaglomerular complex, normal blood pressure and dwarfism: report of a case. *Pediatrics*, **37**, 43–50.

Cannon, P. J., Leeming, J. M., Sommers, S. C., Winters, R. W. & Laragh, J. H. (1968). Juxtaglomerular cell hyperplasia and secondary hyperaldosteronism (Bartter's syndrome). A reevaluation of the pathophysiology. *Medicine*, **47**, 107–131.

Chaimovitz, C., Levi, J., Better, O. S., Oslander, L. & Benderli, A. (1973). Studies on the site of the renal salt loss in a patient with Bartter's syndrome. *Pediatric Research*, **7**, 89–94.

Dehart, H. S., Bath, N. M., Glenn, J. F. & Gunnells, J. C. (1974). Urologic considerations in Bartter's syndrome. *Journal of Urology*, **111**, 420–424.

Erkelens, D. W. & Statius van Eps, L. W. (1973). Bartter's syndrome and erythrocytosis. *American Journal of Medicine*, **55**, 711–719.

Fichman, M. P., Litenburg, G., Brooker, G. & Horton, R. (1972). Effect of prostaglandin A_1 on renal and adrenal function in man. *Circulation Research*, (suppl.) **32**, 11–19.

Fichman, M. P., Telfer, N. & Zia, P. (1976). Role of prostaglandins in the pathogenesis of Bartter's syndrome. *American Journal of Medicine*, **60**, 785–797.

Fleischer, N., Brown, H., Graham, D. Y. & Delena, S. (1969). Chronic laxative-induced hyperaldosteronism and hypokalemia simulating Bartter's syndrome. *Annals of Internal Medicine*, **70**, 791–798.

France, R., Shelley, W. M. & Gray, M. E. (1974). Abnormal intracellular granules of the renal papilla in a child with potassium depletion (Bartter's syndrome) and renal tuberous sclerosis. *Johns Hopkins Medical Journal*, **135**, 274–285.

Gall, G., Vaitukaitis, J., Haddow, J. E. & Klein, R. (1971). Erythrocyte sodium flux in a patient with Bartter's syndrome. *Journal of Clinical Endocrinology and Metabolism*, **32**, 562–567.

Galvez, O. G., Bay, W. H., Roberts, B. W. & Ferris, T. F. (1977). The hemodynamic effects of potassium deficiency in the dog. *Circulation Research* (suppl. 1), **40**, I11–16.

Gardner, J. D., Simopoulos, A. P., Lapey, A. & Shibolet, S. (1972). Altered membrane sodium transport in Bartter's syndrome. *Journal of Clinical Investigation*, **51**, 1565–1571.

Gill, J. R., Frolich, J. C., Bowden, R. E., Taylor, A. A., Keiser, H. R., Seyberth, W. W., Oates, J. A. & Bartter, F. C. (1976). Bartter's syndrome: a disorder characterized by high urinary prostaglandins and a dependence of hyperreninemia on prostaglandin synthesis. *American Journal of Medicine*, **61**, 43–51.

Golub, N., Zia, P., Matsumo, M. & Horton, R. (1975). Metabolism of prostaglandins A_1 and E_1 in man. *Journal of Clinical Investigation*, **56**, 1404–1410.

Goodman, A. D., Vagnucci, A. H. & Hartroft, P. M. (1969). Pathogenesis of Bartter's syndrome. *New England Journal of Medicine*, **281**, 1435–1439.

Grantham, J. J. & Orloff, J. (1968). Effect of prostaglandin E_1 on the permeability response of the isolated collecting tubule to vasopressin, adenosine 3'5'-monophosphate, and theophylline. *Journal of Clinical Investigation*, **47**, 1154–1161.

Greenberg, A. J., Arboit, J. M., New, M. I. & Worthen, H. G. (1966). Normotensive secondary hyperaldosteronism. *Journal of Pediatrics*, **69**, 719–727.

Hall, B. D. (1971). Preponderance of Bartter's syndrome among blacks. *New England Journal of Medicine*, **285**, 581.

Halushka, P. V., Wohltmann, H., Privitera, P. J., Hurwitz, G. & Margolius, H. S. (1977). Bartter's syndrome: urinary prostaglandin E-like material and kallikrein; indomethacin effects. *Annals of Internal Medicine*, **87**, 281–286.

Hollenberg, N. K., Chenitz, W. R., Adams, D. F. & Williams, G. H. (1974). Reciprocal influence of salt intake on adrenal glomerulosa and renal vascular responses to angiotensin II in normal man. *Journal of Clinical Investigation*, **54**, 34–42.

Imai, M., Yabuta, K., Murata, H., Takita, S., Ohbe, Y. & Sokabe, H. (1974). A case of Bartter's syndrome with abnormal renin response to salt load. *Journal of Pediatrics*, **74**, 738–749.

Kaplan, N. M. & Silah, J. G. (1964). The effect of angiotensin II on the blood pressure in humans with hypertensive disease. *Journal of Clinical Investigation*, **43**, 659–669.

Kono, T., Oseko, F. & Shimbo, S. (1976). Blood pressure fall by angiotensin II antagonist in patients with Bartter's syndrome. *Journal of Clinical Endocrinology and Metabolism*, **43**, 692–695.

Kurtzman, N. A. & Gutierrez, L. E. (1975). The pathophysiology of Bartter's syndrome. *Journal of the American Medical Association*, **234**, 758–759.

Lecchi, A., Covi, G., Lecchi, C., Mantero, F. & Scuro, L. A. (1976). Urinary kallikrein excretion in Bartter's syndrome. *Journal of Clinical Endocrinology and Metabolism*, **43**, 1175–1178.

Mace, J. W., Hambridge, K. M., Gotlin, R. W., Dubois, R. S., Solomon, C. S. & Katz, F. H. (1973). Magnesium supplementation in Bartter's syndrome. *Archives of Disease in Childhood*, **48**, 485–487.

Meyer, W. J., III, Gill, J. R., Jr. & Bartter, F. C. (1975). Gout as a complication of Bartter's syndrome. A possible role for alkalosis in the decreased clearance of uric acid. *Annals of Internal Medicine*, **83**, 56–59.

Modlinger, R. S., Nicolis, G. L., Krakoff, L. R. & Gabrilove, J. L. (1973). Pathogenesis of Bartter's syndrome. *New England Journal of Medicine*, **289**, 1022–1024.

Norby, L., Flamenbaum, W., Lentz, R. & Ramwell, P. (1976). Prostaglandins and aspirin therapy in Bartter's syndrome. *The Lancet*, **2**, 604–606.

Norby, L., Mark, A. L. & Kaloyanides, G. J. (1976). On the pathogenesis of Bartter's syndrome: report of studies in a patient with this disorder. *Clinical Nephrology*, **6**, 404–413.

Orloff, J., Handler, J. S. & Bergstrom, S. (1965). Effect of prostaglandin (PGE_1) on the permeability response of toad bladder to vasopressin, theophylline and adenosine 3'5'-monophosphate. *Nature*, **205**, 397–398.

Pasternack, A., Perheentupa, J., Launiala, K. & Hallman, N. (1967). Kidney biopsy findings in familial chloride diarrhoea. *Acta Endocrinologica*, **55**, 1–10.

Pronove, P., MacCardle, R. C. & Bartter, F. C. (1960). Aldosteronism, hypokalemia and a unique renal lesion in a five year old boy. *Acta Endocrinologica* (suppl.), **51**, 167.

Relman, A. & Schwartz, W. (1958). The kidney in potassium depletion. *American Journal of Medicine*, **24**, 764–773

Rubini, M. (1961). Water excretion in potassium deficient man. *Journal of Clinical Investigation*, **40**, 2215–2224.

Sasaki, H., Okumura, M., Ikeda, M., Kawasaki, I. & Fukiyama, K. (1976). Hypotensive response to angiotensin II analogue in Bartter's syndrome. *New England Journal of Medicine*, **294**, 611–612.

Schwartz, G. J. & Cornfeld, D. (1975). Bartter's syndrome: clinical study of its treatment with salt loading and propanolol. *Clinical Nephrology*, **4**, 45–51.

Simopoulos, A. P. & Bartter, F. C. (1972). Growth characteristics and factors influencing growth in Bartter's syndrome. *Journal of Pediatrics*, **81**, 56–65.

Solomon, R. J. & Brown, R. S. (1975). Bartter's syndrome: new insights into pathogenesis and treatment. *American Journal of Medicine*, **58**, 575–583.

Tannen, R. L. (1970). The effect of uncomplicated potassium depletion on urine acidification. *Journal of Clinical Investigation*, **49**, 813–827.

Tarm, F., Juncos, L. L., Anderson, C. F. & Donadio, J. V. (1973). Bartter's syndrome: an unusual presentation. *Proceedings of the Mayo Clinic*, **48**, 280–283.

Tomko, D. K., Yeh, B. P. & Falls, W. F., Jr. (1976). Bartter's syndrome: study of a 52 year old man with evidence for a defect in proximal tubular sodium reabsorption and comments on therapy. *American Journal of Medicine*, **61**, 111–118.

Trygstad, C. W., Mangos, J. A., Bloodworth, M. D., Jr. & Lobeck, C. C. (1969). A sibship with Bartter's syndrome: failure of total adrenalectomy to correct the potassium wasting. *Pediatrics*, **44**, 234–242.

Vander, A. J. (1970). Direct effects of potassium on renin secretion and renal function. *American Journal of Physiology*, **219**, 455–459.

Verberckmoes, R., Vandamme, B., Clement, J., Amery, A. & Michielsen, P. (1976). Bartter's syndrome with hyperplasia of renal medullary cells: successful treatment with indomethacin. *Kidney International*, **9**, 302–307.

Wald, M. K., Perrin, E. V. & Bolande, R. P. (1971). Bartter's syndrome in early infancy. *Pediatrics*, **47**, 254–263.

Weber, P. C., Larsson, C., Anggard, E., Hamberg, M., Corey, E. J., Nicolaou, K. C. & Samuelsson, B. (1975). Stimulation of renin release from rabbit renal cortex by arachidonic acid and prostaglandin endoperoxides. *Circulation Research*, **39**, 868–874.

White, M. G. (1972). Bartter's syndrome: a manifestation of renal tubular defects. *Archives of Internal Medicine*, **129**, 41–47.

Index

(Note: Page numbers in *italics* represents illustrations, page numbers followed by (t) represent tables.)

Acetazolamide administration, in hydrogen ion secretion, 11
Acetyl CoA, 69, 94
Acid(s)
  excretion of, 2, 108, 109
    and bicarbonate reabsorption, 32–34, *33*
    in hypocapnia, 156
    in metabolic alkalosis, 111
    in potassium deficiency, 125, 212
    vs. acid secretion, *31*, 32
  metabolic production of, vs. renal excretion of, 30
  metabolic removal of, 72, 73
  noncarbonic, production of, 108
Acid anions. See *Anions.*
Acid-base balance
  and lactate turnover, 71–73
  and potassium levels, 23, 170, 174
  effect of CO_2 transport on, 141
  in renal tubular acidosis, 51
  in stable uremic acidosis, 36, 36(t)
  renal regulation of, 108–111, *109*
  respiratory regulation of, disturbances in, 103–106
Acid-base disorders, mixed, in respiratory acidosis, 151–153
  mixed, in respiratory alkalosis, 161–162
  of respiratory origin, 137–162
  potassium-induced, 210
Acid-base response, to gastric drainage, in metabolic alkalosis, *116*, 117
  to uncomplicated dietary potassium depletion, 121–123, *123*
Acid-disequilibrium pH, in hypercapnia, 13
  in tubular fluid, 5, 6
Acidemia, hypercapnia and, 149
  hypocapnia and, 159
Acidification of urine. See *Urine acidification.*
Acidity, plasma, and carbonic acid concentration, 137
Acidosis(es)
  chronic, parathyroid hormone in, 37
  hyperchloremic, 35, 43, 47, 50, 256
  in hyporeninemic hypoaldosteronism, 256

Acidosis (*continued*)
  intracellular, 60
    in potassium depletion, 211
  lactic. See *Lactic acidosis.*
  metabolic. See *Metabolic acidosis.*
  mixed, 35, 151–153
  potassium levels in, 23, 170, 189
  renal, 30–61
    introductory terms and concepts, 30–35
    pathogenesis and effects of, *31*
  renal tubular. See *Renal tubular acidosis.*
  respiratory. See *Respiratory acidosis.*
  uremic. See *Uremic acidosis.*
Aciduria, paradoxical, 210
ACTH, and aldosterone secretion, 255
ACTH syndromes, ectopic, 180, 245
ACTH-induced paralysis, 176
Addison's disease, 192, 252
Adenoma, adrenal. See *Adrenal adenoma.*
  villous, and hypokalemia, 186, 208
Adrenal adenoma, and hypokalemia, 178
  and primary aldosteronism, 241, *242*
Adrenal enzyme defects, 238(t), 243–244
  in mineralocorticoid deficiency, 253–254
Adrenal hyperplasia, 179, 180, 241, *242*
Adrenocortical insufficiency, 192, 252–253
Adrenogenital syndrome, 253
AG. See *Anion gap.*
Airway obstruction, and hypercapnia, 151
Alanine, 76
  transamination of, 68, 72
Alcohol, and lactic acidosis, 81
  synergism with phenformin, 82
Aldosterone
  administration, and bicarbonate reabsorption, 16
  and potassium secretion, 22
  concentration of, and ion transport, 233, *234*, 235, 237
  deficiency of, 249
    hypokalemia-induced, 211
  effect on intracellular potassium, 172
  excretion of, in hypokalemia, 220
  hypersecretion of, 50. See also *Hyperaldosteronism.*
  impaired metabolism of, 194
  in mineral homeostasis, 232

Aldosterone (*continued*)
  production of, excessive, See *Hyperaldosteronism*.
  isolated deficiency of, 254
  reduced. See *Hypoaldosteronism*.
  secretion of, in Bartter's syndrome, 269, 276, 277
    in potassium restriction, 122
Aldosteronism, primary, 178–179, 238(t), 239–243, 239
  metabolic alkalosis in, 124, 125
Alkalemia, hypercapnia and, 149
  hypocapnia and, 159
  in treatment of lactic acidosis, 97
Alkali, fractional excretion of, 59
  reclamation of, from glomerular filtrate, 31
  removal of, from cell to blood, 11
Alkali leak, 39, 44
Alkali synthesis, 32–34, 33
Alkali therapy, 46–47
  for hyperkalemia, 196
  in distal RTA, 60–61
  in lactic acidosis, 96
  potassium supplementation in, 60
Alkaline urine, in renal failure, 40
Alkaline-disequilibrium pH, in tubular fluid, 6
Alkalosis
  and potassium secretion, 23, 170, 176
  contraction, 101, 107, 107, 130
  metabolic. See *Metabolic alkalosis*.
  mixed, 161–162
  posthypercapnic, 147, 152
  respiratory. See *Respiratory alkalosis*.
  Altitude, and hypocapnia, 156, 160
Amino acid(s), plasma, in lactic acidosis, 76
Amino acid storage disorders, 55
Aminoaciduria, 51
  in hypokalemia, 214
  in proximal RTA, 54
Ammonia
  excretion of, diminished, 57
    in hypercapnia, 144
    in hyporeninemic hypoaldosteronism, 257
    in metabolic alkalosis, 111
    in mineralocorticoid deficiency, 250, 251
    in potassium deprivation, 121
    in chronic renal failure, 45–46. 45(t)
  production of, 45(t), 59
    and potassium homeostasis, 125, 212
  urinary, 33
Ammonia "trapping," impairment of, 213
Ammoniagenesis, 45(t). See also *Ammonia, production of*.
Amphotericin-B, and distal RTA, 53
Androgen deficiency, 244

Anemia, hemolytic, 190
Anesthesia overdosage, and hypercapnia, 151
Angiotensin II, in Bartter's syndrome, 269
  in hypokalemia, 221
  in hyporeninemic hypoaldosteronism, 255
  vascular insensitivity to, 279, 280, 281, 282–283
Anion(s), acid, in acidoses, 31
Anion gap, in differentiation of renal acidoses, 34–35
  of metabolic acidosis, 58(t), 59
Anion reabsorption, selectivity of, in metabolic alkalosis, 118
Antidiuretic hormone, in glucocorticoid deficiency, 252
Anxiety, 160
Arrhythmias, in hypokalemia, 219
Aspirin, in Bartter's syndrome, 285
Asterixis, 150
Azotemia, in acidoses, 35

Backleak hypothesis, 48, 49, 114
Barium poisoning, redistribution of potassium and, 176
Bartter's syndrome, 181, 248
  differential diagnosis of, 284
  histologic changes in, 273–275
  laboratory findings in, 272, 272(t)
  pathogenesis of, 271–272, 279–284, 282
  pathophysiology of, 269–285
  radiographic findings in, 273
  renal biopsy in, 275
  symptoms of 270–275, 270(t)
  treatment of, 284–285
Base. See also *Acid-base* and *Alkali* entries.
  endogenous, gain of, and metabolic alkalosis, 107
  excretion in stool, 107, 108
Bence Jones proteinuria, 56
Bicarbonate
  control of, in plasma and extracellular fluid, 1
  concentration of, and hyperkalemia, 189
    in hypocapnia, 153
    in potassium regulation, 171
  increase in, 101
  luminal, changes in, 8, 10
  normalization in chronic respiratory acidosis, 105
  excretion of, in metabolic alkalosis, 111
  filtered, reclamation of, 3
  generation of, in potassium depletion, 211
  reabsorption of, 2
    and calcium, 42
    and carbonic anhydrase inhibition, 7
    and parathyroid hormone, 41, 42
    and pCO_2, 13
    in hypocapnia, 156

INDEX 291

Bicarbonate (continued)
  in potassium deprivation, 121
  in uremic acidosis, 39–44, 40
  ionic, 5
  reclamation of, 31
  recycling of, 8, 9
  regeneration of, 1, 3, 31
  secretion of. See Hydrogen ion secretion.
  tubular, 114–115
  serum, renal regulation of, 31
Bicarbonate leaks, 39, 44
Bicarbonate therapy, in renal tubular acidosis, 51
Bicarbonate wasting, 47
  in proximal RTA, 55, 55(t)
Bicarbonaturia, after bicarbonate infusion, 113
  in fructose intolerance, 15
  in renal failure, 40, 44
  self-limited, 48
Biliary cirrhosis, 52
Blood, lactate in, 74
Blood flow, and lactic acidosis, 79
Blood pressure, in hypokalemia, 220
Blood volume. See also Hypovolemia.
  arterial, and urinary sodium excretion, 112
Bone, and hydrogen ion absorption, 36
  in acidosis, 47
Bone carbonates, in acidosis, 37
  in CO_2 storage, 138
Bone disease, acid-base balance in, 51, 54
Brain, in hyperventilation, 74
  lactate accumulation in, 74
Brain lesions, in pyruvate carboxylase deficiency, 94
Brain syndrome, acute, 188
Buffer(s), availability of, in chronic renal failure, 45–46
  bone carbonates as, 36
  excretion of, low, 57
  hemoglobin as, 139
  in hypercapnia, 143
  nonbicarbonate, 4, 5, 154
  urinary, titration of, 33
Burn patients, hypokalemia in, 187

Calcinosis, 51
Calcium
  administration of, and renal acidification, 131
  and bicarbonate reabsorption, 41
  and hydrogen ion secretion, 14
  chelation of, 51
  in renal tubular acidosis, 51
  metabolism of, disorders of, 53
Calcium carbonates of bone, as buffers, 36
Calcium wasting, 47
Carbamino compounds, in CO_2 transport, 139

Carbenoxolone, 245
Carbohydrate metabolism, potassium depletion and, 221–223
Carbon dioxide
  diffusion of, tubular, 6
  excretion of, 140–141
  hydration of, 2, 4, 139
  input of, 138
  storage of, 138
  transport of, 139
Carbon dioxide dissociation curve, 139, 140
Carbon dioxide tension. See pCO_2.
Carbon dioxide therapy, in lactic acidosis, 98
Carbon monoxide poisoning, and lactic acidosis, 79
Carbonic acid
  buffering of, 141
  concentration of, and plasma acidity, 137
  dissociation of, 2, 139
  formation of, 33
  recycling of, 8, 9
Carbonic anhydrase
  and potassium excretion, 210
  catalytic effect of, 4
  in acid-base regulation, 110
  inhibition of, 7
Cardiac abnormalities, in hypokalemia, 187, 218
Cardiac arrest, and hypercapnia, 151
Cardiac arrhythmias, in hypercapnia, 149
  in hypocapnia, 159
Cardiac conduction, in hyperkalemia, 195
Catecholamines, in lactic acidosis, 78, 87
Cation exchange resin, in hyperkalemia, 196
Cell membrane, in acid-base homeostasis, 170
Central nervous system lesions, and hypocapnia, 160
Cerebrospinal fluid, acidity of, in respiratory acidosis, 147
  in respiratory alkalosis, 156–158
  lactate accumulation in, 74
Chemotherapy, and hyperkalemia, 191
Chloride
  dietary, retention of, 31
  in correction of alkalosis, 115–119, 116
  concentration of, relative to bicarbonate concentration, 102
  conservation of, in Bartter's syndrome, 277
  depletion of, in hypokalemia, 211
  excretion of, in hypercapnia, 145, 152
    in mineralocorticoid excess, 237
  reabsorption of, in response to mineralocorticoids, 126
    inhibition of, 127–130
Chloride diarrhea, congenital, 107

Chloride ion, in hydrogen ion secretion, 12
  in respiratory acidosis, 146
Chloride-resistant metabolic alkalosis, 120, 122
Chloruretic agents, in metabolic alkalosis, 129
Choline HCO_3, 11
Chronic obstructive lung disease, 151
  and hypercapnia, 144–147, *146*
  and metabolic alkalosis, 104
Circulation, effect of hypercapnia on, 148
Circulatory collapse, and lactic acidosis, 78
Circulatory reflexes, depression of, in hypokalemia, 220
Cirrhosis, and distal RTA, 52, 53
Cirrhosis, potassium deficiency in, 213, 222
Citrate metabolism, in renal tubular acidosis, 51
Citraturia, 51
Clay ingestion, and potassium loss, 208
Collecting duct, cystic dilatation of, 53
  potassium transport in, 19–20
Collecting tubule, potassium transport in, 19–20
Colon, sodium absorption in, 175
Coma, hepatic, hypokalemia and, 213
Conn's syndrome, 178–179
Contraction alkalosis, 101, 107, *107*, 130
Copper toxicity, 56
Cori cycle, *71*
Corticosterone, 179, 243
Cortisol, 244, 252, 253
Cortisone, and metabolic alkalosis, 130
Cushing's syndrome, 130, 180, 238(t), 244–245
Cyclic-AMP generation, in hypokalemia, 214
Cystinosis, in proximal RTA, 54

Dehydration constant, 4
Dehydroxylase, deficiency of, 253
Deoxycorticosterone, 179, 243
  and hydrogen ion secretion, 16
  in potassium depletion, 124
Dexamethasone therapy, 243
Diabetes insipidus, nephrogenic, 54
Diabetes mellitus, and hyperkalemia, 195
  and lactic acidosis, 90
  in hyporeninemic hypoaldosteronism, 255, 256
Diabetic ketoacidosis, and hypokalemia, 209
  vs. lactic acidosis, 77
Dialysis, peritoneal, in hyperkalemia, 196
  in lactic acidosis, 96
Diarrhea, and potassium excretion, 186
Dichloroacetate, in lactic acidosis, 97, 98
Diet, potassium-deficient, 177
  sodium-deficient, 23

Digitalis toxicity, and potassium depletion, 219
Diuresis, osmotic, in renal failure, 40–41
Diuretics
  and carbohydrate handling, 222
  and hydrogen ion secretion, 129
  and hypokalemia, 208, 209, 216, 224
  and kaliuresis, 174, 177
  chronic administration of, and chloride reabsorption, 127
  for proximal RTA, 61
  in lactic acidosis, 96
Drugs, and distal RTA, 53
  and lactic acidosis, 77, 80–87, 80(t)
  and proximal RTA, 55(t), 56
Duct, collecting. See *Collecting duct.*
Dysproteinemia, 52
  and proximal RTA, 55(t), 56

ECF. See *Extracellular fluid.*
Edema, in hypokalemia, 214
  pulmonary, and lactic acidosis, 80
Edema-forming states, and distal RTA, 53
EKG, in hyperkalemia, 195
  in hypokalemia, 219
Electrogenicity, of hydrogen ion secretion, 10
Electrolytes, in renal acidoses, 34(t)
  pattern in serum, *31*
  response to gastric drainage in metabolic alkalosis, *116*, 117
Embden-Meyerhof pathway, 67, 71, 72
Embolism, pulmonary, and lactic acidosis, 89
End-organ unresponsiveness, and hyperkalemia, 193
Enzyme deficiency states, in mineralocorticoid deficiency, 253
  in mineralocorticoid excess, 243
Epinephrine, effect on potassium concentration, 171
  in lactic acidosis, 87
Epithelium, tubular, transport characteristics of, 114
Erythrocyte. See *Red cell.*
Ethacrynic acid, and kaliuresis, 177
  and net acid excretion, 128
Ethanol, and lactic acidosis, 81
Ethylene glycol, and lactic acidosis, 86
Extracellular fluid, control of bicarbonate in, 1
Extracellular fluid volume
  and hydrogen ion secretion, 12–13
  changes in, and sodium and chloride reabsorption, 113
  depletion of, in Addison's disease, 252
  expansion of, and renal bicarbonate reabsorption, 112, *113*
  "isometric," 115–119, *116*

Extracellular fluid volume (*continued*)
  in mineralocorticoid excess, 238, 239
  protection of, in renal failure, 44

Fanconi syndrome, 54, 56
Fludrocortisone, in hyporeninemic hypoaldosteronism, 257, *258*, *259*
Fructose, and lactic acidosis, 87, 93
Fructose intolerance, and bicarbonaturia, 15
Furosemide, 37, 223
  and aldosterone, 255, 257
  and increased net acid excretion, 128
  in kaliuresis, 177

Gastric drainage, and acid-base response to metabolic alkalosis, *116*, *117*
Gastric fluid, loss of, 185
Gastrointestinal abnormalities, in Bartter's syndrome, 270
  in hyperkalemia, 195
Gastrointestinal loss of potassium, 185–187
Geophagia, and potassium loss, 208
Glomerular filtration, insufficiency of, and acidoses, 31
Glomerular hypercellularity, in Bartter's syndrome, 275
Glomerular insufficiency, chronic, 254
Glucocorticoids, and ammoniagenesis, 130
  deficiency of, 252
  vs. mineralocorticoid activity, 237
Gluconeogenesis, 69
  defects in, 92(t)
  hepatic, ethanol and, 81
Glucose, conversion to lactate, *68*, *71*
  for hyperkalemia, 196
  for lactic acidosis, 93
Glucose intolerance, in Bartter's syndrome, 273
Glucose tolerance, potassium depletion and, 222
Glucose-6-phosphatase, deficiency of, and lactic acidosis, 92
Glutamine, in ammonia production, 45
  in renal gluconeogenesis, 69
Glycogen storage, diseases of, 92
  potassium depletion and, 223
Glycolytic pathway, 67
Glycyrrhizic acid, 185
Gout, 57, 271
Growth retardation, 37, 54
  in Bartter's syndrome, 271

Haldane effect, 139
HCO_3^-, reabsorption of, 32
  in nephron, 110
  titration of, in renal failure, 40, *41*
H_2CO_3. See also *Bicarbonate*.
  in lumen, dehydration of, 4
  synthesis of, 32

Heart, effect of potassium on, 195
Heart failure, congestive, and potassium depletion, 207, 218
Hemodialysis, and potassium reduction, 196
Hemodynamics, in hypokalemia, 220
  in respiratory acidosis, 148
  in respiratory alkalosis, 158
Hemoglobin, as buffer, 139
Hemolysis, and hyperkalemia, 190
Henle's loop, potassium transport in, 17–18
Hepatic coma, hypokalemia and, 213
Hepatic encephalopathy, in potassium depletion, 213
Hepatic insufficiency, and hypocapnia, 160, 162
Hormones, effect of potassium depletion on, 222
  in proximal RTA, 55(t), 56
Hydration constant, 4
Hydrocortisone, and metabolic alkalosis, 130
Hydrogen ion(s)
  absorption of, by bone, 36
  and control of lactate production, 73
  backleak of, 48, *49*, 114
  concentration of, in hyperkalemia, 189
    in hypocapnia, 153
    in respiratory acidosis, 143, *145*
    intracellular, in hypocapnia, 158
  distribution of, 4
  excretion of, 1, 2
  loss of, and metabolic alkalosis, 106
  production of, 2
    endogenous, 108
    failure of, 48, *49*
  secretion of
    active vs. passive, 9–11
    and calcium, 14
    and extracellular fluid volume, 12–13
    and HCO_3^- reabsorption, 32
    and mineralocorticoids, 15
    and parathyroid hormone, 14
    and pCO_2, 13–14
    and phosphorus, 14
    and potassium, 15, 210
    diuretics and, 129
    electrogenicity of, 10
    hormone-stimulated sodium-nondependent, 127
    in nephron, 110
    in relation to other ions, 11–12
    in tubule, evidence of, 4–7
    in urinary acidification, 7–8
    ineffective, 47
    rate of, 8–9, *10*
  transport of, 1–16
    aldosterone and, 235, *236*
Hydrogen ion balance, and potassium distribution, 170

Hydrogen ion concentration gradient, maintenance of, 110
Hydrogen ion pump, 48, *49*
Hydronephrosis, 54
Hydroxylase deficiency, 179, 180, 243
Hyperaldosteronism, 50, 241
  primary, 178–179, 238(t), 239–243, *239*
  and Liddle's syndrome, 185
  secondary, 183, 248–249
Hypercalcemia, 131
  and hydrogen ion secretion, 15
  in Bartter's syndrome, 273
Hypercalciuria, 51, 53
  chronic acidosis and, 37
Hypercapnia, 102, 138, 141
  acute, 142–144, *145*
  and bicarbonate reabsorption, 13
  chronic, 144–147, *146*
Hyperchloremia. See *Acidosis, hyperchloremic.*
Hyperemia, muscular, 217, 221
Hyperglobulinemic states, 52
  in proximal RTA, 55(t), 56
Hyperglucocorticoidism, 130
Hyperkalemia, 169, 189–197, 249, 260. See also *Potassium.*
  and periodic paralysis, 190
  associated with potassium retention, 191–195
  associated with redistribution of potassium, 189
  causes of, 189(t)
  artificial, 197
  in uremic acidosis, 35
  manifestations of, 195
  therapy for, 195–197
Hyperlactatemia, 65
Hypermineralocorticoidism, and metabolic alkalosis, 120–127. See also *Mineralocorticoids.*
Hypernatremia. See *Sodium.*
Hyperparathyroidism, 14, 15
  bicarbonate reabsorption in, 42
Hyperphosphatemia, 37, 131
  alkali therapy in, 46
  hypercapnia and, 149
Hyplerplasia, adrenal, 179, 180, 241, *242*
  medullary interstitial cell, 275, *275*
  of juxtaglomerular apparatus, 269, 273, *274*
Hyperreninemia, 248, 260
Hypertension
  and hypokalemia, 178
  in aldosteronism, 239, *240*
  in Cushing's syndrome, 244
  low-renin essential, 245–248, *247*
  malignant, 183
  renal vascular, 182
Hyperuricemia, in Bartter's syndrome, 271

Hyperventilation, alveolar, 141
  and bicarbonate reabsorption, 13
  and hypocapnia, 153
  in lactic acidosis, 79, 89
  of pregnancy, 161
Hypervitaminosis D, 131
Hypoaldosteronism, 54, 192. See also *Aldosterone.*
  hyporeninemic, 194–195, 254–259
Hypocalcemia, alkali therapy in, 46
Hypocalciuria, 37
Hypocapnia, 138, 153
  acute, 154–156, *155*
  and bicarbonate reabsorption, 14
  and high altitudes, 156, 160
  chronic, 156
  in pregnancy, 161
  in uremic acidosis, 38
Hypoglycemic biguanides, and lactic acidosis, 82
Hypokalemia, 169, 175–189, 175(t). See also *Potassium.*
  and alkalosis, 15
  and periodic paralysis, 51, 176, 215–217
  associated with potassium deficiency, 177–185
  associated with potassium redistribution, 176
  causes of, 175(t)
  cortisone and, 130
  dialysis-induced, *283*
  histologic alterations in, 215
  in Bartter's syndrome, 269, 271, 276
  in Cushing's syndrome, 244
  in distal RTA, 51
  in mineralocorticoid excess, 238
  in proximal RTA, 54
  manifestations of, 187–188
  postburn, 187
  treatment of, 188–189
Hypokalemic paralysis, 51, 176, 215, 217
Hypomagnesemia, 184, 272, 285
Hyponatremia, 214, 249. See also *Sodium.*
Hypoparathyroidism, 131
Hypophosphatemia, 159
Hypophosphaturia, 49
Hyporeninemic hypoaldosteronism, 194–195, 254–259
Hyporeninism, 54
Hypotension, 260
Hypovolemia, 249
  potassium deficit in, 122
Hypoxemia, 142, 160
  arterial, and lactic acidosis, 80
  in respiratory acidosis, 150
Hypoxia, 142
  cerebral, 158
  tissue, and lactic acidosis, 78–80, 80(t)

INDEX 295

Immunologic disorders, and tubular dysfunction, 52
Indomethacin, 278, 279
  in Bartter's syndrome, 285
Insulin, and serum potassium concentration, 171
  defects in, and hyperkalemia, 195
  effect of potassium depletion on, 222
  secretion of, defect in, 256
Insulin therapy, in hyperkalemia, 196
  in lactic acidosis, 97
Integument, potassium losses from, 187
Interstitial renal disease, and acidosis, 35
Intestinal excretion of potassium, 174–175
Intracranial pressure, increased, 150
Intrathoracic processes, and hypocapnia, 160
Ion transport, mineralocorticoids and, 232
Ionic bicarbonate reabsorption, 5. See also *Bicarbonate; Hydrogen ion(s)*.
Ischemia, and hypokalemia, 207
Isotope, potassium, 205, 206
Juxtaglomerular apparatus, hyperplasia of, 269, 273, 274

Kaliopenia, 213
Kaliuresis, 172–174
  diuretics and, 177
  in mineralocorticoid excess, 239
  in $NaHCO_3$ therapy, 50
  in potassium-loading, 21
Ketoacidosis, 84–85
Ketones, urine test for, 84
$KHCO_3$ therapy, in distal RTA, 61
Kidney
  abnormalities in, in hypokalemia, 187
  effects of potassium depletion on, 209–215
  in regulation of serum bicarbonate, 31
  lactate metabolism in, 70, 73
  round cell infiltration of, 52, 53

Lactate
  accumulation of, in blood, 65
  biochemistry of, 66–69, *68*
  concentration of, 67
  conversion to glucose, *68*
  metabolism of, 66–75
  organ physiology of, 70–71, *71*
  production of, during exercise, 70
  synthesis of, hypocapnia, 158
  utilization of, in proton removal, 73
Lactate dehydrogenase, 66
Lactic acid, conversion to glucose, *71*
Lactic acidosis, 65–98
  causes of, 78–90, 80(t)
  clinical, 75–98
    characteristics of, 77
  congenital, 91–95, 92(t)
  definitions of, 75–77

Lactic acidosis (*continued*)
  drug-induced, 77, 80–87, 80(t)
  idiopathic acute, 90
  prognosis of, 77, 85
  treatment of, 95–98
Laxative abuse, 186
  and potassium loss, 208
  vs. Bartter's syndrome, 284
Left ventricular failure, and lactic acidosis, 80
Leukemia, 56
  lactic acidosis in, 89
Leukocytes, potassium content of, 207
Licorice ingestion, and hypokalemia, 185, 216, 245
Liddle's syndrome, 185, 240
Lipoic acids, in lactic acidosis, 97
Lithium, and distal RTA, 53
Liver, disease of, and lactic acidosis, 88
  metabolism of lactate in, 70, 73
Lumen, bicarbonate concentration of, changes in, 8, *10*
Lung. See *Pulmonary* entries.
Lysozymuria, 56

Magnesium deficiency, 184, 272
  in Bartter's syndrome, 285
Malignancy, and lactic acidosis, 89
  in proximal RTA, 55(t) 56
Malignant hypertension, 183
Medullary interstitial cell hyperplasia, 275, *275*
Medullary sponge kidney, 53
Mental retardation, in Bartter's syndrome, 271
Metabolic acidosis, 249. See also *Uremic acidosis*.
  and respiratory acidosis, 152
  bicarbonate loss in, 31
  hyperlactatemia and, 65
  in hypokalemia, 211
  steady-state, ammonium excretion in, *251*
Metabolic alkalosis, 101–131
  and respiratory acidosis, 152
  chloride-resistant, 120, 122
  in Bartter's syndrome, 269, 276
  in mineralocorticoid excess, 238
  loss of gastric fluid and, 185
  of extrarenal origin, renal response to, 111–120
  of renal origin, 120–131
  pathogenesis of, 106–108, *107*, 119(t)
  potassium depletion and, 15
  saline-resistant, *123*
Metals, exposure to, 55(t), 56
Metformin, and lactic acidosis, 82
Methanol, and lactic acidosis, 86
Methylene blue, in lactic acidosis, 97
Mineral homeostasis, 232

Mineralocorticoids
  and hydrogen ion secretion, 15
  and metabolic alkalosis, 123
    effect of potassium depletion on, 124–126
  and potassium transport, 22
  deficiency of, experimental states of, 249–251
  effect on potassium concentration, 172, 173
    syndromes of, 249–261, 250(t)
  effect on sodium concentration, 173
  excessive production of, 178, 179
    and hypokalemia, 208, 216
    experimental states of, 237–239
    syndromes associated with, 237–249, 238(t)
  in distal RTA, 54
  physiologic effects of, 232–237
  site of action of, 174
  unknown, syndromes associated with, 238(t), 245–248
Mitochondrial oxidative reaction, impairment of, 68
Muscle, skeletal. See *Skeletal muscle.*
Muscle weakness, in Bartter's syndrome, 269
  in distal RTA crisis, 60
  in hypokalemia, 215
Myeloma, 56
Myocardial abnormalities, in hypokalemia, 187, 218–219
Myocardial contractility, in hypocapnia, 158
  in hypokalemia, 219
  in respiratory acidosis, 148

NaCl in diet, and kaliuresis, 239, 241
NaHCO$_3$, 11. See also *Sodium bicarbonate.*
Na$^+$K$^+$ATPase activity, and kaliuresis, 21, 22
  and mineralocorticoids, 235
  and potassium concentration, 170, 219
Necrosis, myocardial, 218
  tissue, and hyperkalemia, 190
Neoplastic disease. See *Malignancy.*
Nephritis, interstitial, and hyperkalemia, 193
  and hypokalemia, 215
Nephrocalcinosis, 51, 53
Nephrolithiasis, 51, 57
Nephron, acid-base regulation in, 108–111
  site of mineralocorticoid action in, 174
Nephron "challenge," 43–44
Nephropathy, kaliopenic, 213
Neurologic damage, in pyruvate carboxylase deficiency, 94
Neuromuscular abnormalities, in Bartter's syndrome, 270

Neuromuscular abnormalities (*continued*)
  in hyperkalemia, 195
  in hypokalemia, 187
Neutron activation, and potassium measurement, 205
NH$_3$. See *Ammonia.*
NH$_4$ excretion, 111
Nitrogen metabolism, and hypokalemia, 223
Norepinephrine, and lactic acidosis, 87
  effect of hypokalemia on, 220

Oat cell carcinoma, 181
Oliguria, 192
Osmotic diuresis, in renal failure, 40–41
Osteodystrophy, renal, 37
Oxaloacetate, 68
Oxidase, mixed function, deficiency in, 254
Oxidative phosphorylation, defects in, 95
Oxyhemoglobin dissociation curve, in hypercapnia, 149
  in hypocapnia, 158

Pacemaker, transvenous, in hyperkalemia, 197
PaCO$_2$, and hydrogen ion concentration, 155
  changes in, 138, 140, 142, *143*
  in hypocapnia, 158
Papilledema, 150
Para-aminohippurate, clearance of, in hypokalemia, 214
Paralysis, periodic, hyperkalemic, 190
  hypokalemic, 51, 176, 215, 217
Parathyroid hormone, and bicarbonate reabsorption, 41
  and hydrogen ion secretion, 14
  in chronic acidosis, 37
Parathyroidectomy, and bicarbonate reabsorption, 131
Pars recta, potassium transport in, 17
Pasteur effect, 74
pCO$_2$
  and hydrogen ion secretion, 13–14
  elevation of, 141
  in metabolic alkalosis, 103
  in plasma, 140
  in respiratory alkalosis, 59
  in tissues, 139
  in tubule vs. plasma, 6
  in uremic acidosis, 38
  of cerebrospinal fluid, in hypocapnia, 157
  physiologic regulation of, 138–141
  reduction in, 153
  urinary, in renal tubular acidosis, 49
Peritoneal dialysis, in hyperkalemia, 196
  in lactic acidosis, 96
pH, acid-disequilibrium, in hypercapnia, 13
pH, alkaline-disequilibrium, in tubular fluid, 6

pH, alterations in, potassium-induced, 210
pH, intracellular, in respiratory acidosis, 147
pH, lactate in regulation of, 73
pH, of CSF, in hypocapnia, 157
pH, of cerebrospinal fluid, in respiratory acidosis, 147
pH, of tubular urine, 4
pH, in tubular fluid, acid-disequilibrium, 5, 6
pH, urinary, and potassium excretion, 23
pH, urinary, in incomplete RTA, 59
pH, urinary, inmetabolic alkalosis, 111
pH, urinary, mineralocorticoids and, 124
pH, urinary, relative to hydrogen ion concentration, 34
Phenformin intoxication, and lactic acidosis, 77, 82–86
Phosphate
  and bicarbonate reabsorption, 41
  and hydrogen ion secretion, 14
  depletion of, 57
  in chronic renal failure, 45
  in renal tubular acidosis, 51
  serum, in lactic acidosis, 76
  urinary, 33
Plasma acid-base equilibrium, in respiratory alkalosis, *154*
Plasma bicarbonate concentration, control of, 1
Plasma carbon dioxide tension, 140
Polyuria, in hypokalemia, 214
Potassium
  and hydrogen ion secretion, 15
  concentration of, extracellular, 170
    normal, 206
  deficiency of. See also *Hypokalemia.*
    conditions associated with, 177–185
    consequences of, 205–224
    etiology of, 224(t)
    in Bartter's syndrome, 271
    metabolic factors in, 221–223
    therapy for, 188–189
  depletion of, and alkalotic effect of mineralocorticoids, 124–127
    and metabolic alkalosis, 102, 120–127
    effects on kidney, 209–215
    etiology of, 207–209
    physiologic effects of, 215–221
  distribution of, 168–175, *169*
  effect of mineralocorticoids on, 172, 173
  excessive tissue release of, 189
Potassium excretion, 168–175, 208
  and acid-base balance, 23
  carbonic anhydrase and, 210
  intestinal, 174–175
  reduced, 192
  regulation of, 172–175
  renal, 172–174. See also *Kaliuresis.*

Potassium excretion (*continued*)
  homeostasis of, and ammonia production, 212
  changes in, 169
  disorders of, 168–197
  in plasma, 171
  intake of, and transport, 21
    excessive, 191
    inadequate, 177
    reduced, 208
  intracellular, aldosterone and, 172
    insulin and, 171
  losses of, from integument, 187
    gastronintestinal, 185–187
    in urine, 177
  redistribution of, 176, 189–191
  retention of, 249. See also *Hyperkalemia.*
    syndromes associated with, 191–195
  secretion of, defect in, 260
  serum, in lactic acidosis, 76
  supplementing alkali therapy, 60
  total body, measurement of, 205–207, 206(t)
    reduction of, 196
  transport of, 16–25, 207
    and mineralocorticoids, 22, 236
    and potassium intake, 21
    and sodium, 20
    in collecting tubule and duct, 19–20
    in distal convoluted tubule, 18–19, *19*
    in Henle's loop, 17–18
    in pars recta, 17
    in proximal convoluted tubule, 16–17
    urine flow rate and, 24
Potassium ions, concentration gradient of, 207
Potassium loading, and kaliuresis, 21
Potassium wasting, in Bartter's syndrome, 281, *282*
  in proximal RTA, 54
  in renal tubular acidosis, 50
Potassium-hydrogen exchange system, 210
Pregnancy, chronic hypocapnia in, 161
Progesterone, and hypercapnia, 161
Propranolol, in Bartter's syndrome, 285
Prostaglandins excretion of, in hypokalemia, 220
  urinary, 281, *283*
  in Bartter's syndrome, 278, *279*
  synthesis of, in hypokalemia, 182, 214
Protein synthesis, impaired, in hypokalemia, 223
Proton removal, 73
Pseudohyperkalemia, 197
Pseudohypermineralocorticoidism, 249
Pseudohypoaldosteronism, 260
Pseudohypomineralocorticoidism, 260–261
Pseudotumor cerebri, 150
PTH. See *Parathyroid hormone.*

Pulmonary compensation, in uremic acidosis, 38
Pulmonary disease, in metabolic alkalosis, 104
Pulmonary embolism, and lactic acidosis, 89
Pump-leak system, 9
Pyelonephritis, 215
Pyruvate
 concentration of, 67
 oxidation of, 69
  defects in, 92(t), 94
 production of, 67
 reduction of, 66
 removal of, 68
Pyruvate amino-transferase reaction, 72
Pyruvate carboxylase deficiency, and lactic acidosis, 93
Pyruvate dehydrogenase, 68
Pyruvate dehydrogenase pathway, 94

Red cell, potassium content of, 207
 rapid destruction of, and hyperkalemia, 190
 transport of sodium, in Bartter's syndrome, 273
Renal failure. See also *Uremic acidosis*.
 chronic, pathophysiology of acidosis in, 39–46
 hyperkalemia in, 191
Renal sodium handling, in Bartter's syndrome, 277
Renal transport, of hydrogen, 1–16
 of potassium, 16–25
Renal tubular acidosis, 47–61
 acquired, 211, 212
 complete, 59–60
 diagnosis of, 57–60, 58(t)
 distal, 51–54, 52(t)
  primary, 52
  therapy for, 60–61
 hypokalemia in, 184, 208
 incomplete, 57–59
 pathophysiology of, 48–51
 proximal, 54–57, 55(t)
  therapy for, 61
 proximal vs. distal, 48
Renin
 excessive production of, 182–184
 in hyperkalemia, 194
 in primary aldosteronism, 239
 increased levels of, 181
 secretion of, impairment of, 254
  in Bartter's syndrome, 269, 277
  renal, 54
Renin-angiotensin-aldosterone system, in hyperaldosteronism, 248
 in hypokalemia, 221
 in potassium depletion, 209
Renin-secreting tumor, 183

Respiration, effects of metabolic acidoses on, 38
Respiratory acidosis, 138, 141–153
 acute and chronic, 153
 causes of, 150–151, 150(t)
 CSF and intracellular acidity in, 147
 diagnosis and clinical manifestations of, 149–150
 in renal tubular acidosis, 51
 mixed acid-base disorders in, 151–153
 pathophysiology of, 141–142
 physiologic and biochemical effects of, 148–149
 secondary physiologic responses, 142–147
 treatment of, 151
Respiratory alkalosis, 59, 138, 153–162
 and lactic acidosis, 88
 and metabolic acidosis, 162
 and metabolic alkalosis, 162
 causes of, 159–160, 160(t)
 CSF acidity in, 156–158
 diagnosis and clinical manifestations of, 159
 in lactic acidosis, 78
 intracellular acidity in, 158
 mixed acid-base disorders in, 161–162
 pathophysiology of, 153
 physiologic and biochemical effects of, 158
 secondary physiologic responses to, 153–156, *154*
 treatment of, 161
Respiratory compensation, in metabolic alkalosis, 102–103, *103, 105*
Respiratory muscle paresis, potassium depletion and, 60
Rhabdomyolysis, in hypokalemia, 215
RTA. See *Renal tubular acidosis*.

Salicylate(s), and lactic acidosis, 86
Salicylate intoxication, 160, 162
Saline-resistant metabolic alkalosis, *123*
Salt, in diet, and kaliuresis, 239, 241
Salt retention, renal, 50
Salt wasting, in Bartter's syndrome, 280
Salt-wasting nephropathy, 260
Saralasin, 280
Sedative overdosage, and hypercapnia, 151
Sepsis, and hypocapnia, 162
 lactic acidosis in, 89
Septicemia, and hypocapnia, 161
Sexual development, in aldosteronism, 244
Shock, and lactic acidosis, 77, 78
Shohl's solution, 61
Sickle cell anemia, 54
Sjögren's syndrome, 52
 in proximal RTA, 56
Skeletal muscle, analysis of, to measure potassium, 207, 215
 in hypokalemia, 221

Sodium, absorption of, in colon, 175
　　and potassium transport, 20
　　conservation of, in Bartter's syndrome, 277
　　effect of mineralocorticoids on, 173
　　excretion of, in mineralocorticoid excess, 237
　　　volume constraints on, 115
　　metabolism of, altered, effect on nephrons, 43–44
　　reabsorption of, and hydrogen ion secretion, 11
　　　in potassium depletion, 209, 210
　　retention of, in hyperaldosteronism, 248
　　retention of, in hypokalemia, 214
　　transport of, and potassium transport, 207
　　　in Liddle's syndrome, 185
　　urinary losses of, and hypokalemia, 181
Sodium bicarbonate, excretion of, and ECF bicarbonate, 112
　　and volume contraction, 50
　　reabsorption of, 33
Sodium bicarbonate therapy, and kaliuresis, 50
　　in distal RTA, 60, 61
Sodium chloride, administration of, in metabolic alkalosis, 115–119, *116*
　　in diet, in kaliuresis, 239, 241
Sodium citrate solution, for distal RTA, 61
Sodium diuresis, in renal failure, 40–41
Sodium nitroprusside, in lactic acidosis, 98
Sodium salt, formation of, 1
Sodium wasting, 249
Sodium-cation exchange, in severe potassium depletion, 123
Sodium-hydrogen exchange, 11, 110
　　in hypercapnia, 144
　　in hypocapnia, 156
Sodium-potassium exchange pump, and mineralocorticoids, 234
Sorbitol, and lactic acidosis, 87
Spinal fluid, lactate accumulation in, 74
Spironolactone, 130
　　and hyperkalemia, 194
　　in aldosteronism, 239
　　in Bartter's syndrome, 284
Steady state, acute, 143
　　in hypocapnia, 154
Steroids, 237
Stones. See *Nephrolithiasis*.
Stool, excretion of base in, 107, 108
　　potassium losses in, 186
Stress, physiological, 74
Sweat, and hypokalemia, 187

Tetany, alkali therapy and, 46
Tetracycline, 56
Thiamin, in lactic acidosis, 97
Thiazides, 223

Tissue analysis, to measure potassium, 205, 207
Tissue necrosis, and hyperkalemia, 190
Toxins, and lactic acidosis, 80(t), 86
Transplantation, renal, and distal RTA, 53
　　and hyperkalemia, 193
　　and proximal RTA, 55(t), 56
Transtubular electrical potential difference, 127, 128
Triamcinolone, 130
Triamterene, 249
　　and hyperkalemia, 194
Tubular epithelial cells, vacuolization of, in hypokalemia, 215
Tubular fluid, acid-disequilibrium pH in, 5, 6
　　alkaline-disequilibrium pH in 6
　　flow rate of, and potassium transport, 24
Tubular resistance to mineralocorticoids, 260
Tubular urine, pH of, and distribution of hydrogen ions, 4
Tubule, acidification of, components of, 7(t)
　　bicarbonate secretion by, 114–115
　　collecting, potassium transport in, 19–20
　　distal convoluted, potassium transport in, 18–19, *19*
　　　properties of, *19*
　　hydrogen ion secretion in, evidence for, 4–7
　　defects of, hypokalemia and, 214
　　dysfunction of, dysproteinemias and, 52
　　in acid-base regulation, 108–111
　　in renal acidoses, 31
　　proximal convoluted, potassium transport in, 16–17
　　sodium reabsorption in, 112
Tumors, ACTH-producing, 180
　　renin-secreting, 183

Urate, serum, in lactic acidosis, 76
Urema. See *Renal failure*.
Uremic acidosis, 32, 35–47
　　bicarbonate reabsorption in, 39–44, *40*
　　buffer availability in, 45–46
　　clinical signs and symptoms of, 37(t)
　　failure to acidify urine in, 44
　　pathophysiology of, 39–46
　　stable, 36, 36(t)
　　therapy for, 46–47
Urinary abnormalities, in Bartter's syndrome, 270
Urinary buffers, in renal failure, 45–46
Urinary tract obstruction, 54
Urine
　　acid-disequilibrium pH in, 5, 6
　　acidification of, failure of, 44
　　　in renal tubular acidosis, 48
　　normal, 7(t), 108, *109*

Urine (continued)
  quantitative role of hydrogen ions in, 7–8
  regulation of, 1
  alkaline, in renal failure, 40
    inappropriate, 53
  alkaline-disequilibrium pH in, 6
  ketones in, 84
  lactate in, 74
  potassium loss in, 177
Urine concentrating defect, in Bartter's syndrome, 271, 283
  in hypokalemia, 213–214
Urine diluting ability, impairment of, 252
Urine flow rate, and potassium transport, 24

Vascular sensitivity to angiotensin II, 279, 280, 281, 282–283
Vascular system, effect of potassium depletion on, 220–221
Vasoconstriction, and lactic acidosis, 87
  in hypercapnia, 148
  in hypocapnia, 158
Vasodilation, in hypercapnia, 148
  in hypocapnia, 158

Venous admixture, 141
Ventilation, alveolar, increased, 153
  reduced, 102
  and carbon dioxide concentration, 140
  in metabolic acidosis, 38, 102
Ventilation-perfusion inequality, 142
Ventilators, and hypercapnia, 151
  and hypocapnia, 161
Villous adenoma, and hypokalemia, 186, 208
Virilism, 244, 253
Vitamin D, antiphosphaturic effect of, 131
  in uremic acidosis, 37
Volume contraction, absence of, and metabolic alkalosis, 119–120, 119(t)
  and gastrointestinal potassium loss, 185
  correction of, in metabolic alkalosis, 115–119, *116*
  in Addison's disease, 192
Vomiting, protracted, and potassium loss, 208
von Gierke's disease, 92

Water molecules, splitting of, 2
Wilson's disease, 56